The Foundations
of Quantum Theory

THE FOUNDATIONS OF QUANTUM THEORY

SOL WIEDER

Fairleigh Dickinson University

ACADEMIC PRESS   New York and London

ACADEMIC PRESS, INC.
111 Fifth Avenue, New York, New York 10003

United Kingdom Edition published by
ACADEMIC PRESS, INC. (LONDON) LTD.
24/28 Oval Road, London NW1

LIBRARY OF CONGRESS CATALOG CARD NUMBER: 72-9422

PRINTED IN THE UNITED STATES OF AMERICA

כבד את אביך ואת אמך ...

TO MY PARENTS
MAYER AND TOBY

Contents

Chapter 8.   **The Theory of Scattering**

Part II   **MANY-PARTICLE SYSTEMS**

Chapter 9.   **Noninteracting Particles**

Part III RELATIVISTIC QUANTUM MECHANICS AND FIELD THEORY

Preface

The undergraduate physics curriculum at many institutions has been revised to include a two-semester junior level course which surveys topics in modern physics and treats quantum mechanics on a semiquantitative basis. The student later takes a course which stresses formal quantum theory. This book provides a smooth transition from early undergraduate work to the more advanced material. The text evolved from lecture notes used in a two-semester course at Fairleigh Dickinson University. It draws on the student's background in mechanics, electricity and magnetism, and modern physics.

From my own experience, I have found two pedagogical approaches to quantum theory useful. The first relies rather heavily on the analogy between the wave equation in physical optics and the Schroedinger equation. This viewpoint is likely responsible for the somewhat restrictive and misleading label "wave mechanics." Since the de Broglie hypothesis is used as a cornerstone in wave mechanics, this approach is understandably consistent with a historical development. In addition the classical limit can easily be explained in terms of an analogy with geometrical optics by illustrating the similarities between the Hamilton–Jacobi equation and the eikonal equation. However, the *formal* correspondence between classical and quantum mechanics is not immediately clear.

In this text, I have chosen the second alternative, sometimes referred to as the "canonical" approach. The importance of first setting up the classical Hamiltonian in terms of the canonical coordinates and their conjugate momenta and then studying the inherent symmetries of the problem is an underlying theme of this book. In this respect the book follows closely the elegant but formal text of Dirac.* This book approaches the subject in a manner more palatable to the average senior. The correspondence between the classical and quantum theories is made via the Poisson bracket-commutator analogy. Ehrenfest's theorem is used as a postulate to connect the classical equations of motion with the Schroedinger equation. After presenting the general form of wave mechanics in bra-ket notation, it is shown that wave mechanics is only one of the many representations of quantum theory.

There is general agreement that where wave mechanics is applicable, it is by far the simplest and most direct way of dealing with a problem. Most of the problems in the text deal with particles in prescribed potentials and so wave mechanics is used extensively and the techniques for solving Schroedinger's differential equation are discussed in detail. However, where feasible, solutions are also obtained using the more general methods of Dirac. The quantum theory is applied to selected problems in modern physics with the expressed purpose of teaching the former and not the latter. It has been my feeling that modern physics is best appreciated after a thorough exposition of quantum mechanics. Since the electromagnetic interactions are understood more completely than are the other interactions, I have deliberately avoided most of the applications to nuclear physics and have limited most of the discussions to the atomic and molecular domain.

The text is divided into three parts—One-Particle Systems (Chapters 1–8), Many-Particle Systems (Chapters 9 and 10), and Relativistic Quantum Mechanics and Field Theory (Chapters 11 and 12). In the first part, Chapters 3 and 7 are crucial and will probably be the most difficult for the student to master. Without proper attention to this material, what follows may be meaningless. In Part II, Chapter 9 deals with noninteracting indistinguishable particles and the material covered is fundamental to almost all branches of physics. Chapter 10 covers interacting particle systems and if necessary certain sections may be omitted. Part III contains somewhat more advanced material and if time is short this part may be neglected altogether. If all twelve chapters are to be covered, at least half of Chapter 7 should be completed by the end of the first semester.

---

* P. A. M. Dirac, "The Principles of Quantum Mechanics," 4th ed., Oxford University Press, London, 1958.

I am very much indebted to my colleagues, students, and editors who have contributed their time and effort to make this text a better one. My thanks to my Honors student Mr. Alan Blumberg who scanned the original manuscript and to Professor W. Arthur whose careful scrutiny of the galleys has reduced the number of errors significantly. I am especially grateful to Professors D. Flory and R. Zeidler for the many interesting discussions and invaluable suggestions on quantum theory which have improved this book immeasurably. Any remaining errors or shortcomings are entirely my own fault. My wife Suzanne patiently typed and retyped the manuscript making many valuable comments while at the same time caring for our young sons Ari, Jonah, and Jeremy. For this and for her tolerance throughout the course of this work, she deserves my most special thanks.

I | ONE-PARTICLE SYSTEMS

1 | Historical Aspects

The development of physics in the twentieth century has been marked by two great discoveries. The first, special relativity (Einstein, 1905), corrects the equations of classical dynamics when the characteristic speed of matter becomes comparable to that of light. The second, quantum mechanics (Schroedinger, Heisenberg, Born, Dirac, 1925–1928), provides us with a more accurate picture of the dynamics of microscopic systems than do Newton's laws. By the end of the nineteenth century, experimental evidence had gradually accumulated to suggest that the classical theories of Newton and Maxwell were not adequate to explain many phenomena associated with matter and radiation. As a first step in our study of quantum mechanics, we examine some of the problems that faced the physicist at the turn of the century.

I Black–Body Radiation

Matter is constantly emitting and absorbing radiation. A material emits radiation due to thermal agitation. For example, a metal may become "red hot" when heated to a few thousand degrees Kelvin. Thus, when any material at a temperature T is fashioned into a cavity to enclose a region in space, the cavity will contain electromagnetic radiation. In equilibrium, it is this radiation which is known as "black-body" radiation and is found experimentally to contain a characteristic mixture of frequencies (that is, color) which depends only on the Kelvin temperature T and not on the chemical composition, contents, or shape of the enclosure.

We define the spectral density of the radiation $\rho(\omega, T)$ as the energy (per unit volume) of that radiation lying in the frequency (radians/sec) range between ω and $\omega + d\omega$. Figure 1-1 gives the spectral density at two different

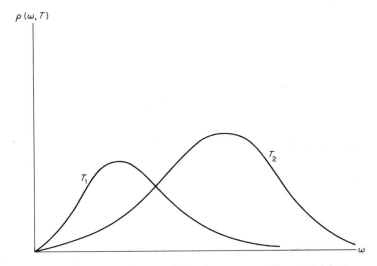

Figure 1-1 The black-body spectral density at two different Kelvin temperatures $(T_2 > T_1)$.

temperatures. It is the curves, of which these two are typical, that we shall try to derive from classical theory. Our failure in this regard will give us a better perspective on the need for quantum mechanics and at the same time will provide us with an exercise in classical physics. Mathematically we have

$$d\mathscr{E} = \rho(\omega, T)\, d\omega. \qquad (1\text{-}1)$$

The total energy density contained by all possible frequencies is

$$\mathscr{E}(T) = \frac{E}{\mathscr{V}} = \int_0^\infty \rho(\omega, T)\, d\omega \qquad (1\text{-}2)$$

where E is the total energy and $\mathscr{V}$ is the volume of the cavity.

Various attempts were made to explain the origin of the black-body spectrum. Wien suggested that a general form for the spectral density could be derived by performing a thermodynamic process (Carnot cycle) on the radiation in the cavity. By taking the radiation as the working substance in the Carnot engine, he concluded that $\rho(\omega, T)$ must be of the general form

$$\rho(\omega, T) = \omega^3 F\left(\frac{\omega}{T}\right) \tag{1-3}$$

where F is some function of the variable $x = \omega/T$. Thermodynamics alone could not determine the function F, but it would eliminate those theories that did not conform to Wien's law (1-3).

Using Wien's law and (1-2), the total energy density within the cavity becomes

$$\mathscr{E}(T) = \int_0^\infty \omega^3 F\left(\frac{\omega}{T}\right) d\omega = T^4 \int_0^\infty x^3 F(x)\, dx \tag{1-4}$$

where $x = \omega/T$.

The integral (if it converges) implies that the total energy density within the cavity is proportional to T^4, that is,

$$\mathscr{E}(T) = \sigma' T^4 \tag{1-4a}$$

where the constant is $\sigma' = \int_0^\infty x^3 F(x)\, dx$. Equation (1-4a) relates the electromagnetic energy density to the Kelvin temperature. For isotropic radiation, the radiative flux J (energy crossing unit area per unit time) can in turn be related to the energy density using

$$J = \tfrac{1}{4} c \mathscr{E}$$

where c is the speed of light (see Problem 1-2). Equation (1-4a) becomes

$$J = \tfrac{1}{4} c \sigma' T^4 = \sigma T^4. \tag{1-5}$$

This is the Stefan–Boltzmann law where the Stefan–Boltzmann constant is

$$\sigma = \frac{c}{4} \int_0^\infty x^3 F(x)\, dx.$$

While σ was experimentally known to be $\sigma = 0.567 \times 10^{-4}$ (cgs), $F(x)$ remained to be determined theoretically.

A second consequence of (1-3) is the "displacement" law which suggests that the dominant color (that is, the frequency at which $\rho(\omega, T)$ is a maximum) within the cavity is proportional to the temperature, that is,

$$\omega_0 \propto T. \tag{1-6}$$

This shift in frequency with temperature was also confirmed experimentally. Increasing the temperature of a substance produces a shift from "red hot" to "white hot," the whiteness indicating the presence of a bluish component. The frequency ω_0 at which ρ is a maximum can be derived from (1-3) by differentiation. Using the variable $x = \omega/T$ we find

$$\left.\frac{\partial \omega^3 F(\omega/T)}{\partial \omega}\right|_{\omega=\omega_0} = T^4\left\{\frac{d}{dx}x^3F(x)\right\}_{x=x_0} = 0$$

$$= T^4\{x_0{}^3F'(x_0) + 3x_0{}^2F(x_0)\} = 0$$

or

$$x_0\,F'(x_0) - 3F(x_0) = 0. \tag{1-7}$$

For any "reasonable" function $F(x)$ in Wien's law, (1-7) represents an ordinary equation which can be solved for x_0 giving

$$x_0 = \frac{\omega_0}{T} = \text{const.}$$

as the required displacement law (1-6). Of course, the value of the constant depends on the choice of F. In fact for certain functions, (1-7) has no solutions and the displacement law fails.

    Summarizing, we observe that Wien's law leads to a spectral density which agrees with both the Stefan–Boltzmann law and the displacement law. The constants associated with these laws depend on the particular function F, which cannot be determined using thermodynamics alone. It was in fact the search for the function F that led Planck to the discovery of quantum mechanics. We shall next apply the laws of mechanics and electromagnetism to various models in the hope of obtaining the function $F(\omega/T)$.

II   Characteristic Modes within a Cavity

    In classical theory, electromagnetic radiation is composed of vibrating electric and magnetic fields **E** and **B**. Since black-body radiation within a cavity is independent of the shape and composition of the enclosure, no generality is lost and some mathematical simplification is gained by assuming the enclosure to be a large metallic cube. The laws of electromagnetism require that the fields satisfy the wave equations

$$\nabla^2\mathbf{E} - \frac{1}{c^2}\frac{\partial^2\mathbf{E}}{\partial t^2} = 0$$

$$\nabla^2\mathbf{B} - \frac{1}{c^2}\frac{\partial^2\mathbf{B}}{\partial t^2} = 0 \tag{1-8}$$

where c is the speed of light in vacuo. Furthermore the fields must satisfy certain boundary conditions at the surface of the metal. For example, the tangential components of **E** and the normal components of **B** must vanish at the metallic boundary. This implies that only certain characteristic modes (standing waves) can be supported by the cavity. The modes and their characteristic frequencies can be deduced by making an analogy with a vibrating string.

The natural modes of the vibrating string with fixed ends are represented by

$$y_n = \exp(-i\omega_n t) \sin \frac{n\pi}{L} x \qquad (n = 1, 2, 3, \ldots) \tag{1-9}$$

where L is the length of the string. These solutions satisfy both the wave equation (1-8) and the boundary conditions (that is, $y_n(0, t) = y_n(L, t) = 0$). It may be shown directly that the natural frequencies are given by

$$\omega_n = \frac{v n \pi}{L} \tag{1-10}$$

where v is the phase velocity of transverse waves along the string. Each mode is characterized by a single integer n and each has a unique frequency.

The modes for radiation within a metallic cavity have a form similar to (1-9) with the exception that the modes are characterized by *three* non-negative integers l, m, and n. Furthermore, the characteristic frequencies are given, in analogy with (1-10), by

$$\omega_{l, m, n} = \frac{c\pi}{L} (l^2 + m^2 + n^2)^{1/2}. \tag{1-11}$$

Here L is the length of the edge of the cube. Note that not all the modes have distinct frequencies. From (1-11) it follows, for example, that the three modes $(l = 1, m = 2, n = 1)$, $(l = 1, m = 1, n = 2)$, and $(l = 2, m = 1, n = 1)$ have the same frequency, that is, $\omega = (c\pi/L)\sqrt{6}$. The number of modes having a common frequency is called the mode degeneracy $N(\omega)$ and is equal to *twice* the number of ways in which the integers in (1-11) can be chosen to give the same value of ω. The factor of *two* comes from the fact that light is a transverse wave and has two states of polarization. As ω becomes large, more ways are possible in which to rearrange the integers so that $N(\omega)$ increases with increasing ω.

Let us suppose that the (thermal) average energy of a mode at a given temperature is determined at most by its frequency, that is, let

$$\bar{\varepsilon}_{l, m, n} = \bar{\varepsilon}(\omega). \tag{1-12}$$

This supposition will be justified below. If we wish to calculate the total energy carried by the various modes, we may write

$$E = \sum_{\omega} \bar{\varepsilon}(\omega)N(\omega).\qquad(1\text{-}13)$$

The energy contained by the modes having a frequency ω is just the energy of each mode $\bar{\varepsilon}(\omega)$ multiplied by the number of modes with that frequency $N(\omega)$. The total energy is obtained by summing over all frequencies. To carry out this summation, let us look at a purely mathematical representation of the physical system.

The modes in a cavity may be graphically displayed by plotting their representative integers as points in the first octant of an "integer" space (Figure 1-2). The points l, m, and n are plotted on the x, y, and z axes, respec-

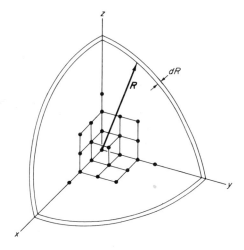

Figure 1-2 Plotting the modes as points in the first octant of an integer space.

tively. Note that the frequency of the mode is determined by the distance from the origin to the mode point, that is, $\omega = (c\pi/L)R$ where

$$R = (l^2 + m^2 + n^2)^{1/2}.$$

The mode degeneracy $N(\omega)$ is equal to twice the number of mode points situated on a sphere of radius R, in the first octant.

For thermal radiation, the dominant mode points of interest are situated far from the origin which of course corresponds to the higher frequencies, that is, those in or above the infrared region. These modes are so closely spaced that we can regard them as being part of a continuum and speak of the number of modes in the interval between ω and $\omega + d\omega$ as given by

$$dN = G(\omega) \, d\omega$$

where $G(\omega)$ is the *mode density* which we shall evaluate next. The total energy within the cavity (1-13) becomes simply

$$E = \int \bar{\varepsilon}(\omega) \, dN = \int \bar{\varepsilon}(\omega) G(\omega) \, d\omega. \qquad (1\text{-}14)$$

To find $G(\omega)$ we note that dN represents the volume of the "integer" space within a shell of radius R and thickness dR in the first octant, that is,

$$dN = 2 \cdot \tfrac{1}{8} \cdot 4\pi R^2 \, dR = \pi R^2 \, dR. \qquad (1\text{-}15)$$

The factor of 2 comes from the physics rather than our mathematical model in that we must include the two degrees of polarization. Using $\omega = (c\pi/L)R$ we have

$$R^2 \, dR = \left(\frac{L}{c\pi}\right)^3 \omega^2 \, d\omega = \frac{\mathscr{V}}{c^3 \pi^3} \omega^2 \, d\omega \qquad (1\text{-}16)$$

where $\mathscr{V} = L^3$ is the volume of the cavity. Finally we find

$$dN = \frac{\mathscr{V}}{c^3 \pi^2} \omega^2 \, d\omega \qquad (1\text{-}17)$$

and the mode density is

$$G(\omega) = \frac{\mathscr{V}}{c^3 \pi^2} \omega^2. \qquad (1\text{-}18)$$

The total energy density within the cavity follows from (1-14) as

$$\mathscr{E} = \frac{E}{\mathscr{V}} = \frac{1}{c^3 \pi^2} \int_0^\infty \bar{\varepsilon}(\omega) \omega^2 \, d\omega = \int_0^\infty \rho(\omega) \, d\omega \qquad (1\text{-}19)$$

from which we may identify the spectral density as

$$\rho(\omega) = \frac{1}{c^3 \pi^2} \bar{\varepsilon}(\omega) \omega^2. \qquad (1\text{-}20)$$

For thermal (black-body) radiation, the average energy $\bar{\varepsilon}$ per mode is also determined by the Kelvin temperature and (1-20) should be written

$$\rho(\omega, T) = \frac{1}{c^3 \pi^2} \bar{\varepsilon}(\omega, T) \omega^2. \qquad (1\text{-}21)$$

We shall calculate the explicit T dependence below.

To find $\bar{\varepsilon}$ we use the analogy between a mode and a simple harmonic oscillator‡ of natural frequency ω. The assumption that this oscillator is completely classical in nature is the basis of the Rayleigh–Jeans theory.

III   The Rayleigh–Jeans (Classical) Theory

Consider a collection of oscillators all of the same frequency but with varying energies. According to the classical Boltzmann law, the fraction of the oscillators with energy ε is proportional to $e^{-\varepsilon/kT}$ where k is Boltzmann's constant ($k = 1.38 \times 10^{-16}$ erg/deg). The average energy is obtained using the following scheme of classical statistical mechanics§:

$$\bar{\varepsilon}(\omega, T) = \frac{\int_{-\infty}^{\infty} \int_{-\infty}^{\infty} \varepsilon e^{-\varepsilon/kT} \, dp \, dx}{\int_{-\infty}^{\infty} \int_{-\infty}^{\infty} e^{-\varepsilon/kT} \, dp \, dx} \tag{1-22}$$

where the oscillator energy is

$$\varepsilon = \frac{p^2}{2m} + \frac{1}{2} m\omega^2 x^2.$$

The integrals may be performed directly with the result $\bar{\varepsilon}(\omega, T) = kT$. This result is remarkable since it implies that as long as the energy of an oscillator may be varied continuously, the average energy is *independent* of the frequency of the oscillators being considered.

Substituting $\bar{\varepsilon} = kT$ into (1-21) we find

$$\rho_{\text{RJ}}(\omega, T) = \frac{1}{c^3 \pi^2} kT\omega^2 \tag{1-23}$$

which is the Rayleigh–Jeans spectral density.

Comparing this with Wien's law, we obtain consistency with (1-3) by setting

$$F_{\text{RJ}}\left(\frac{\omega}{T}\right) = \frac{k}{c^3 \pi^2} \left(\frac{\omega}{T}\right)^{-1} \tag{1-24}$$

because then (1-23) reads

$$\rho_{\text{RJ}} = \omega^3 F_{\text{RJ}}\left(\frac{\omega}{T}\right).$$

---

‡ The notion of the equivalence of a radiation mode and an oscillator is a rigorous one which will be verified in Chapter 12.

§ See, for example, C. Kittel, "Elementary Statistical Physics," Chapter 11. Wiley, New York, 1958.

Although the Rayleigh–Jeans result is consistent with Wien's law it is unacceptable as a black-body spectral density for the following reasons:

a. A simple plot of $\rho_{RJ}(\omega, T)$ reveals that this density does not have a maximum; rather it tends to infinity as ω increases. Thus the displacement law is violated.

b. The total energy density $\mathscr{E} = \int_0^\infty \rho(\omega)\, d\omega$ is *infinite* in direct violation of Stefan's law, that is, $J = \sigma T^4$.

The second difficulty occurs because the integral diverges at the upper limit (high frequencies). This divergence underlies the famous "ultraviolet catastrophe" which, as we shall see, will be remedied by Planck's theory.

IV Planck's (Quantum) Theory

Planck's analysis closely followed the Rayleigh–Jeans theory with the exception that he assumed the energy of an oscillator of frequency ω to be restricted or "quantized" according to the rule

$$\varepsilon_n = n\hbar\omega \tag{1-25}$$

where $\hbar = h/2\pi$ (read "h-bar"), h being a universal (Planck's) constant and $n = 0, 1, 2, \ldots$. The thermal average energy in analogy with (1-22) is given by the quantum Boltzmann law

$$\bar{\varepsilon}(\omega, T) = \frac{\sum_{n=0}^\infty \varepsilon_n \exp(-\varepsilon_n/kT)}{\sum_{n=0}^\infty \exp(-\varepsilon_n/kT)}. \tag{1-26}$$

The sums are evaluated easily if we observe that (1-26) may be expressed as

$$\bar{\varepsilon}(\omega, T) = -\frac{\partial}{\partial\beta} \ln \sum_{n=0}^\infty \exp(-\beta\varepsilon_n) \tag{1-27}$$

where $\beta = 1/kT$. Using $\varepsilon_n = n\hbar\omega$, the sum becomes

$$\sum_{n=0}^\infty \exp(-\beta n\hbar\omega) = \sum_{n=0}^\infty [\exp(-\beta\hbar\omega)]^n. \tag{1-28}$$

Setting $y = e^{-\beta\hbar\omega}$, the sum is that of a geometric series and we find

$$\sum_{n=0}^\infty y^n = \frac{1}{1-y}$$

so that

$$\sum_{n=0}^\infty \exp(-\beta\varepsilon_n) = \sum_{n=0}^\infty \exp(-\beta n\hbar\omega) = \frac{1}{1-e^{-\beta\hbar\omega}}. \tag{1-29}$$

Taking the derivative as required by (1-27) and simplifying we find

$$\bar{\varepsilon}(\omega, T) = \frac{\hbar\omega e^{-\hbar\omega/kT}}{1 - e^{-\hbar\omega/kT}} = \frac{\hbar\omega}{(e^{\hbar\omega/kT} - 1)}. \tag{1-30}$$

Note that the quantization condition (1-25) leads to an average oscillator energy which depends *both* on T and ω. Using (1-30) in (1-21) we are led to the Planck spectral density

$$\rho_P(\omega, T) = \frac{\omega^2}{c^3\pi^2} \frac{\hbar\omega}{(e^{\hbar\omega/kT} - 1)} = \omega^3 F_P\left(\frac{\omega}{T}\right) \tag{1-31}$$

where

$$F_P\left(\frac{\omega}{T}\right) = \frac{\hbar}{c^3\pi^2} [e^{(\hbar/k)(\omega/T)} - 1]^{-1}. \tag{1-32}$$

The Planck formula is consistent with Wien's law. However now the Planck function $F_P(\omega/T)$ leads to a spectral density which is also consistent with experimental observations (Figure 1-1). Both the displacement law and the Stefan–Boltzmann law follow directly from the Planck formula.

In making a comparison of the theory with experiment, it is convenient to express the spectral density in terms of the wavelengths instead of the frequencies. By definition, $\rho(\lambda, T)$ represents the energy carried by the wavelengths in the interval between λ and $\lambda + d\lambda$. We therefore set

$$-\rho(\lambda, T)\, d\lambda = \rho(\omega, T)\, d\omega. \tag{1-33}$$

Using

$$\lambda = \frac{2\pi c}{\omega} \quad \text{and} \quad d\lambda = -\frac{2\pi c}{\omega^2}\, d\omega$$

we find

$$\rho(\lambda, T) = \frac{\omega^2}{2\pi c}\rho(\omega, T)$$

or

$$\rho_P(\lambda, T) = \frac{8\pi hc}{\lambda^5} \left[\exp\left(\frac{hc}{\lambda kT}\right) - 1\right]^{-1}. \tag{1-34}$$

Excellent agreement with experiment is obtained by setting $h = 6.63 \times 10^{-27}$ erg-sec.

In resolving the difficulties with the classical theory all that was necessary was to restrict or quantize the energies of an oscillator according to $\varepsilon_n = n\hbar\omega$. Additional evidence will be presented to suggest that quantization must occur in other systems as well.

V The Photoelectric Effect

All matter is composed of atoms which in turn contain electrons, protons, and neutrons. In a metal some of the electrons are free to move within the interior of the metal from atom to atom. The minimum energy W required to free a single electron from the metal itself is known as the *work function* of the metal. Energy to produce electron emission may be supplied thermally (thermionic emission), by impact with energetic particles (secondary emission), or by electromagnetic radiation (photoemission). We shall examine the last process, the photoelectric effect (Figure 1-3), in some detail since it sheds further light on the early developments of quantum mechanics.

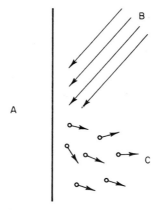

Figure 1-3 The photoelectric effect. A, metal; B, light; C, electrons.

Classical theory gives the energy of an electromagnetic wave as being proportional to the square of its amplitude. Consequently any frequency with sufficient amplitude should be able to supply the necessary energy to eject electrons. Yet, experimentally, an increase in amplitude, or in the case of light, brightness, does not initiate photoemission. Contrary to classical thought, photoemission is produced by raising the *frequency* beyond a threshold value. This threshold frequency v_0 is found experimentally to be directly proportional to the work function W.

An explanation of the photoelectric effect was first offered by Einstein (1905). In his quantum theory of light, he conjectured that light had particle-like characteristics and was composed of light quanta or *photons* having the following properties:

$$\text{speed} = c \qquad \text{energy} = hv$$

$$\text{mass} = 0 \qquad \text{momentum} = \frac{h}{\lambda} \qquad \text{where} \quad v = \frac{\omega}{2\pi}.$$

According to Einstein, an electron acquires energy by absorbing a *single* photon. The minimum photon energy required to liberate one electron is therefore

$$\varepsilon_{min} = h\nu_0 = W. \tag{1-35}$$

If the incoming photon energy is greater than $h\nu_0$, the excess energy will go elsewhere, possibly into kinetic energy of the ejected electron. Intensifying the beam merely increases the number of incident photons and results in multiplying the number of photoelectrons. The maximum kinetic energy available to each electron is

$$T_{max} = h\nu - W = h(\nu - \nu_0). \tag{1-36}$$

This is Einstein's photoelectric equation. If we plot the maximum observed kinetic energy T_{max} versus the incident frequency for a variety of metals, we do in fact obtain a linear relation (Figure 1-4). All the curves have a common slope h but differ in their intercepts according to the particular work function of the metal. The experimentally obtained value of h is found to be in perfect agreement with the one obtained from black-body radiation data.

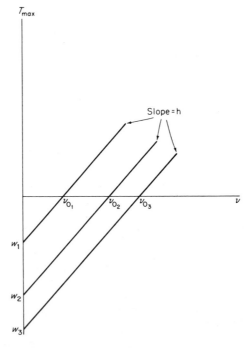

Figure 1-4 A plot of the photoelectric equation for three metals illustrating the work functions and the threshold frequencies.

A complete justification of Einstein's theory requires a detailed quantum-dynamical investigation of the photoemission process. A rather complicated formula has been derived to explain the subtle features of this effect.‡

VI The Compton Effect

Einstein's quantum theory of light may also be applied to the scattering of X-ray photons from free electrons. In practice the electrons are bound to a metal; however the energies associated with the X rays are much greater than the work function W and the electrons may be regarded as free. This effect when treated classically is called *Thomson scattering*. Classically, the incident and scattered radiation have the same frequency and the scattered intensity varies with the scattering angle according to

$$I \propto I_0(1 + \cos^2 \theta) \tag{1-37}$$

where I_0 is the incident intensity. When X rays are scattered from electrons in a metal foil neither prediction holds true. The scattered X rays suffer a drop in frequency and their intensity is given by a far more complicated formula than (1-37).

The decrease in the frequency of the scattered X rays was first explained by Compton (1923) using Einstein's quantum theory of light. Compton viewed the scattering as a collision between a photon and a free electron, the latter being initially at rest (Figure 1-5).

The conservation requirements on energy and momentum read:

$$h\nu - h\nu' = T \qquad \text{(energy)}$$

$$\frac{h}{\lambda} = \frac{h}{\lambda'} \cos \theta - p \cos \phi \qquad \text{(momentum in the } x \text{ direction)} \quad (1\text{-}38)$$

$$\frac{h}{\lambda'} \sin \theta = p \sin \phi \qquad \text{(momentum in the } y \text{ direction).}$$

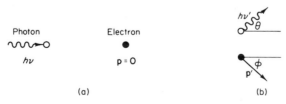

Photon   Electron

$h\nu$ $p = 0$

(a)   (b)

Figure 1-5 The kinematics of the Compton effect (a) before and (b) after the collision between a photon and an electron.

‡ For a discussion of the photoelectric formula, see W. Heitler, "The Quantum Theory of Radiation," 3rd ed., pp. 204–211. Oxford Univ. Press, London and New York, 1954.

Here T and p are, respectively, the kinetic energy and the linear momentum of the scattered electron. Using the wave formula

$$c = v\lambda = v'\lambda'$$

and the relativistic relation between momentum and kinetic energy

$$T^2 + 2Tmc^2 = p^2c^2$$

we may systematically eliminate p, T, v, v', and ϕ in (1-38) and obtain the Compton formula

$$\Delta\lambda = \lambda' - \lambda = \frac{h}{mc}(1 - \cos\theta). \tag{1-39}$$

This formula relates the scattered wavelength λ' to the photon scattering angle θ. The quantity

$$\lambda_C = h/mc \simeq 0.024 \times 10^{-8} \quad \text{cm} \tag{1-40}$$

is known as the Compton wavelength. The maximum shift occurs for back scattering ($\theta = 180°$) in which case $\Delta\lambda = 2\lambda_C$.

The intensity pattern of the scattered X rays fits the so-called Klein–Nishina formula (instead of the Thomson formula); the formula is based on relativistic quantum electrodynamics and is too complicated to be discussed here.‡

VII   The Quantum Theory of Matter

Although quantization was originally introduced in connection with radiation, it soon became apparent that matter too must be quantized. Even the characteristics of the simplest atom, hydrogen, could not be explained in terms of classical mechanics alone. Classically, the hydrogenic electron should move in an elliptical orbit about the proton in a manner resembling planetary motion with the electrostatic attraction providing the central force.

It is well known, classically, that when charged particles accelerate they emit radiative energy. Classical electrodynamics shows the emission rate to be

$$\frac{\text{energy}}{\text{time}} = \frac{2e^2a^2}{3c^3} \tag{1-41}$$

where e is the charge, a the acceleration, and c the speed of light. Since the hydrogenic electron is always accelerating toward the proton, it should lose

---

‡ For a derivation of the Klein–Nishina formula see W. Heitler, "The Quantum Theory of Radiation," 3rd ed., p. 215. Oxford Univ. Press, London and New York, 1954.

energy via the radiation process and spiral into the proton. Yet, this atomic collapse does not occur in nature; the hydrogen electron is never observed (on the average) closer to the proton than $d \sim \frac{1}{2} \times 10^{-8}$ cm.

Another consequence of the classical theory is that as the electron spirals inward, its frequency of revolution should change smoothly and the emission spectrum should be continuous, that is, all colors should be constantly emitted. Experiment clearly indicates that hydrogen's spectrum is discrete, that is, it contains characteristic colors.

Rydberg observed that the wavelength of the spectral lines in hydrogen's emission spectrum fit the following empirical formula:

$$\frac{1}{\lambda_{m,n}} = R\left(\frac{1}{n^2} - \frac{1}{m^2}\right) \qquad (n, m = 1, 2, \ldots; \quad m > n) \qquad (1\text{-}42)$$

where $R = 109677.576$ cm^{-1} is Rydberg's constant. An explanation of Rydberg's observations is offered below.

VIII The de Broglie Hypothesis and the Davisson–Germer Experiment

When light of a given wavelength passes through a series of closely spaced apertures, it interferes with itself and produces a diffraction pattern visible on a screen as variations in optical intensity. This phenomenon is not restricted to light but occurs in all systems with wave properties (for example, sound, water, etc.). As the wavelength tends to zero, wavelike properties (wave optics) disappear and the waves behave as rays (geometrical optics). As $\lambda \to 0$, instead of producing interference patterns, the light produces geometrical shadows of the apertures. Equivalently, in this geometric limit the light travels in straight lines.

Louis de Broglie, in his doctoral dissertation (1924), suggested that matter also had wavelike properties. He conjectured that with every free particle we could associate a wavelength

$$\lambda = \frac{h}{p} \qquad (1\text{-}43)$$

where p is the linear momentum of the particle and h is Planck's constant. Since h is extremely small, macroscopic particles ($p = mv \sim 1$) have such short "de Broglie wavelengths" that wave effects are not discernible. For microscopic particles, the wavelengths may be sufficiently long to observe interference effects.

Davisson and Germer (1927) passed a monoenergetic beam of electrons through a crystal and detected them on a photographic film. They found the intensity pattern to be strikingly similar to the one observed in optical diffraction, and they conjectured that the electron beam was "diffracted" by the regular atomic arrangement of the crystalline lattice. The experimental analysis confirmed that de Broglie's hypothesis (1-43) was indeed correct. In fact, this wavelike behavior of electrons is fundamental to the operation of the electron microscope. The availability of extremely small wavelengths permits greater resolution than is possible with conventional optical micro-scopes. Focusing is accomplished with electric and magnetic fields rather than with ordinary lenses.

The quantum nature of the de Broglie relation is evident from the presence of Planck's constant. If h were zero, then wavelike properties would be absent in matter. Evidently we can regard $h \to 0$ as the "classical limit." Remarkably, while quantum theory attributes particlelike properties to light it also attributes wavelike properties to matter. The wave–particle duality is an integral part of the quantum theory of all forms of energy whether it be matter or radiation.

IX   The Bohr Theory of Hydrogen

Long before the de Broglie hypothesis, Bohr (1913) had offered an explanation of the quantum nature of hydrogen. He made the following assumptions:

(1)   The electrons move in circular orbits.

(2)   The permissible orbits are those for which the angular momentum is quantized according to the rule

$$L_n = mv_n r_n = n\hbar \qquad (n = 1, 2, \ldots). \tag{1-44}$$

(3)   An electron in any one of these orbits does not radiate energy. Rather, radiation is emitted (or absorbed) only when the electron "jumps" from orbits of higher (or lower) to lower (or higher) energy (Figure 1-6). The quantum of energy emitted is equal to the loss of orbital energy.

The second postulate can be understood in terms of de Broglie's hypo-thesis. If we assume that in any orbit the de Broglie wave of the electron must be "in phase" with itself, we must require that the circumference of the orbit be an integral number of wavelengths, that is,

$$2\pi r = n\lambda.$$

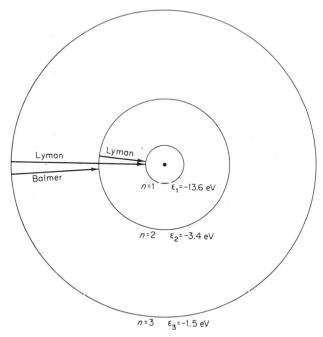

Figure 1-6 The quantized orbits in hydrogen according to the Bohr theory. The first few transitions are shown.

But since $\lambda = h/p = h/mv$, we obtain

$$mvr = n\hbar \tag{1-45}$$

which is Bohr's postulate.

The permissible energies (kinetic plus potential)

$$\varepsilon_n = \frac{1}{2} m v_n^2 - \frac{e^2}{r_n} \tag{1-46}$$

may be found using Newton's law for circular motion:

$$F = ma, \qquad \frac{e^2}{r_n^2} = m \frac{v_n^2}{r_n}. \tag{1-47}$$

Solving for ε_n and r_n from (1-45), (1-46), and (1-47), we obtain

$$\varepsilon_n = \frac{\varepsilon_1}{n^2} \qquad (n = 1, 2, \ldots) \tag{1-48a}$$

and

$$r_n = a n^2 \tag{1-48b}$$

where the *ground state energy* is

$$\varepsilon_1 = -\frac{me^4}{2\hbar^2} \simeq -13.6 \quad \text{eV} \tag{1-49a}$$

and the *first Bohr radius* is

$$a = \frac{\hbar^2}{me^2} \simeq 0.53 \times 10^{-8} \quad \text{cm}. \tag{1-49b}$$

Finally, using the Einstein relation for a photon $\varepsilon = h\nu$, we find the characteristic frequencies of hydrogen's spectrum to be

$$h\nu_{mn} = \varepsilon_m - \varepsilon_n = |\varepsilon_1| \left(\frac{1}{n^2} - \frac{1}{m^2}\right) \quad (n < m). \tag{1-50}$$

The wavelengths are given by

$$\frac{1}{\lambda_{mn}} = \frac{\nu_{mn}}{c}$$

so that

$$\frac{1}{\lambda_{mn}} = \frac{|\varepsilon_1|}{hc} \left(\frac{1}{n^2} - \frac{1}{m^2}\right) \quad (n < m). \tag{1-51}$$

This agrees with Rydberg's formula where Rydberg's constant is given by $R = |\varepsilon_1|/hc$. The hydrogenic emission lines are classified as shown in Table 1-1.

Table 1-1

Classification of Hydrogenic Emission Lines

Series	Transition	Range	Frequency (cyc/sec)
Lyman	$m > 1 \to n = 1$	ultraviolet	$\sim 3 \times 10^{15}$
Balmer	$m > 2 \to n = 2$	visible	$\sim 6 \times 10^{14}$
Paschen	$m > 3 \to n = 3$	infrared	$\sim 2 \times 10^{14}$
Brackett	$m > 4 \to n = 4$	infrared	$\sim 7 \times 10^{13}$
Pfund	$m > 5 \to n = 5$	infrared	$\sim 5 \times 10^{13}$

X   The Correspondence Principle

Bohr's postulates regarding radiation from the hydrogen atom were clearly inconsistent with classical electrodynamics. Yet the classical laws seemed to be adequate for macroscopic electromagnetic phenomena. We must

therefore assume that there exists some limit in which quantum electrodynamics corresponds to classical electrodynamics.

To see how this limit is reached we note that a charged particle moving in
uniform circular motion should, according to classical electrodynamics, emit
radiation of frequency equal to the rotational frequency, that is,

$$v_{rot} = \frac{v}{2\pi r} = \frac{1}{2\pi} \left(\frac{e^2}{mr^3} \right)^{1/2} \qquad (1\text{-}52)$$

or using (1-48b) for the nth orbit

$$v_{rot_n} = \frac{1}{2\pi} \frac{me^4}{\hbar^3} \frac{1}{n^3}. \qquad (1\text{-}53)$$

However quantum theory predicts that the radiation frequency between two
adjacent orbits, n and $n + 1$, is

$$v_n = \frac{\varepsilon_{n+1} - \varepsilon_n}{2\pi\hbar} = \frac{|\varepsilon_1|}{2\pi\hbar} \left[\frac{1}{n^2} - \frac{1}{n+1} \right]^2. \qquad (1\text{-}54)$$

Clearly (1-53) and (1-54) are inconsistent. If however we study (1-54) in the
limit as $n \to \infty$, we find

$$\left[\frac{1}{n^2} - \frac{1}{(n+1)^2} \right] = \frac{2n+1}{n^2(n+1)^2} \xrightarrow[n \to \infty]{} \frac{2}{n^3}$$

so that

$$v_n = \frac{|\varepsilon_1|}{2\pi\hbar} \frac{2}{n^3} = \frac{1}{2\pi} \frac{me^4}{2\hbar^2} \frac{1}{\hbar} \frac{2}{n^3} = v_{rot_n}.$$

In the limit $n \to \infty$ the emission frequencies as given by the classical and
quantum laws agree and we may regard $n \to \infty$ as the classical limit. Thus
classically when a charged particle emits radiation it is actually making
transitions between adjacent quantum levels of very large quantum numbers.
However in these regions the fractional energy separation $\Delta\varepsilon/\varepsilon$ of the levels is
so small that quantization goes unnoticed.

XI Summary

The developments discussed above formed the basis for the old quantum
theory. While this theory resolved some of the problems associated with the
classical theory of matter and radiation, it was lacking in many respects. Some
of the shortcomings were:

(1) In the form presented, it could only be applied to a limited number of
systems.

(2) It gave unconvincing reasons for quantization.

(3) It did not evolve formally as a generalization of classical mechanics and the principle of correspondence was not always clear.

The older theory was to be corrected, generalized, and given a rigorous theoretical foundation by Schroedinger, Heisenberg, Born, and Dirac (1925–1929). While some results of the old and new quantum theories agreed, many did not. For example, the quantized energies in hydrogen given by the modern quantum theory were in agreement with Bohr's result (1-48a). However, the concept of well-defined orbits used in the Bohr theory clearly violated one of the fundamental postulates of the newer theory, Heisenberg's uncertainty principle.

In the next chapter, we give a brief review of classical mechanics and illustrate how it can be used to develop the theory of quantum mechanics.

Suggested Reading

Born, M., "Atomic Physics." Hafner, New York, 1957.
Eisberg, R. M., "Fundamentals of Modern Physics." Wiley, New York, 1961.
McGervey, J. D., "Introduction to Modern Physics." Academic Press, New York, 1971.
Richtmyer, F. K., Kennard, E. H., and Lauritsen, T., "Introduction to Modern Physics." McGraw-Hill, New York, 1955.

Problems

1-1. Which of the following functions for the black-body spectral density are consistent with Wien's law:

(a) $\rho(\omega, T) = A\omega^3 [\ln \omega - \ln T]/\sin^2 \dfrac{T}{\omega}$

(b) $\rho(\omega, T) = \omega^3 T \cos \dfrac{\omega}{T}$

(c) $\rho(\omega, T) = \omega^2 [\omega^{1/\omega} e^{1/T}]^\omega$

(d) $\rho(\omega, T) = \dfrac{\omega^4}{T} + \omega^3$

Express $F(x)$ for the acceptable cases.

1-2. For an isotropic gas of photons of energy density $\mathscr{E}$, show that the energy crossing unit area per unit time along any direction is given by

$$\bar{J} = \tfrac{1}{4} c \mathscr{E}.$$

(Hint: Evaluate the average flux along the z direction by integrating $J_z = J \cos \theta$ over a hemisphere. Note that only half the energy density flows along a given direction.)

1-3. Show that the Rayleigh–Jeans spectral density violates Stefan's law as well as the displacement law. Explain why the Planck spectral density remedies the "ultraviolet catastrophe."

1-4. Evaluate the integrals in Eq. (1-22) and thereby show that $\bar{\varepsilon} = kT$.

1-5. Set up the integral for the Stefan–Boltzmann constant using the Planck spectral density.

1-6. Derive Compton's formula, Eq. (1-39), as suggested in Section VI.

1-7. Estimate the de Broglie wavelength of an electron in the ground state of hydrogen. Compare this to the de Broglie wavelength of the earth in its orbit about the sun.

1-8. (a) Estimate the de Broglie wavelength of a thermal neutron (at room temperature) by assuming $\varepsilon \simeq kT$.
(b) Assume that a beam of thermal neutrons is incident on a set of atomic planes in a crystalline lattice. According to the Bragg theory, reflective reinforcement occurs when

$$2d \sin \theta = n\lambda \qquad \text{(Figure 1-7)}$$

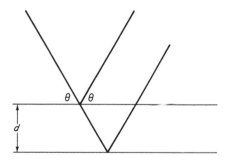

Figure 1-7

where d is the interplane spacing, θ is the angle between the beam and the planes, n is an integer representing the order of the reflection, and λ is the wavelength of the incident beam. In first order, $n = 1$, the observed reflection occurs at $\theta = 30°$. Find the spacing between the atomic planes.

1-9. Verify that the Lyman and Balmer series reside respectively in the ultraviolet and visible parts of the electromagnetic spectrum.

1-10. According to the correspondence principle, what is a charged classical oscillator actually doing when it emits light of frequency equal to its natural frequency v?

1-11. Compare the theoretical and experimental values of Rydberg's constant.

1-12. Use Eq. (1-41) to estimate the time it should take for the hydrogenic electron to spiral onto the proton from the first Bohr radius, according to the classical theory.

2 | Classical Mechanics

The subject of quantum mechanics can be developed in a variety of ways. Historically, an important approach is one based on de Broglie's hypothesis of the wave–particle duality of matter. In this approach, a wave equation is developed for de Broglie waves in a manner similar to that for electromagnetic waves. Using physical analogies, the characteristics of a particle may be related to its wave properties, with diffraction, interference, and resonance effects becoming an integral part of the quantum mechanics of matter. The relationships between classical and quantum mechanics are not immediately obvious in this treatment.

A second approach views quantum theory as a direct generalization of classical dynamics, with the wave–particle duality being a natural consequence of the theory. Its advantages are that it establishes the correspondence with the classical theory and that it permits a description of both fields and particles in a unified way. The connection between the quantum and classical theories is, however, not to be found in Newtonian dynamics but rather in the Hamiltonian formulation of classical mechanics.

The formalization of classical mechanics, that is, the transition from Newtonian to Lagrangian and finally to Hamiltonian dynamics, is discussed below.

I   The Newtonian Form of Mechanics (Nonrelativistic)

We will concern ourselves at this point with the description of the dynamics of a single point particle under the influence of external forces. The particle is assumed to be nonrelativistic, that is, its speed is much less than that of light. The net force on the particle is assumed irrotational ($\nabla \times \mathbf{F} = \mathbf{0}$) and is therefore derivable from a potential energy function, $V(\mathbf{r}, t)$,‡ using

$$\mathbf{F} = -\nabla V. \tag{2-1}$$

Velocity-dependent forces such as those associated with charged particles in magnetic fields will be discussed presently. Friction, which is not irrotational, does not occur in microscopic phenomena and will be omitted from our present discussion.

The object of theoretical mechanics is to describe the general motion of the particle when

(1)   its initial position and velocity are given
and
(2)   the net force on the particle $\mathbf{F}(\mathbf{r}, t)$ is indicated.

The solution is mathematically contained in the position vector $\mathbf{r}(t)$ which locates the particle at all times along its trajectory (Figure 2-1). The instan-

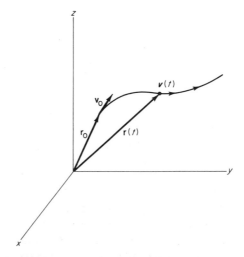

Figure 2-1   The trajectory and the position and velocity vectors of a particle.

---

‡ We shall use the notation $V(\mathbf{r}, t) = V(x, y, z, t)$.

taneous velocity and acceleration of the particle may be obtained by the differentiation,

$$v(t) = \frac{d\mathbf{r}}{dt} \quad \text{and} \quad a(t) = \frac{d\mathbf{v}}{dt} = \frac{d^2\mathbf{r}}{dt^2}. \tag{2-2}$$

The net force on the particle is related to its acceleration by Newton's second law,

$$-\nabla V = \mathbf{F} = m\mathbf{a} = m\frac{d^2\mathbf{r}}{dt^2} \tag{2-3}$$

or more generally,

$$\mathbf{F} = \frac{d\mathbf{p}}{dt} \tag{2-4}$$

where $\mathbf{p} = m\mathbf{v}$ is the linear momentum of the particle. Equation (2-3) is a vector differential equation and may be solved for

$$\mathbf{r} = \mathbf{r}(t) \tag{2-5}$$

once the initial conditions

$$\mathbf{r}_0 = \mathbf{r}(0) \quad \text{and} \quad \mathbf{v}_0 = \mathbf{v}(0)$$

have been specified.

In Cartesian coordinates, the components of (2-3) give

$$m\frac{d^2x}{dt^2} = F_x = -\frac{\partial V}{\partial x}(\mathbf{r}, t)$$

$$m\frac{d^2y}{dt^2} = F_y = -\frac{\partial V}{\partial y}(\mathbf{r}, t) \tag{2-6}$$

$$m\frac{d^2z}{dt^2} = F_z = -\frac{\partial V}{\partial z}(\mathbf{r}, t).$$

Solving these coupled differential equations, we obtain the kinematic solution

$$\mathbf{r}(t) = x(t)\mathbf{i} + y(t)\mathbf{j} + z(t)\mathbf{k} \tag{2-7}$$

where $\mathbf{i}$, $\mathbf{j}$, and $\mathbf{k}$ are unit vectors along the x, y, and z axes respectively.

If the kinematic solution is known, it is possible to eliminate the time dependence between $x(t)$, $y(t)$, and $z(t)$ and to obtain the trajectory

$$y = y(x) \quad \text{and} \quad z = z(x) \tag{2-8}$$

representing the path along which the particle will travel.

It is instructive to construct and study certain dynamical functions of the variables **v** and **r** throughout the course of a particle's motion. For example, the components of the angular momentum

$$\mathbf{L} = m\mathbf{r} \times \mathbf{v} \qquad (2\text{-}9)$$

are functions of interest in problems involving planetary motion. The total energy

$$E = \tfrac{1}{2}mv^2 + V(\mathbf{r}, t) \qquad (2\text{-}10)$$

is also important in a variety of problems. Any dynamical function is called a *constant of the motion* if it does not change in time as the particle progresses along its trajectory. We can also say that the function is *conserved*.

When a particle is under the influence of conservative (irrotational and time-independent) forces, its energy is always a constant of the motion, that is, energy is conserved. From (2-10), we obtain, using the chain rule for differentiation,

$$\frac{dE}{dt} = m\mathbf{v} \cdot \frac{d\mathbf{v}}{dt} + \mathbf{v} \cdot \nabla V = \mathbf{v} \cdot (m\mathbf{a} + \nabla V).$$

Using Newton's law, the term in parentheses is zero and E is a constant of the motion.

In order to illustrate the theory above, we shall apply Newton's laws to the isotropic oscillator, that is, to a particle attracted to the origin by a linear restoring force,

$$\mathbf{F} = -k\mathbf{r} \qquad (2\text{-}11)$$

where k is the elastic coefficient. The potential energy associated with this force is

$$V = \tfrac{1}{2}kr^2. \qquad (2\text{-}12)$$

The equations of motion take the form

$$m\ddot{x} = -\frac{\partial V}{\partial x} = -kx$$

$$m\ddot{y} = -\frac{\partial V}{\partial y} = -ky \qquad (2\text{-}13)$$

and

$$m\ddot{z} = -\frac{\partial V}{\partial z} = -kz.$$

We choose the initial conditions to be

$$x(0) = x_0, \qquad y(0) = y_0, \qquad z(0) = 0,$$
$$v_x(0) = v_{x_0}, \qquad v_y(0) = v_{y_0}, \qquad \text{and} \qquad v_z(0) = 0.$$

The solution for the z coordinate is

$$v_z \equiv z \equiv 0 \tag{2-14}$$

so that the motion remains in the xy plane. The x and y solutions are

$$x = A_x \cos(\omega t + \phi_x), \qquad y = A_y \cos(\omega t + \phi_y), \tag{2-15}$$

where $\omega = (k/m)^{1/2}$,

$$A_x = \left(x_0^2 + \frac{v_{x_0}^2}{\omega^2}\right)^{1/2}, \qquad A_y = \left(y_0^2 + \frac{v_{y_0}^2}{\omega^2}\right)^{1/2},$$

$$\phi_x = \tan^{-1}\frac{v_{x_0}}{\omega x_0}, \qquad \phi_y = \tan^{-1}\frac{v_{y_0}}{\omega y_0}.$$

The velocity is obtained by differentiation of (2-15).

As was already demonstrated, the energy

$$E = \tfrac{1}{2}mv^2 + \tfrac{1}{2}kr^2 = \tfrac{1}{2}m(v_x^2 + v_y^2 + v_z^2) + \tfrac{1}{2}k(x^2 + y^2 + z^2) \tag{2-16}$$

is a constant of the motion since the oscillator is a conservative system. It is possible to verify that the nonvanishing component of angular momentum,

$$L = L_z = m(xv_y - yv_x) \tag{2-17}$$

is also a constant of the motion. Both constants of the motion may be evaluated under the initial conditions and expressed in terms of $\mathbf{r}_0$ and $\mathbf{v}_0$.

We simplify the problem somewhat by assuming that $v_{x_0} = 0$ and $y_0 = 0$. The solutions (2-15) become

$$x = x_0 \cos \omega t, \qquad y = \frac{v_{y_0}}{\omega} \sin \omega t \tag{2-18}$$

and the constants of the motion may be written

$$E = \tfrac{1}{2}mv_{y_0}^2 + \tfrac{1}{2}kx_0^2 \tag{2-19a}$$

$$L = mx_0 v_{y_0}. \tag{2-19b}$$

The trajectory is obtained by eliminating the time variable from (2-18) giving

$$\frac{x^2}{a^2} + \frac{y^2}{b^2} = 1 \tag{2-20}$$

where $a = x_0$ and $b = v_{y_0}/\omega$. This orbit is an ellipse in normal form, with the major and minor axes determined by the initial conditions. These conditions may be expressed in terms of E and L using (2-19a) and (2-19b).

II Lagrange's Equations

There is no reason to restrict Newton's laws to Cartesian coordinates. In fact, many problems which are insoluble in Cartesian coordinates can be solved using other coordinate systems. For example, the Kepler problem which concerns itself with the motions of particles attracted to a force center by

$$\mathbf{F} = -\frac{k\mathbf{r}}{r^3} \quad \begin{array}{ll} (k = GMm, & \text{gravitation}) \\[2mm] (k = qq', & \text{electrostatics}) \end{array} \tag{2-21}$$

involves equations of motion of the form

$$m\ddot{x} = -\frac{kx}{(x^2 + y^2 + z^2)^{3/2}} = F_x$$

$$m\ddot{y} = -\frac{ky}{(x^2 + y^2 + z^2)^{3/2}} = F_y \tag{2-22}$$

$$m\ddot{z} = -\frac{kz}{(x^2 + y^2 + z^2)^{3/2}} = F_z.$$

Unlike the equations of motion for the isotropic oscillator, these equations are coupled, that is, the x equation also involves y and z, etc. The decoupling process is usually not a simple matter. If, however, the problem were formulated in spherical polar coordinates using r, θ, and ϕ as variables, the equations of motion would readily separate as we show shortly.

One could imagine the convenience of having a formalism equivalent to Newton's laws which is independent of any special system of coordinates. This generalization, known as *Lagrange's equations*, may be expressed in terms of a set of generalized coordinates; these new coordinates need have no simple geometrical relationship to Cartesian coordinates. The only restriction is that the coordinates be independent and that they uniquely determine the position of the particle.

A set of generalized coordinates may be defined in terms of the Cartesian variables using a transformation of the form

$$q_j = q_j(\mathbf{r}) \quad (j = 1, 2, 3). \tag{2-23}$$

The number of independent coordinates (in this case three) required to locate the particle is said to be the *number of degrees of freedom*. For example, the transformation to spherical polar coordinates is (Figure 2-2)

$$r = (x^2 + y^2 + z^2)^{1/2}$$

$$\theta = \tan^{-1} \frac{(x^2 + y^2)^{1/2}}{z} \tag{2-24}$$

$$\phi = \tan^{-1} \frac{y}{x}.$$

Hence $q_1 = r, q_2 = \theta, q_3 = \phi$ is one particular set of generalized coordinates.

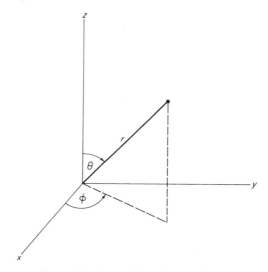

Figure 2-2 Spherical polar coordinates.

While Newton's laws deal directly with forces, Lagrange's equations introduce a quantity called the *Lagrangian* defined as

$$\mathscr{L} = T - V \tag{2-25}$$

where $T(q_j, \dot{q}_j)$ is the kinetic energy expressed in terms of generalized coordinates and velocities and $V(q_j)$ is the potential energy. In terms of the Lagrangian, Lagrange's equations of motion for the generalized coordinates take the form

$$\frac{d}{dt} \frac{\partial \mathscr{L}}{\partial \dot{q}_j} - \frac{\partial \mathscr{L}}{\partial q_j} = 0 \qquad (j = 1, 2, 3, \ldots). \tag{2-26}$$

A derivation of (2-26) can be found in standard texts on classical mechanics.‡ When (2-26) has been solved for the $q_j(t)$, the time dependence of the Cartesian variables can be established using the transformation inverse to (2-23).

In Cartesian coordinates, the Lagrangian is simply

$$\mathscr{L} = \tfrac{1}{2}m(\dot{x}^2 + \dot{y}^2 + \dot{z}^2) - V(x, y, z, t). \tag{2-27}$$

An application of Lagrange's equations to the variable x gives

$$\frac{d}{dt}\left(\frac{\partial \mathscr{L}}{\partial \dot{x}}\right) - \frac{\partial \mathscr{L}}{\partial x} = 0 \quad \text{or} \quad m\ddot{x} = -\frac{\partial V}{\partial x} = F_x. \tag{2-28}$$

An application to other Cartesian variables completely reproduces Newton's second law.

The correspondence between Lagrange's equations and Newton's law becomes more transparent if we define the generalized momentum associated with q_j as

$$p_j = \frac{\partial \mathscr{L}}{\partial \dot{q}_j} \tag{2-29}$$

and the generalized force as

$$F_j = -\frac{\partial V}{\partial q_j}. \tag{2-30}$$

Further, we regard

$$``F_j" = \frac{\partial T}{\partial q_j} \tag{2-31}$$

as a kinetic or *inertial* force. Lagrange's equations now take the form

$$\dot{p}_j = F_j + ``F_j" \tag{2-32}$$

which may be compared with (2-4).

In plane polar coordinates the Lagrangian is

$$\mathscr{L} = \tfrac{1}{2}m(\dot{r}^2 + r^2\dot{\phi}^2) - V(r, \phi, t). \tag{2-33}$$

Using Lagrange's equations, we obtain

$$\frac{d}{dt}p_r = \frac{d}{dt}(m\dot{r}) = -\frac{\partial V}{\partial r} + mr\dot{\phi}^2 \tag{2-34}$$

---

‡ See, for example, K. R. Symon, "Mechanics," 3rd ed., Chapter 9. Addison-Wesley, Reading, Massachusetts, 1971.

and

$$\frac{d}{dt} p_\phi = \frac{d}{dt} (mr^2 \dot{\phi}) = -\frac{\partial V}{\partial \phi}. \tag{2-35}$$

The first equation equates the change in radial momentum to the component of the force along the radius plus the inertial (centrifugal) force. The second equation relates the change in angular momentum to the generalized force associated with an angular variable, that is, the torque.

Lagrange's equations generally lead to a set of coupled second-order differential equations for $q_j(t)$. When the solutions have been obtained subject to specific initial conditions, the classical problem has been solved.

An important example utilizing polar coordinates is the Kepler problem which we introduced in connection with our discussion of Newtonian mechanics. The potential associated with this problem is

$$V = -\frac{k}{r}. \tag{2-36}$$

It may be verified without solving the entire problem that as with the isotropic oscillator the motion here is also planar. Setting the motion in the xy plane, the Lagrangian in polar coordinates becomes

$$\mathscr{L} = \frac{1}{2} m(\dot{r}^2 + r^2 \dot{\phi}^2) + \frac{k}{r} \tag{2-37}$$

and Lagrange's equations give

$$m\ddot{r} - mr\dot{\phi}^2 = -\frac{k}{r^2} \tag{2-38}$$

and

$$\dot{p}_\phi = \frac{d}{dt} (mr^2 \dot{\phi}) = 0 \qquad \text{or} \qquad p_\phi = mr^2 \dot{\phi} = L = \text{const.} \tag{2-39}$$

Since ϕ does not appear in the Lagrangian, the corresponding generalized momentum is a constant of the motion. Coordinates absent in the Lagrangian are termed *cyclic* or *ignorable*. Notice how the equations of motion are found immediately from the Lagrangian, whereas in Newtonian mechanics all the forces must first be found and then often resolved. But, once the differential equations of motion are found, it matters little how they were obtained in finding their solution.

For the Kepler problem, both the energy,

$$E = \frac{1}{2} m(\dot{r}^2 + r^2 \dot{\phi}^2) - \frac{k}{r} \tag{2-40}$$

and the angular momentum,

$$L = mr^2\dot{\phi} \qquad (2\text{-}41)$$

are constants of the motion. Using (2-41) we may substitute $\dot{\phi} = L/mr^2$ into (2-40) and solve for $\dot{r}$, obtaining

$$\dot{r} = \frac{dr}{dt} = \left[\frac{2}{m}\left(E + \frac{k}{r} - \frac{L^2}{2mr^2}\right)\right]^{1/2}$$

or

$$t = \int_0^t dt = \int_{r_0}^r \frac{dr}{\left[\frac{2}{m}\left(E + \frac{k}{r} - \frac{L^2}{2mr^2}\right)\right]^{1/2}} = I(r). \qquad (2\text{-}42)$$

While the integral is not a trivial one as it stands, it can be transformed to a more familiar form and evaluated or it can be treated numerically. In either case, (2-42) may be inverted giving

$$r = r(t). \qquad (2\text{-}43)$$

Finally, we may integrate (2-41) to give

$$\phi(t) = L \int_0^t \frac{dt}{mr^2(t)} + \phi_0. \qquad (2\text{-}44)$$

This integral may be evaluated once $r(t)$ is known.

While it is not possible to obtain $r(t)$ and $\phi(t)$ in simple form, it is possible to indirectly eliminate the time variable and establish the trajectory

$$r = r(\phi). \qquad (2\text{-}45)$$

The result is well known to be

$$r = \frac{a(1 - \varepsilon^2)}{1 + \varepsilon \cos(\phi - \phi_0)} \qquad (2\text{-}46)$$

where the geometrical constants are determined by

$$\varepsilon = \left(1 + \frac{2EL^2}{mk^2}\right)^{1/2} \qquad \text{(eccentricity)} \qquad (2\text{-}47)$$

and

$$a = -\frac{k}{2E}. \qquad (2\text{-}48)$$

The trajectory or orbit is a conic section with the center of force at one focus.

In particular, it will be an ellipse if $\varepsilon < 1$ ($E < 0$), a parabola if $\varepsilon = 1$ ($E = 0$), and a hyperbola if $\varepsilon > 1$ ($E > 0$). In the case of the ellipse, a represents the semimajor axis and depends on E but not on L. Many ellipses are possible for a fixed energy; all have the same major axis. Their eccentricities differ according to the angular momentum L of the particle. The existence of different orbits of the same energy also occurs in quantum mechanics where it is referred to as a *degeneracy* (Figure 2-3).

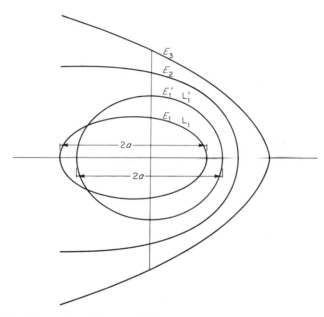

Figure 2-3 Orbits of a $1/r$ potential for various energies. Note that the two bound orbits have been chosen to be of equal energies.

It is possible to include velocity-dependent potentials,

$$U(q_j, \dot{q}_j, t) \tag{2-49}$$

in the Lagrangian formulation. To include these potentials, the generalized forces must be derivable from them by the relation

$$F_j = \frac{d}{dt} \frac{\partial U}{\partial \dot{q}_j} - \frac{\partial U}{\partial q_j}. \tag{2-50}$$

Indeed this is the case for electromagnetic forces acting on charged particles, as we show below. We need merely use U as we would V in the Lagrangian, that is, set $\mathscr{L} = T - U$, and proceed with Lagrange's equations.

In the case of electrostatic and magnetostatic forces, we require that the force on a charge be

$$\mathbf{F} = q\mathbf{E} + \frac{q\mathbf{v}}{c} \times \mathbf{B} \quad \text{(Lorentz force)}. \qquad (2\text{-}51)$$

Introducing the vector and scalar potentials **A** and Φ, such that

$$\mathbf{E} = -\nabla\Phi \quad \text{and} \quad \mathbf{B} = \nabla \times \mathbf{A} \qquad (2\text{-}52)$$

we may construct a velocity-dependent potential function‡

$$U = q\Phi(\mathbf{r}) - \frac{q}{c}\mathbf{A}(\mathbf{r}) \cdot \mathbf{v}. \qquad (2\text{-}53)$$

An application of (2-50) to (2-53) gives for the x component

$$F_x = -\frac{d}{dt}\frac{q}{c}A_x - q\frac{\partial\Phi}{\partial x} + \frac{q}{c}\frac{\partial}{\partial x}(\mathbf{v}\cdot\mathbf{A}). \qquad (2\text{-}54)$$

Since the rate of change of A_x, as seen by the particle, is

$$\frac{d}{dt}A_x = \frac{\partial A_x}{\partial x}\frac{dx}{dt} + \frac{\partial A_x}{\partial y}\frac{dy}{dt} + \frac{\partial A_x}{\partial z}\frac{dz}{dt} = \frac{\partial A_x}{\partial x}v_x + \frac{\partial A_x}{\partial y}v_y + \frac{\partial A_x}{\partial z}v_z$$

(2-54) becomes

$$F_x = -\frac{q}{c}\left[\frac{\partial A_x}{\partial x}v_x + \frac{\partial A_x}{\partial y}v_y + \frac{\partial A_x}{\partial z}v_z\right] - q\frac{\partial\Phi}{\partial x} + \frac{q}{c}\frac{\partial}{\partial x}(\mathbf{v}\cdot\mathbf{A}). \qquad (2\text{-}55)$$

This is just the x component of

$$\mathbf{F} = -\frac{q}{c}\mathbf{v}\cdot\nabla\mathbf{A} - q\nabla\Phi + \frac{q}{c}\nabla(\mathbf{v}\cdot\mathbf{A}). \qquad (2\text{-}56)$$

For **v** independent of **r**, we may use the identity

$$\nabla(\mathbf{v}\cdot\mathbf{A}) = \mathbf{v}\cdot\nabla\mathbf{A} + \mathbf{v}\times(\nabla\times\mathbf{A}) \qquad (2\text{-}57)$$

and (2-56) finally becomes

$$\mathbf{F} = q(-\nabla\Phi) + \frac{q}{c}\mathbf{v}\times(\nabla\times\mathbf{A})$$

$$= q\mathbf{E} + \frac{q}{c}\mathbf{v}\times\mathbf{B}$$

which is the desired result. The electromagnetic Lagrangian is therefore

$$\mathscr{L} = T - U = \frac{1}{2}mv^2 + \frac{q}{c}\mathbf{A}\cdot\mathbf{v} - q\Phi. \qquad (2\text{-}58)$$

---

‡ This result is also valid for time-varying fields.

III Hamilton's Equations

The Lagrangian formalism developed in the preceding section is inadequate for the purpose of developing the subject of quantum mechanics. As we shall see in Chapter 3, quantum mechanics requires that the generalized coordinates play an equal role with their corresponding *canonically conjugate momenta*. The required classical formalism must formally involve equations of motion which are symmetric in q_j and p_j. Lagrange's equations (and Newton's law) treat the coordinates differently from their momenta in that only time derivatives of momenta appear in (2-32). Hamilton's formalism, to be discussed below, provides the symmetry in p_j and q_j that we need.

We construct a function from the Lagrangian using the transformation

$$\mathcal{H} = \sum_{i=1}^{3} p_i \dot{q}_i - \mathcal{L} \tag{2-59}$$

where

$$p_i = \frac{\partial \mathcal{L}}{\partial \dot{q}_i} = p_i(q_1, \ldots, \dot{q}_1, \ldots). \tag{2-60}$$

Equation (2-59) is an example of a *Legendre* transformation which in this case generates a function known as the *Hamiltonian*, which contains q_j and p_j as its natural variables, rather than q_j and $\dot{q}_j$ which are the variables of the Lagrangian. The construction of the Hamiltonian is accomplished using (2-60) to express the $\dot{q}_i$ in terms of the coordinates and momenta where they appear on the right-hand side of (2-59). *The Hamiltonian is always expressed in terms of q_i and p_i.*

It can be shown that for most systems with velocity-independent potentials the Hamiltonian is merely the total energy of the system expressed in terms of the coordinates and momenta. For such systems, we may write

$$\mathcal{H} = \frac{p^2}{2m} + V(\mathbf{r}, t) = T(p_i, q_i) + V(q_i, t) \tag{2-61}$$

where T is the kinetic energy. We obtain the Hamiltonian for a particle in the presence of electromagnetic forces by performing a Legendre transformation on (2-58), giving

$$\mathcal{H} = \frac{(\mathbf{p} - q\mathbf{A}/c)^2}{2m} + q\Phi. \tag{2-62}$$

It is possible to show, using Lagrange's equations, that the equations of motion are obtained from the Hamiltonian using

$$\dot{p}_j = -\frac{\partial \mathcal{H}}{\partial q_j}, \qquad \dot{q}_j = \frac{\partial \mathcal{H}}{\partial p_j}, \qquad \frac{d\mathcal{H}}{dt} = \frac{\partial \mathcal{H}}{\partial t} = -\frac{\partial \mathcal{L}}{\partial t}. \tag{2-63}$$

Equations (2-63) are known as Hamilton's canonical equations of motion. They reflect the symmetry‡ in p_j and q_j and are completely equivalent to Lagrange's equations (2-26). From (2-63) it follows that if $\mathscr{L}$ or $\mathscr{H}$ has no explicit time dependence, then the Hamiltonian (and in most cases the energy) is a constant of the motion.§

In contrast to the three second-order differential Lagrange equations, Hamilton's equations (2-63) lead to six first-order coupled differential equations. The latter are to be solved for $p_j(t)$ and $q_j(t)$. To effect a separation of these equations, it is permissible to introduce transformations to new coordinates and momenta of the form

$$p_j' = p_j'(p_1, \ldots, q_1, \ldots), \qquad q_j' = q_j'(p_1, \ldots, q_1, \ldots) \qquad (2\text{-}64)$$

with a new Hamiltonian

$$\mathscr{H}'(p_i', q_i') = \mathscr{H}(p_i, q_i). \qquad (2\text{-}65)$$

Equations (2-64) are not necessarily simple coordinate transformations since they mix the p_i and q_i. We are however limited to those transformations for which the equations

$$\dot{p}_j' = -\frac{\partial \mathscr{H}'}{\partial q_j'} \quad \text{and} \quad \dot{q}_j' = \frac{\partial \mathscr{H}'}{\partial p_j'} \qquad (2\text{-}66)$$

and those of (2-63) give equivalent differential equations of motion. Such transformations are termed *canonical* because they maintain the form of Hamilton's equations. A simple change of coordinates (a point transformation) is always canonical.

In certain cases it is possible to find a canonical transformation that will trivially decouple Hamilton's equations. For example, the one-dimensional oscillator Hamiltonian

$$\mathscr{H} = \frac{p^2}{2m} + \frac{1}{2} m\omega^2 x^2, \qquad \omega = \left(\frac{k}{m}\right)^{1/2} \qquad (2\text{-}67)$$

leads to the coupled equations of motion

$$\dot{p} = -m\omega^2 x, \qquad \dot{x} = \frac{p}{m}. \qquad (2\text{-}68)$$

A simpler set of differential equations is obtained if we use the canonical transformation

$$p = (2m\omega p')^{1/2} \cos x', \qquad x = \left(\frac{2p'}{m\omega}\right)^{1/2} \sin x'. \qquad (2\text{-}69)$$

---

‡ Except for a minus sign the symmetry is intact.
§ We shall assume $\partial \mathscr{H}/\partial t = d\mathscr{H}/dt = 0$ unless otherwise noted.

The Hamiltonian (2-67) becomes

$$\mathscr{H}' = \omega p' \cos^2 x' + \omega p' \sin^2 x' = \omega p' = E \qquad (2\text{-}70)$$

and generates the decoupled equations of motion

$$\dot{p}' = 0 \qquad \text{and} \qquad \dot{x}' = \omega. \qquad (2\text{-}71)$$

The solutions are

$$p' = \text{const} = \frac{E}{\omega} \qquad \text{and} \qquad x' = \omega t + \phi. \qquad (2\text{-}72)$$

Expressing the results in terms of the old coordinates and momenta we finally obtain

$$p = (2mE)^{1/2} \cos(\omega t + \phi), \qquad x = \left(\frac{2E}{m\omega^2}\right)^{1/2} \sin(\omega t + \phi). \qquad (2\text{-}73)$$

These are the familiar results for an oscillator, except that the amplitude has been expressed in terms of a constant of the motion, that is, the energy.

Of course the ability to decouple and trivially solve a given problem rests with a knowledge of the appropriate canonical transformation. Obtaining that transformation is the subject of Hamilton–Jacobi theory and will not be presented here.‡

IV Poisson Brackets

The equations of motion for any canonical function $G(p_i, q_i, t)$ can be written, using Hamilton's equations, as

$$\frac{dG}{dt} = \sum_{i=1}^{3} \frac{\partial G}{\partial q_i} \dot{q}_i + \frac{\partial G}{\partial p_i} \dot{p}_i + \frac{\partial G}{\partial t}$$

$$= \sum_{i=1}^{3} \frac{\partial G}{\partial q_i} \frac{\partial \mathscr{H}}{\partial p_i} - \frac{\partial G}{\partial p_i} \frac{\partial \mathscr{H}}{\partial q_i} + \frac{\partial G}{\partial t}. \qquad (2\text{-}74)$$

We define the *Poisson bracket* of two functions A and B by§

$$\{A, B\} = \sum_{i=1}^{3} \frac{\partial A}{\partial q_i} \frac{\partial B}{\partial p_i} - \frac{\partial B}{\partial q_i} \frac{\partial A}{\partial p_i} \qquad (2\text{-}75)$$

so that (2-74) may be abbreviated

$$\frac{dG}{dt} = \{G, \mathscr{H}\} + \frac{\partial G}{\partial t}. \qquad (2\text{-}76)$$

---

‡ See for example, H. Goldstein, "Classical Mechanics," Chapter 9. Addison-Wesley, Reading, Massachusetts, 1950.

§ Some authors prefer the symbol [ ] for the Poisson bracket. We use { } to avoid any confusion with the *commutator* to be introduced in Chapter 3.

For canonical functions with no explicit time dependence ($\partial G/\partial t = 0$), we observe that if $\{G, \mathcal{H}\} = 0$, then G is a constant of the motion. In fact, (2-76) actually includes all of Hamilton's equations as can be verified by setting $G = p_i$, $G = q_i$, or $G = \mathcal{H}$.

It is trivial to demonstrate that

$$\{q_i, p_j\} = \delta_{ij} \qquad \text{and} \qquad \{q_i, q_j\} = \{p_i, p_j\} = 0 \qquad (2\text{-}77)$$

where the *Kronecker* delta is defined by

$$\begin{aligned} \delta_{ij} &= 1 \qquad \text{for} \quad i = j \\ \delta_{ij} &= 0 \qquad \text{for} \quad i \neq j. \end{aligned} \qquad (2\text{-}78)$$

Any relation involving Poisson brackets must be invariant under a canonical transformation. This is in fact another *definition* of a canonical transformation. Equations (2-77) are fundamental Poisson bracket relations for any coordinates and their canonically conjugate momenta. It is not very difficult to verify that the components of the angular momentum

$$\mathbf{L} = \mathbf{r} \times \mathbf{p} \qquad (2\text{-}79)$$

satisfy

$$\{L_i, L_j\} = L_k \qquad (i, j, k = x, y, z, \text{ in cyclic order})$$

and   (2-80)

$$\{L_i, L^2\} = 0.$$

It follows from (2-75) that the derivative of a canonical function with respect to a coordinate q_j is equal to the Poisson bracket of that function with the canonically conjugate momentum p_j, that is $\partial F/\partial q_j = \{F, p_j\}$. In particular, we have

$$\frac{\partial F}{\partial x} = \{F, p_x\} \qquad \text{or} \qquad F(x + dx, y, z) = F(x, y, z) + \{F, p_x\}\, dx. \qquad (2\text{-}81)$$

The canonical momentum p_x is said to be the *generator* of infinitesimal translations along the x direction.

V Relativistic Dynamics

We conclude by generalizing some of the previous results so that they may be applied to relativistic particles moving at speeds comparable to that of light ($v \lesssim c$). Newton's second law must be modified and written as

$$\mathbf{F} = \dot{\mathbf{p}} \qquad (2\text{-}82)$$

where **p** is now the *relativistic momentum*

$$\mathbf{p} = \frac{m\mathbf{v}}{(1 - (v^2/c^2))^{1/2}} \tag{2-83}$$

and m is the "rest" mass of the particle. Lagrange's and Hamilton's equations remain unaltered provided that we use the appropriate relativistic functions

$$\mathscr{L} = -mc^2 \left(1 - \frac{v^2}{c^2}\right)^{1/2} - V(\mathbf{r}, t) \tag{2-84}$$

and

$$\mathscr{H} = (p^2 c^2 + m^2 c^4)^{1/2} + V(\mathbf{r}, t). \tag{2-85}$$

The relativistic Hamiltonian for a charged particle in an electromagnetic field may be shown to be

$$\mathscr{H} = \left[\left(\mathbf{p} - \frac{q\mathbf{A}}{c}\right)^2 c^2 + m^2 c^4\right]^{1/2} + q\Phi. \tag{2-86}$$

This last result is obtained using the *minimal* substitution which makes the replacement

$$\mathbf{p} \rightarrow \mathbf{p} - \frac{q\mathbf{A}}{c} \tag{2-87}$$

and uses $q\Phi$ as a conventional potential energy function.

It may be verified directly that an application of Lagrange's equations to (2-84), or equivalently Hamilton's equations to (2-85), reproduces Newton's second law (2-82). In the limit $v/c \ll 1$, a Taylor expansion of (2-84) and (2-85) indicates that the relativistic forms properly reduce to their nonrelativistic counterparts except for the fact that the *rest energy*, mc^2, appears to have been subtracted from $\mathscr{L}$ and added to $\mathscr{H}$. This constant energy, however, does not affect the equations of motion.

The Hamiltonian–Poisson bracket formulation is particularly suited for the purpose of developing a generalization to quantum theory. We shall see that such a generalization will introduce Planck's constant, h, the fundamental constant of quantum theory, and that $h \rightarrow 0$ will represent the classical limit. In the next chapter, we develop the mathematical framework of quantum theory in a manner that will clearly demonstrate its relationship to classical mechanics.

Suggested Reading

Corben, H. C., and Stehle, P., "Classical Mechanics." Wiley, New York, 1960.
Goldstein, H., "Classical Mechanics." Addison-Wesley, Reading, Massachusetts, 1950.
Konopinski, E. J., "Classical Descriptions of Motion." Freeman, San Franscisco, 1969.

Marion, J., "Classical Dynamics of Particles and Systems," 2nd ed. Academic Press, New York, 1970.
Symon, K. R., "Mechanics," 3rd ed. Addison-Wesley, Reading, Massachusetts, 1971.

Problems

2-1. Show directly using (2-15) that $L = m(xv_y - yv_x)$ is a constant of the motion for an isotropic oscillator.

2-2. Express the solution given by (2-18) in terms of the constants of the motion E (energy) and L (angular momentum).

2-3. Consider a system of N particles whose Hamiltonian is

$$\mathcal{H} = \sum_{j=1}^{N} \frac{p_j^2}{2m} + V(\mathbf{r}_j) + \frac{1}{2} \sum_{i=1}^{N} \sum_{j \neq i}^{N} V(\mathbf{r}_i - \mathbf{r}_j).$$

Assume that the coordinates and momenta are periodic functions in time with a period τ.

(a) Prove the relation

$$\left\langle \sum_{j=1}^{N} \mathbf{r}_j \cdot \nabla_j \mathcal{H} \right\rangle = \left\langle \sum_{j=1}^{N} \mathbf{p}_j \cdot \nabla_{p_j} \mathcal{H} \right\rangle$$

where the period average is defined by

$$\langle A \rangle = \int_0^{\tau} \frac{A \, dt}{\tau}.$$

(Hint: Differentiate the function, $G = \sum_{j=1}^{N} \mathbf{p}_j \cdot \mathbf{r}_j$, with respect to time.)

(b) The negative of the left side of the result above is known as the *virial* of Clausius, $\langle \mathfrak{B} \rangle$. Show that the right side is equal to twice the kinetic energy. This establishes the virial theorem

$$-\langle \mathfrak{B} \rangle = 2\langle T \rangle.$$

(c) Show that for a single particle in a potential $V = Ar^n$ the virial theorem takes the form

$$\langle T \rangle = \tfrac{1}{2} n \langle V \rangle.$$

(d) Show by a *direct* calculation that for a one-dimensional simple harmonic oscillator the virial theorem gives $\langle T \rangle = \langle V \rangle$.

2-4. Consider the case of a charged particle in the presence of an electromagnetic field.

(a) Show using the Lagrangian given by (2-58) that the momentum canonically conjugate to **x** is

$$p_x = mv_x + \frac{qA_x}{c}.$$

(b) Show also that the momentum conjugate to the azimuthal angle ϕ is

$$p_\phi = (\mathbf{r} \times m\mathbf{v})_z + \left(\mathbf{r} \times \frac{q}{c}\mathbf{A}\right)_z.$$

(c) From the Lagrangian in (2-58) use a Legendre transformation to obtain the corresponding Hamiltonian.

2-5. (a) Write the Hamiltonian for two electrons in the helium atom, accounting for their attraction to the nucleus and for their mutual repulsion.

(b) Show that the total angular momentum of the electrons, $\mathbf{L} = \mathbf{L}_1 + \mathbf{L}_2$, is conserved by demonstrating that

$$\{\mathcal{H}, \mathbf{L}\} = 0.$$

2-6. Verify that the components of angular momentum satisfy

$$\{L_i, L_j\} = L_k, \qquad \{L^2, L_i\} = 0.$$

2-7. (a) Apply Lagrange's equations and Hamilton's equations respectively to (2-84) and (2-85) and verify that Newton's law follows.

(b) Show that

$$\mathcal{L}_{rel} \xrightarrow[v/c \to 0]{} \mathcal{L}_{nonrel} - mc^2$$

$$\mathcal{H}_{rel} \xrightarrow[v/c \to 0]{} \mathcal{H}_{nonrel} + mc^2.$$

3 | The Formalism of Quantum Mechanics

The early experiments of modern physics suggested that the assignment of only certain permissible values to canonical variables such as energy and angular momentum was necessary to explain the fundamental behavior of matter and radiation. However the origin of quantization was not really understood until 1925. The formal theory of quantum mechanics was to predict more than just the characteristic values of measurable canonical functions or *observables*. It was also to contain the format with which to derive all possible information from a mechanical system. As the theory developed, the correspondence between classical and quantum mechanics became more apparent.

The modern form of quantum theory was formulated in different but equivalent ways by E. Schroedinger, W. Heisenberg, and P. A. M. Dirac. The approach used here is due to Dirac; the relationship with the Schroedinger formulation will be established later. Dirac's representation of quantum theory requires some knowledge of a complex linear vector space. Some of the mathematics that will be required for our discussion is presented below.

I   Vectors in a Complex, N-Dimensional, Linear Space

A vector in a real three-dimensional space is characterized by a magnitude and a direction. To represent the vector it is convenient to introduce an *orthonormal basis* consisting of three mutually orthogonal vectors, each of unit length. It is customary to label the vectors associated with the x, y, and z axes as **i**, **j**, and **k** respectively. These vectors have the orthonormal properties

$$\mathbf{i} \cdot \mathbf{i} = \mathbf{j} \cdot \mathbf{j} = \mathbf{k} \cdot \mathbf{k} = 1$$
$$\mathbf{i} \cdot \mathbf{j} = \mathbf{i} \cdot \mathbf{k} = \mathbf{j} \cdot \mathbf{k} = 0.$$

If the space is linear, then an arbitrary vector may be represented as a linear combination of the basis vectors, that is,

$$\mathbf{a} = a_x \mathbf{i} + a_y \mathbf{j} + a_z \mathbf{k}. \tag{3-1}$$

The quantities a_x, a_y, and a_z are called the components of the vector **a** in the basis **i**, **j**, **k**. It should be stressed that while **a** is unique, a_x, a_y, and a_z are arbitrary varying with the basis used. Using the properties of the basis vectors we find

$$a_x = \mathbf{a} \cdot \mathbf{i}, \qquad a_y = \mathbf{a} \cdot \mathbf{j}, \qquad \text{and} \qquad a_z = \mathbf{a} \cdot \mathbf{k}.$$

In another basis the components have the form

$$a_x' = \mathbf{a} \cdot \mathbf{i}', \qquad a_y' = \mathbf{a} \cdot \mathbf{j}', \qquad \text{and} \qquad a_z' = \mathbf{a} \cdot \mathbf{k}'.$$

The length of the vector is defined by

$$|\mathbf{a}| = (a_x^2 + a_y^2 + a_z^2)^{1/2} = (a_x'^2 + a_y'^2 + a_z'^2)^{1/2}$$

and is the same regardless of the basis. Basis-invariant quantities are termed *scalars*.

The dot or *inner* product of two vectors is also a scalar and may be evaluated in terms of their components using

$$\mathbf{a} \cdot \mathbf{b} = a_x b_x + a_y b_y + a_z b_z = a_x' b_x' + a_y' b_y' + a_z' b_z'.$$

It is possible to display the components of a vector in a basis using a column matrix, that is,

$$\mathbf{a} = \begin{pmatrix} a_x \\ a_y \\ a_z \end{pmatrix}.$$

The dot product $\mathbf{b} \cdot \mathbf{a}$ may be displayed by writing the first vector as a row matrix and the second as a column matrix. Using the rules of matrix multiplication, we find

$$\mathbf{b} \cdot \mathbf{a} = (b_x, b_y, b_z)\begin{pmatrix} a_x \\ a_y \\ a_z \end{pmatrix} = a_x b_x + a_y b_y + a_z b_z. \tag{3-2}$$

It is clear that a vector can be expressed as either a row or a column matrix. We shall call the row display the *adjoint* or *dual* to the column. The inner product involves multiplication of a column matrix by its dual—a row matrix. In column notation the basis vectors always have the form

$$\mathbf{i} = \begin{pmatrix} 1 \\ 0 \\ 0 \end{pmatrix}, \qquad \mathbf{j} = \begin{pmatrix} 0 \\ 1 \\ 0 \end{pmatrix}, \qquad \mathbf{k} = \begin{pmatrix} 0 \\ 0 \\ 1 \end{pmatrix}. \tag{3-3}$$

Suppose now we generalize by assuming that the vectors of interest belong to an N-dimensional space. An arbitrary vector can be expanded in an orthonormal basis in analogy with (3-1) as

$$\mathbf{a} = \sum_{i=1}^{N} a^{(i)} \boldsymbol{\alpha}_i \tag{3-4}$$

where the orthonormality relations are

$$\boldsymbol{\alpha}_i \cdot \boldsymbol{\alpha}_j = \delta_{ij} \qquad (i, j = 1, \ldots, N). \tag{3-5}$$

The N basis vectors are displayed in column form as

$$\boldsymbol{\alpha}_1 = \begin{pmatrix} 1 \\ 0 \\ 0 \\ 0 \\ \vdots \\ 0 \end{pmatrix}, \qquad \boldsymbol{\alpha}_2 = \begin{pmatrix} 0 \\ 1 \\ 0 \\ 0 \\ \vdots \\ 0 \end{pmatrix}, \qquad \boldsymbol{\alpha}_3 = \begin{pmatrix} 0 \\ 0 \\ 1 \\ 0 \\ \vdots \\ 0 \end{pmatrix}, \qquad \boldsymbol{\alpha}_N = \begin{pmatrix} 0 \\ 0 \\ 0 \\ \vdots \\ 0 \\ 1 \end{pmatrix}. \tag{3-6}$$

The vector $\mathbf{a}$ is displayed in terms of its components as

$$\mathbf{a} = \begin{pmatrix} a^{(1)} \\ a^{(2)} \\ a^{(3)} \\ \vdots \\ a^{(N)} \end{pmatrix}. \tag{3-7}$$

Again it should be stressed that the vector $\mathbf{a}$ is fixed by its direction and magnitude but its components vary with the basis. A representation of a vector is meaningful only when the basis as well as the components in that basis are indicated.

Since the *N*-dimensional space is already abstract, we shall make one further generalization useful in quantum theory. We assume that the vectors of the space are complex. Equivalently, the components of a vector in a basis need not be real. It is convenient to redefine the adjoint of a column vector as a row vector with the corresponding elements complex conjugated. Mathematically we have

$$\text{adjoint of } \begin{pmatrix} a^{(1)} \\ a^{(2)} \\ \vdots \\ a^{(N)} \end{pmatrix} = (a^{*(1)}, a^{*(2)}, \ldots, a^{*(N)}). \tag{3-8}$$

We introduce the notation of Dirac and abbreviate the column matrix by

$$\mathbf{a} = |a\rangle = \begin{pmatrix} a^{(1)} \\ a^{(2)} \\ \vdots \\ a^{(N)} \end{pmatrix} \tag{3-9}$$

where $|a\rangle$ is called a *ket* vector. Similarly, the adjoint form of $|a\rangle$ is written

$$\langle a| = (a^{*(1)}, a^{*(2)}, \ldots, a^{*(N)}) \tag{3-10}$$

where $\langle a|$ is a *bra* vector. The inner product $\mathbf{b} \cdot \mathbf{a}$ is a complex scalar obtained by multiplying the bra form of $\mathbf{b}$ by the ket form of $\mathbf{a}$, that is,

$$c = \mathbf{b} \cdot \mathbf{a} = \langle b|a\rangle = (b^{*(1)}, b^{*(2)}, \ldots, b^{*(N)}) \begin{pmatrix} a^{(1)} \\ a^{(2)} \\ \vdots \\ a^{(N)} \end{pmatrix}$$

$$= \sum_{i=1}^{N} b^{*(i)} a^{(i)}. \tag{3-11}$$

The terms bra and ket originate with the fact that the inner product forms a " bra-ket " or bracket. Note that the dot product is no longer a commutative operation since a reversal in the order of multiplication leads to a complex conjugated inner product.

The adjoint of the bra vector is the ket and vice versa; mathematically, we have

$$|a\rangle^{\dagger} = \langle a| \qquad \text{and} \qquad \langle a|^{\dagger} = |a\rangle. \tag{3-12}$$

Using the rules of matrix multiplication, it can be shown that the adjoint of a product of elements is equal to the product, in reverse order, of the adjoints of the individual elements. Therefore,

$$c^{\dagger} = (\langle b|a\rangle)^{\dagger} = \langle a|b\rangle = c^{*}; \tag{3-13}$$

thus the adjoint of a scalar is simply equal to its complex conjugate. The inner product of a vector with itself is called the *norm* of the vector and is always real and positive, namely,

$$\langle a | a \rangle = \sum_{i=1}^{N} a^{*(i)} a^{(i)} = \sum_{i=1}^{N} |a^{(i)}|^2 \geqslant 0.$$

Using Dirac's notation the vector expansion in (3-4) may be written in ket form as

$$|a\rangle = \sum_{i=1}^{N} a^{(i)} |\alpha_i\rangle \qquad (3\text{-}14)$$

or in bra form as

$$\langle a| = \sum_{i=1}^{N} \langle \alpha_i | a^{*(i)}. \qquad (3\text{-}15)$$

The orthonormality relation for the basis vectors takes the form

$$\langle \alpha_i | \alpha_j \rangle = \delta_{ij} \qquad (3\text{-}16)$$

that is, the inner product of two different basis vectors is zero (orthogonality) and each has a norm of unity.

II Linear Operators

An operator in an N-dimensional complex space is defined by its action on the various vectors of the space. For example, in the relation

$$|b\rangle = \hat{A} |a\rangle \qquad (3\text{-}17)$$

the operator $\hat{A}$ is unique when a prescription is specified for obtaining $|b\rangle$ from any $|a\rangle$. We restrict our discussion to *linear* operators, that is, operators which have the property

$$\hat{A}[c|a\rangle + d|b\rangle] = c\hat{A}|a\rangle + d\hat{A}|b\rangle \qquad (3\text{-}18)$$

for arbitrary c, d, $|a\rangle$, and $|b\rangle$.

Expressing the vector $|a\rangle$ in the basis $|\alpha_i\rangle$, the operation $|b\rangle = \hat{A}|a\rangle$ becomes

$$|b\rangle = \hat{A} \sum_{i=1}^{N} a^{(i)} |\alpha_i\rangle = \sum_{i=1}^{N} a^{(i)} \hat{A} |\alpha_i\rangle. \qquad (3\text{-}19)$$

If the components of $|b\rangle$ in the basis are also known, then the effect of the operation is unique. Taking the inner product of $\langle \alpha_j |$ with both sides of (3-19), we obtain

$$\langle \alpha_j | b \rangle = b^{(i)} = \sum_{i=1}^{N} a^{(i)} \langle \alpha_j | \hat{A} | \alpha_i \rangle. \qquad (3\text{-}20)$$

The N^2 elements

$$A_{ji} = \langle \alpha_j | \hat{A} | \alpha_i \rangle$$

are called the *matrix elements* of the linear operator $\hat{A}$ in the α_i basis. Inspection of (3-20) reveals that the components of $|a\rangle$ and $|b\rangle$ satisfy the matrix relation

$$|b\rangle = \hat{A}|a\rangle \rightarrow \begin{pmatrix} b^{(1)} \\ \vdots \\ b^{(N)} \end{pmatrix} = \begin{pmatrix} A_{11} & \cdots & A_{1N} \\ \vdots & \vdots & \vdots \\ A_{N1} & \cdots & A_{NN} \end{pmatrix} \begin{pmatrix} a^{(1)} \\ \vdots \\ a^{(N)} \end{pmatrix}. \qquad (3\text{-}21)$$

Equivalently a linear operator may be displayed as an $N \times N$ matrix in a given N-dimensional basis. Again, while the operator $\hat{A}$ is *absolute*, that is, it is defined by its operation on the vectors of the space, its elements depend on the basis used in the representation.

An operator may operate toward the right on a ket or toward the left on a bra. Successive operations are symbolized by a product of operators. An operator product obeys the *associative* and *distributive* laws but not necessarily the *commutative* law, namely,

$$(\hat{A}\hat{B})\hat{C} = \hat{A}(\hat{B}\hat{C})$$
$$\hat{A}(\hat{B} + \hat{C}) = \hat{A}\hat{B} + \hat{A}\hat{C} \qquad (3\text{-}22)$$
$$\hat{A}\hat{B} \neq \hat{B}\hat{A}.$$

These rules conform to the well-known results for matrix multiplication.

The simplest operator is the *unit* or *identity* operator $\hat{1}$, which has the property

$$\hat{1}|a\rangle = |a\rangle$$

for all vectors of the space. Its matrix form in all bases is

$$\hat{1} = \begin{pmatrix} 1 & 0 & 0 & 0 & \cdots \\ 0 & 1 & 0 & 0 & \cdots \\ 0 & 0 & 1 & 0 & \cdots \\ 0 & 0 & 0 & 1 & \cdots \\ \vdots & \vdots & \vdots & \vdots & \ddots \end{pmatrix}.$$

The *inverse* of an operator A^{-1}, if it exists,‡ is defined by the operation

$$\hat{A}^{-1}\hat{A} = \hat{1}. \qquad (3\text{-}23)$$

In effect, the inverse A^{-1} undoes the original operation $\hat{A}$.

The *adjoint* of an operator $\hat{A}^\dagger$ may be defined in a variety of ways. The

‡ An operator whose determinant [see (3-29)] vanishes is said to be *singular* and does not have an inverse.

definition used here defines $\hat{A}^\dagger$ by the relation

$$|b\rangle = \hat{A}|a\rangle$$
$$\langle b| = \langle a|\hat{A}^\dagger \qquad \text{for arbitrary } |a\rangle \text{ and } |b\rangle.$$

In words, if $\hat{A}$ is the operator that generates the ket $|b\rangle$ from the ket $|a\rangle$, then $\hat{A}^\dagger$ is that operator which generates the bra $\langle b|$ from the bra $\langle a|$. Using this definition, it is possible to verify that

$$\langle \alpha_i|\hat{A}|\alpha_j\rangle = (\langle \alpha_j|\hat{A}^\dagger|\alpha_i\rangle)^*. \qquad (3\text{-}24)$$

In matrix form, the adjoint is obtained by interchanging rows and columns and taking the complex conjugate of the elements.

Two very important operators in quantum mechanics are the Hermitian operator and the unitary operator. An operator is *Hermitian* if it is equal to its own adjoint:

$$\hat{A} = \hat{A}^\dagger$$

or

$$\langle \alpha_i|\hat{A}|\alpha_j\rangle = (\langle \alpha_j|\hat{A}|\alpha_i\rangle)^*. \qquad (3\text{-}25)$$

It is *unitary* if its inverse is equal to its adjoint:

$$\hat{A}^\dagger = \hat{A}^{-1}$$

or using (3-23),

$$\hat{A}\hat{A}^\dagger = \hat{A}^\dagger\hat{A} = \hat{1}. \qquad (3\text{-}26)$$

Using the properties of operators, it is possible to verify the following:

$$(\hat{A}\hat{B})^{-1} = \hat{B}^{-1}\hat{A}^{-1}, \qquad (\hat{A}\hat{B})^\dagger = \hat{B}^\dagger\hat{A}^\dagger. \qquad (3\text{-}27)$$

Let us now examine the effect of a unitary operator $\hat{U}$ on a set of ortho-normal bases vectors $|\alpha_i\rangle$. We write

$$|\beta_i\rangle = \hat{U}|\alpha_i\rangle$$

or in adjoint (bra) form

$$\langle \beta_j| = \langle \alpha_j|\hat{U}^\dagger.$$

Taking the inner product and using the definition of a unitary operator, we find

$$\langle \beta_j|\beta_i\rangle = \langle \alpha_j|\hat{U}^\dagger\hat{U}|\alpha_i\rangle = \langle \alpha_j|\alpha_i\rangle = \delta_{ij};$$

thus the vectors $|\beta_i\rangle$ constitute another orthonormal basis. We may therefore associate $\hat{U}$ with a rotation and conclude that under a unitary operation all vectors maintain their lengths and relative orientations (inner products). More generally, all scalars are invariant under unitary operations (rotations).

The *trace* of an operator, written Tr $\hat{A}$, is a scalar and is therefore invariant under a change of basis. It may be evaluated in a basis by summing the diagonal elements of the matrix associated with $\hat{A}$:

$$\text{Tr } \hat{A} = \sum_{i=1}^{N} \langle \alpha_i | \hat{A} | \alpha_i \rangle = \sum_{i=1}^{N} A_{ii}. \tag{3-28}$$

The *determinant* of an operator is also a scalar and is obtained by evaluating the determinant of the elements:

$$|\hat{A}| = \det \hat{A} = \begin{vmatrix} \langle \alpha_1 | \hat{A} | \alpha_1 \rangle & \cdots & \langle \alpha_1 | \hat{A} | \alpha_N \rangle \\ \vdots & \vdots & \vdots \\ \langle \alpha_N | \hat{A} | \alpha_1 \rangle & \cdots & \langle \alpha_N | \hat{A} | \alpha_N \rangle \end{vmatrix}. \tag{3-29}$$

As we have already pointed out, the product of two operators is not in general commutative. The difference between the direct and commuted product is called the *commutator* and is mathematically expressed as

$$[\hat{A}, \hat{B}] = \hat{A}\hat{B} - \hat{B}\hat{A}. \tag{3-30}$$

It follows that if two operators commute, then their commutator is zero. We shall see below that the commutator of two operators plays a fundamental role in the mathematical development of quantum mechanics.

III Eigenvalues and Eigenvectors

Generally, the action of an operator on a vector changes its magnitude and direction. We would like to confine our attention for the moment to the class of vectors that change only in magnitude under the action of the given operator. This change in magnitude amounts to multiplying the vector by a constant; this leads us to write

$$\hat{A}|a_i\rangle = a_i|a_i\rangle. \tag{3-31}$$

The subscript is added to emphasize that there is more than one vector with this property. It will become clear that there are N linearly independent vectors which satisfy (3-31). The vectors $|a_i\rangle$ and the scalars a_i are termed respectively the eigenvectors (or eigenkets) and eigenvalues of the operator $\hat{A}$. The prefix "*eigen*" means "self" or "own" in German. Hence these quantities are *self* or *characteristic* vectors and values of the operator. The set of eigenvalues is called the *spectrum* of the operator. Since (3-31) does not restrict the magnitudes of the eigenvectors, it is convenient to normalize them, that is, to require

$$\langle a_i | a_i \rangle = 1.$$

If the elements of an operator are given in some orthonormal basis, then the eigenvectors and eigenvalues may be readily calculated. For example, consider the following matrix representation of the Hermitian operator $\hat{A}$ in a two-dimensional space:

$$\hat{A}^\dagger = \hat{A} = \begin{pmatrix} 0 & i \\ -i & 0 \end{pmatrix}.$$

The eigenvalue equation $\hat{A}|a_j\rangle = a_j|a_j\rangle$ $(j = 1, 2)$ may be expressed in matrix form as

$$\begin{pmatrix} 0 & i \\ -i & 0 \end{pmatrix}\begin{pmatrix} a_j^{(1)} \\ a_j^{(2)} \end{pmatrix} = a_j\begin{pmatrix} a_j^{(1)} \\ a_j^{(2)} \end{pmatrix} \qquad (j = 1, 2). \qquad (3\text{-}32)$$

We wish to find the two components of each of the eigenvectors along with their corresponding eigenvalues. Equation (3-32) is equivalent to a pair of homogeneous equations for the components $a_j^{(1)}$ and $a_j^{(2)}$. Nontrivial solutions $(a_j^{(1)} \neq 0 \neq a_j^{(2)})$ exist only for values of a_j which are the roots of the algebraic equation

$$\det(\hat{A} - \hat{1}a_j) = \begin{vmatrix} 0 - a_j & i \\ -i & 0 - a_j \end{vmatrix} = a_j^2 - 1 = 0.$$

The roots of this *secular* equation, $a_j = \pm 1$, determine the eigenvalues of the operator $\hat{A}$. Setting $a_1 = +1$ and $a_2 = -1$, the components of the corresponding eigenvectors $|a_1\rangle$ and $|a_2\rangle$ are obtained by direct substitution in (3-32). As we already mentioned, (3-31) determines only the direction of the eigenvectors, that is, the ratio of their components. We are at liberty to set any one (nonzero) component equal to unity. Using the first eigenvalue $a_1 = +1$ in (3-32) and setting $a_1^{(1)} = 1$, we find $a_1^{(2)} = -i$ and write

$$|a_1\rangle = \begin{pmatrix} 1 \\ -i \end{pmatrix} \qquad \text{for} \quad a_1 = +1.$$

The second eigenvector is found in a similar manner to be

$$|a_2\rangle = \begin{pmatrix} 1 \\ i \end{pmatrix} \qquad \text{for} \quad a_2 = -1.$$

These vectors have been represented in the same basis as the original matrix. Each of them has a norm of 2 so that multiplication by $1/\sqrt{2}$ effects normalization. It is simple to verify that the two eigenvectors are mutually orthogonal. This result along with the fact that the eigenvalues are real is no coincidence, but follows from the Hermitian character of the original operator, as we shall show below.

The above example may be somewhat misleading in that it implies that an operator must be represented as a matrix in some basis before its eigen-

values can be found. It must be stressed that the eigenvalues and the eigenvectors are characteristic to the operator and are *independent* of the basis used
to calculate them. In certain cases where a matrix display is not feasible, it
is still possible to deduce the eigenvalues and eigenvectors without using a
specific basis. This is especially important when the dimensionality of the
space N is large.

IV   Eigenvalue–Eigenvector Algebra for Hermitian Operators

The theorems to be discussed below are quite independent of any basis
that might be used in a representation. They are of the utmost importance in
quantum theory and merit special consideration on the part of the student.

Theorem I   *The eigenvalues of a Hermitian operator are all real.*

Proof   Let $\hat{A}$ be Hermitian and let $|a_j\rangle$ and a_j denote its eigenvectors and
eigenvalues. The eigenvalue equation in ket form is

$$\hat{A}|a_j\rangle = a_j|a_j\rangle \qquad (j = 1, \ldots, N), \tag{3-33}$$

and in bra form,

$$\langle a_k|\hat{A}^\dagger = \langle a_k|a_k^* \qquad (k = 1, \ldots, N). \tag{3-34}$$

Taking the inner product of (3-33) with $\langle a_k|$ and of (3-34) with $|a_j\rangle$, and then
subtracting the results, we obtain

$$\langle a_k|\hat{A}^\dagger|a_j\rangle - \langle a_k|\hat{A}|a_j\rangle = (a_k^* - a_j)\langle a_k|a_j\rangle.$$

Since the operator has been assumed Hermitian ($\hat{A} = \hat{A}^\dagger$), the left-hand side
vanishes and we find

$$(a_k^* - a_j)\langle a_k|a_j\rangle = 0. \tag{3-35}$$

Setting $k = j$ and noting that the norm of an eigenvector cannot be zero, we
obtain

$$(a_j^* - a_j) = 0 \qquad \text{or} \qquad a_j^* = a_j$$

which can only be satisfied if a_j is real.

Definition   *An operator is **degenerate** if two or more of its eigenvalues are equal;
otherwise it is **nondegenerate**.*

Theorem II   *The eigenvectors of any nondegenerate Hermitian operator are mutually orthogonal.*

Proof   Using (3-35), we set $k \neq j$ and obtain

$$(a_k - a_j)\langle a_k | a_j \rangle = 0. \tag{3-36}$$

The eigenvalues a_k have already been proven real $(a_k^* = a_k)$. Using the assumption that $\hat{A}$ is nondegenerate $(a_k \neq a_j$ for $k \neq j)$ we obtain the desired result,

$$\langle a_k | a_j \rangle = \delta_{kj}.$$

We have used the normalization convention for the eigenvectors. We are led to the conclusion that the eigenvectors of a nondegenerate Hermitian operator are unique and constitute an orthonormal basis.

If the Hermitian operator is degenerate, the proof in Theorem II fails. The orthogonality of two eigenvectors is assured only if the corresponding eigenvalues are unequal. In the case of a degeneracy, we need only consider those eigenvectors associated with the equal (degenerate) eigenvalues; the properties of the other eigenvectors are covered by Theorem II.

Lemma 1   *The eigenvectors corresponding to the equal eigenvalues of a degenerate Hermitian operator are not unique.*

Proof   We illustrate this lemma with a twofold degeneracy. Assume that

$$\hat{A}|a_1\rangle = a_1|a_1\rangle \tag{3-37a}$$

and

$$\hat{A}|a_2\rangle = a_2|a_2\rangle \tag{3-37b}$$

with $a_1 = a_2 = a$. It may be verified that the linear combinations

$$|a_1'\rangle = \alpha|a_1\rangle + \beta|a_2\rangle \tag{3-38a}$$

and   (arbitrary $\alpha, \beta, \gamma, \delta$)

$$|a_2'\rangle = \gamma|a_1\rangle + \delta|a_2\rangle \tag{3-38b}$$

are new eigenvectors, both with the eigenvalue a. Direct substitution in (3-37) gives

$$\hat{A}|a_1'\rangle = \alpha\hat{A}|a_1\rangle + \beta\hat{A}|a_2\rangle = a(\alpha|a_1\rangle + \beta|a_2\rangle) = a|a_1'\rangle.$$

A similar result is obtained with $|a_2'\rangle$. By varying α, β, γ, and δ we generate new eigenvectors associated with the degenerate eigenvalue a.

Lemma 2 *Not all eigenvectors associated with equal eigenvalues need be orthogonal.*

Proof This follows from Lemma 1 as it is always possible to choose α, β, γ, and δ so that

$$\langle a_1'|a_2'\rangle \neq 0.$$

It is therefore possible to find N linearly independent eigenvectors of a degenerate Hermitian operator which are not all mutually orthogonal.

Lemma 3 *There exists at least one set of N orthogonal eigenvectors for any degenerate Hermitian operator.*

Proof The proof is essentially by construction. We merely point out that if $|a_1\rangle$ and $|a_2\rangle$ were not originally orthogonal in (3-37), that is, $\langle a_1|a_2\rangle = c \neq 0$, then it is always possible to choose α, β, γ, and δ in (3-38) in such a way that

$$\langle a_1'|a_2'\rangle = 0.$$

The construction of an orthonormal set of eigenvectors is known as the Schmidt orthogonalization procedure.

The results of the previous theorems and lemmas are summarized by the following theorem:

Theorem III *Every Hermitian operator has at least one set of N orthonormal eigenvectors. If the operator is nondegenerate, this set is unique. Otherwise there are many distinct orthonormal sets. (Only those eigenvectors corresponding to nondegenerate eigenvalues are unique.) In the degenerate case, we must specify which orthonormal set we are considering.*

Since every Hermitian operator in an N-dimensional space has at least one set of N mutually orthogonal eigenvectors, we may use that set as a basis with which to represent all the other vectors and operators of the space. The basis composed of an orthonormal set of eigenvectors of the operator $\hat{A}$ is referred to as the $\hat{A}$ *eigenbasis*. The eigenvectors become basis vectors and have the matrix form

$$|a_1\rangle = \begin{pmatrix} 1 \\ 0 \\ 0 \\ 0 \\ 0 \\ \vdots \end{pmatrix}, \qquad |a_2\rangle = \begin{pmatrix} 0 \\ 1 \\ 0 \\ 0 \\ 0 \\ \vdots \end{pmatrix}, \qquad |a_3\rangle = \begin{pmatrix} 0 \\ 0 \\ 1 \\ 0 \\ 0 \\ \vdots \end{pmatrix}, \qquad \text{etc.}$$

The $\hat{A}$ matrix in its own eigenbasis is

$$\hat{A} = \begin{pmatrix} a_1 & 0 & 0 & \cdots & \\ 0 & a_2 & 0 & \cdots & \\ 0 & 0 & a_3 & \cdots & \\ \vdots & \vdots & \vdots & & \ddots & \\ & & & & & a_N \end{pmatrix}.$$

This form of $\hat{A}$ assures that the relation $\hat{A}|a_i\rangle = a_i|a_i\rangle$ will hold in the basis. The transformation from an arbitrary basis to the $\hat{A}$ eigenbasis always results in the *diagonalization* of the $\hat{A}$ matrix. In the eigenbasis, the matrix contains only diagonal elements, each of which is a different eigenvalue of $\hat{A}$.

V   The Commutator and the Eigenvalue Problem

Consider the eigenbases of two Hermitian operators $\hat{A}$ and $\hat{B}$. We show that the relative orientations of the $\hat{A}$ and $\hat{B}$ eigenbases are fundamentally determined by the commutator of the operators. The following theorems for Hermitian operators will establish that relationship.

Theorem IV   *If two operators have a common set of eigenvectors, then their commutator is zero, that is, they commute.*

Proof   Let a set of N orthonormal vectors, denoted by $|a_i, b_i\rangle$, be simultaneous eigenvectors of the operators $\hat{A}$ and $\hat{B}$ with respective eigenvalues a_i and b_i, that is,

$$\begin{aligned} \hat{A}|a_i, b_i\rangle &= a_i|a_i, b_i\rangle \\ \hat{B}|a_i, b_i\rangle &= b_i|a_i, b_i\rangle. \end{aligned} \tag{3-39}$$

Operating on the first equation with $\hat{B}$ and on the second with $\hat{A}$ and subtracting, we obtain

$$(\hat{B}\hat{A} - \hat{A}\hat{B})|a_i, b_i\rangle = (b_i a_i - a_i b_i)|a_i, b_i\rangle = 0.$$

Since the left-hand side must vanish for each of the eigenvectors, $|a_i, b_i\rangle$, we have

$$(\hat{B}\hat{A} - \hat{A}\hat{B}) = [\hat{B}, \hat{A}] = 0. \tag{3-40}$$

Corollary   *If two operators **do not** commute, they **cannot** have a common eigenbasis.*

Theorem V *If two nondegenerate operators commute, then they have a common set of eigenvectors, that is, any eigenvector of the first must also be an eigenvector of the second.*

Proof Let the respective eigenvalue equations of $\hat{A}$ and $\hat{B}$ be

$$\hat{A}|a_i\rangle = a_i|a_i\rangle, \qquad \hat{B}|b_i\rangle = b_i|b_i\rangle$$

and let

$$[\hat{A}, \hat{B}] = (\hat{A}\hat{B} - \hat{B}\hat{A}) = 0.$$

Operating with $\hat{B}$ on the left side of the first equation and using the commutation relation, we obtain

$$\hat{B}\hat{A}|a_i\rangle = \hat{A}\hat{B}|a_i\rangle.$$

This may be written, using the first eigenvalue equation, as

$$\hat{A}\{\hat{B}|a_i\rangle\} = \hat{B}\{\hat{A}|a_i\rangle\} = a_i\{\hat{B}|a_i\rangle\}.$$

Comparing the right and left sides we conclude that the vector in brackets must at worst be proportional to the normalized eigenvector of $\hat{A}$ with the eigenvalue a_i. We therefore write

$$\hat{B}|a_i\rangle = b_i|a_i\rangle \tag{3-41}$$

where b_i is introduced as a proportionality constant. Since both operators have been assumed to be nondegenerate, (3-41) implies that $|a_i\rangle$ is also an eigenvector of $\hat{B}$ with eigenvalue b_i, and we set

$$|a_i\rangle = |b_i\rangle = |a_i, b_i\rangle.$$

Thus we have established that the unique eigenbases of two nondegenerate, commuting, Hermitian operators are identical. If either operator is degenerate, the proof in Theorem V breaks down in the last step. In that case, the vector $\hat{B}|a_i\rangle$ may be a linear combination of the degenerate eigenvectors of $\hat{A}$.

Theorem VI *Any two commuting operators have at least one common set of orthonormal eigenvectors, even if they are degenerate.*

The proof of this theorem is by construction and will not be given here. It suffices to say that if $\hat{A}$ and $\hat{B}$ commute and are degenerate, then each has many distinct eigenbases and that not all eigenbases of $\hat{A}$ coincide with those of $\hat{B}$. *However, there must exist at least one eigenbasis common to both operators if they commute.*

Theorems IV–VI imply that any set of commuting operators share at least one common eigenbasis. The matrices for these operators appear simultaneously diagonal in this common basis.

VI The Projection Operator

If an arbitrary vector can be represented in a given orthonormal basis, then we say the basis is *complete*. For example, the unit vectors **i** and **j** are *incomplete* for representing vectors in a real three dimensional space; the vector **k** is required for representing the z component. It seems plausible that if a vector belongs to an N-dimensional space, any set of N mutually orthogonal unit vectors constitute a complete basis, that is, the vectors *span* the space. It is however useful to express this completeness property in a mathematical form. With each vector $|\alpha_j\rangle$ of an orthonormal basis, we introduce a projection operator $\hat{P}_j$. The effect of $\hat{P}_j$ on an arbitrary vector $|a\rangle$ in the space is to produce a new vector whose direction is along the basis vector $|\alpha_j\rangle$ and whose magnitude is $\langle\alpha_j|a\rangle$ (Figure 3-1).

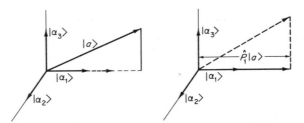

Figure 3-1 Effect of the projection operator $\hat{P}_1$ (associated with the basis vector $|\alpha_1\rangle$) on a vector $|a\rangle$.

Mathematically the effect of the projection operation is

$$\hat{P}_j|a\rangle = (\langle\alpha_j|a\rangle)|\alpha_j\rangle = |\alpha_j\rangle(\langle\alpha_j|a\rangle). \qquad (3\text{-}42)$$

Using the associative law, the right and left sides are related by

$$\hat{P}_j|a\rangle = (|\alpha_j\rangle\langle\alpha_j|)|a\rangle.$$

However, since $|a\rangle$ is completely arbitrary, we find

$$\hat{P}_j = |\alpha_j\rangle\langle\alpha_j|. \qquad (3\text{-}43)$$

The multiplication of a bra by a ket from the left is termed a *dyad* product and corresponds to multiplication of a row by a column matrix, from the left. If the vector $|\alpha_j\rangle$ is expressed in a second basis as

$$|\alpha_j\rangle = \begin{pmatrix} \alpha_j^{(1)} \\ \vdots \\ \alpha_j^{(N)} \end{pmatrix} \quad \text{and} \quad \langle\alpha_j| = (\alpha_j^{*(1)} \cdots \alpha_j^{*(N)}),$$

then $\hat{P}_j$ is represented by the Hermitian matrix

$$\hat{P}_j = |\alpha_j\rangle\langle\alpha_j| \rightarrow \begin{pmatrix} \alpha_j^{(1)} \\ \vdots \\ \alpha_j^{(N)} \end{pmatrix}(\alpha_j^{*(1)} \cdots \alpha_j^{*(N)}) = \begin{pmatrix} \alpha_j^{(1)}\alpha_j^{*(1)} \cdots \alpha_j^{(1)}\alpha_j^{*(N)} \\ \alpha_j^{(N)}\alpha_j^{*(1)} \cdots \alpha_j^{(N)}\alpha_j^{*(N)} \end{pmatrix}. \quad (3\text{-}44)$$

Note that when $\hat{P}_j$ operates on $|\alpha_j\rangle$, it reproduces that vector. On the other hand, since the other basis vectors are orthogonal to $|\alpha_j\rangle$ a projection operation on them gives zero. The basis vectors are therefore eigenvectors of $\hat{P}_j$ with the property

$$\hat{P}_k|\alpha_j\rangle = \delta_{kj}|\alpha_j\rangle \quad (j, k = 1, \ldots, N).$$

Since the N projection operators have a common eigenbasis $(|\alpha_j\rangle)$, they mutually commute (Theorem IV). In this basis all are diagonal and have the matrix form

$$\hat{P}_1 = \begin{pmatrix} 1 & 0 & 0 & 0 & \cdots \\ 0 & 0 & 0 & 0 & \cdots \\ 0 & 0 & 0 & 0 & \cdots \\ 0 & 0 & 0 & 0 & \cdots \\ \vdots & \vdots & \vdots & \vdots & \ddots \end{pmatrix}, \quad \hat{P}_2 = \begin{pmatrix} 0 & 0 & 0 & 0 & \cdots \\ 0 & 1 & 0 & 0 & \cdots \\ 0 & 0 & 0 & 0 & \cdots \\ 0 & 0 & 0 & 0 & \cdots \\ \vdots & \vdots & \vdots & \vdots & \ddots \end{pmatrix},$$

$$\hat{P}_N = \begin{pmatrix} 0 & 0 & 0 & 0 & \cdots \\ 0 & 0 & 0 & 0 & \cdots \\ 0 & 0 & 0 & 0 & \cdots \\ 0 & 0 & 0 & 0. & \cdots \\ \vdots & \vdots & \vdots & \vdots & \cdot 1 \end{pmatrix}.$$

Finally, the basis is complete if any arbitrary vector of the space can be represented as

$$|a\rangle = \sum_{j=1}^{N} (\langle\alpha_j|a\rangle)|\alpha_j\rangle$$

$$= \sum_{j=1}^{N} |\alpha_j\rangle\langle\alpha_j||a\rangle = \sum_{j=1}^{N} \hat{P}_j|a\rangle. \quad (3\text{-}45)$$

Since the expansion is valid for arbitrary $|a\rangle$, the completeness (or closure) relation becomes

$$\sum_{j=1}^{N} \hat{P}_j = \sum_{j=1}^{N} |\alpha_j\rangle\langle\alpha_j| = \hat{1} \quad \text{(completeness relation)}. \quad (3\text{-}46)$$

We shall always assume that the N orthonormal eigenvectors of a Hermitian operator are complete and satisfy (3-46).

VII   The Postulates of Quantum Mechanics

Having reviewed the necessary properties of operators and vectors in a linear space, we can now develop the foundations of quantum mechanics in Dirac notation.

Postulate 1‡   With every dynamical observable (for example, energy, linear momentum, angular momentum), we associate a linear Hermitian operator, in a complex space, whose properties dictate the quantum nature of that observable. (The dimensionality of the space will be discussed presently.)

Thus with the canonical Cartesian variables $\mathbf{r}$ and $\mathbf{p}$, we make the associations

$$\mathbf{r} \to \hat{\mathbf{r}} = \hat{\mathbf{r}}^{\dagger}, \qquad \mathbf{p} \to \hat{\mathbf{p}} = \hat{\mathbf{p}}^{\dagger}. \tag{3-47}$$

With any algebraic function of these variables we associate the corresponding operator function. For example, the operators associated with the kinetic energy and the components of angular momentum of a particle become

$$\hat{T} = \frac{(\hat{p}_x^{\,2} + \hat{p}_y^{\,2} + \hat{p}_z^{\,2})}{2m} \qquad (\hat{p}_x^{\,2} = \hat{p}_x \hat{p}_x)$$

and

$$\hat{L}_x = \hat{y}\hat{p}_z - \hat{z}\hat{p}_y, \qquad \hat{L}_y = \hat{z}\hat{p}_x - \hat{x}\hat{p}_z, \qquad \hat{L}_z = \hat{x}\hat{p}_y - \hat{y}\hat{p}_x.$$

Note that any transcendental operator function must be defined by its Taylor series. For example, the exponential operator function is *defined* by the infinite series

$$e^{\hat{x}} = \sum_{n=0}^{\infty} \frac{1}{n!} \hat{x}^n.$$

Postulate 2   The commutation relations between operators associated with two classical observables is deduced from the Poisson bracket of those observables using the correspondence

$$\{A, B\} \to \frac{[\hat{A}, \hat{B}]}{i\hbar}. \tag{3-48}$$

---

‡ The postulates listed are not necessarily in order of fundamental importance but rather in an order that is believed to be pedagogically best.

We shall see below that this correspondence leads to quantum equations of motion that are consistent with those derived in classical mechanics. Note that this correspondence formally introduces Planck's constant, the fundamental constant of quantum theory. The relationship between the Poisson bracket and the commutator however also introduces the imaginary number i. As we shall see, much of the mathematics of quantum theory involves complex functions.

From the Poisson bracket relations for canonical variables (2-77), we obtain the fundamental commutation relations for the position and momentum operators:

$$[\hat{x}_i, \hat{p}_j] = i\hbar \hat{1} \, \delta_{ij}$$

$$[\hat{x}_i, \hat{x}_j] = [\hat{p}_i, \hat{p}_j] = 0 \qquad (x_1 = x, \quad x_2 = y, \quad x_3 = z). \qquad (3\text{-}49)$$

It follows that if the Poisson bracket of two observables vanishes, then the corresponding commutator is zero and the operators commute.

Postulate 3 The quantum values allowed to any observable are determined by the eigenvalues of the corresponding operator. Any measurement of the system yields one of these values.

For example, let the operator $\mathcal{H}$ correspond to the Hamiltonian (energy) of the system. Any measurement of energy can result in one of the eigenvalues of $\mathcal{H}$ which we call ε_i. Thus once the properties of the operator $\hat{A}$ associated with an observable have been established, the eigenvalues a_i determine the quantized values of that observable. Since these eigenvalues are directly measurable, they must be real. The Hermitian nature of operator observables assures the reality of these eigenvalues (Theorem I).

Postulate 4 The state of any physical system is characterized by a *state vector* of *unit length* in a complex space. Futhermore, if the system is characterized by a vector $|\beta\rangle$ and a measurement of an observable $\hat{A}$ is performed, then the *probability* of observing the system with the characteristic value a_i is given by

$$\mathscr{P}_\beta(a_i) = |\langle a_i|\beta\rangle|^2 \leqslant 1 \qquad (\langle\beta|\beta\rangle = 1). \qquad (3\text{-}50)$$

Postulate 4 is quite remarkable in that it restricts the precision with which we can describe the state of a mechanical system. Equation (3-50) implies that the probabilistic picture of the system is determined by the relative orientation of the state vector with respect to the various eigenvectors of operator observables. Note that the probabilities involve scalars (inner products) and are therefore independent of any basis used in the representation. Summing over all possible values of a_i, we write

$$\sum_i \mathscr{P}_\beta(a_i) = \sum_i |\langle a_i|\beta\rangle|^2$$

$$= \sum_i \langle\beta|a_i\rangle\langle a_i|\beta\rangle = \langle\beta|\left\{\sum_i |a_i\rangle\langle a_i|\right\}|\beta\rangle.$$

Assuming the $\hat{A}$ eigenbasis to be complete, $\sum_i |a_i\rangle\langle a_i| = \hat{1}$, we find

$$\sum_i \mathscr{P}_\beta(a_i) = \langle\beta|\beta\rangle = 1.$$

The normalization requirement on the state vector $|\beta\rangle$ amounts therefore to a normalization of probability,

An alternative way of interpreting the probabilistic aspects of quantum theory is to assume that the system being studied is actually composed of many identical microscopic quantum subsystems. If the system is characterized by $|\beta\rangle$, then

$$\mathscr{P}_\beta(a_i) = |\langle a_i|\beta\rangle|^2 \tag{3-51}$$

can be interpreted as the fraction of subsystems with the characteristic value a_i.

The average or *expectation* value of many measurements of an observable $\hat{A}$, when the system is characterized by $|\beta\rangle$, is given by

$$\overline{A_\beta} = \sum_i \mathscr{P}_\beta(a_i)a_i \Big/ \sum_i \mathscr{P}_\beta(a_i). \tag{3-52}$$

Since the probability is assumed normalized, the denominator is equal to unity. Note that this expectation value is always real for real a_i, that is, for Hermitian operators.

It is convenient to rewrite the expectation value using

$$\overline{A_\beta} = \sum_i \mathscr{P}_\beta(a_i)a_i = \sum_i (\langle\beta|a_i\rangle)(\langle a_i|\beta\rangle)a_i = \sum_i \langle\beta|a_i|a_i\rangle\langle a_i|\beta\rangle.$$

However since $\hat{A}|a_i\rangle = a_i|a_i\rangle$, we find

$$\overline{A_\beta} = \sum_i \langle\beta|\hat{A}|a_i\rangle\langle a_i|\beta\rangle = \langle\beta|\hat{A}\left\{\sum_i |a_i\rangle\langle a_i|\right\}|\beta\rangle.$$

Assuming the completeness of the $\hat{A}$ eigenbasis, $\sum_i |a_i\rangle\langle a_i| = \hat{1}$, the expectation value becomes finally,

$$\overline{A_\beta} = \langle\beta|\hat{A}|\beta\rangle.$$

We adopt the new notation

$$\overline{A_\beta} = \langle A\rangle_\beta = \langle\beta|\hat{A}|\beta\rangle. \tag{3-53}$$

The expectation value of an observable $\hat{A}$ in the state $|\beta\rangle$ is obtained by making a "sandwich" with the bra and ket state vectors on the outside and the operator in the middle.

To describe the degree to which a random measurement deviates from the expectation value, we define the *uncertainty* or root-mean-square deviation of a measurement of A (when the system is characterized by $|\beta\rangle$) by

$$\Delta A_\beta = [\langle\beta|(\hat{A} - \langle A\rangle_\beta)^2|\beta\rangle]^{1/2}$$
$$= [\langle\beta|\hat{A}^2|\beta\rangle - 2\langle A\rangle_\beta\langle\beta|\hat{A}|\beta\rangle + \langle A\rangle_\beta^2]^{1/2}$$
$$= [\langle A^2\rangle_\beta - \langle A\rangle_\beta^2]^{1/2}. \tag{3-54}$$

The uncertainty is the square root of the difference between the average square and the square of the average of $\hat{A}$.

Postulate 5‡   If the state vector of a system is coincident with an eigenvector $|a_i\rangle$ of an operator $\hat{A}$, then the system is said to be in an *eigenstate* of the operator. The probability that a measurement will yield the value a_i for the observable $\hat{A}$ is unity, as can be seen from

$$|\langle a_i|\beta\rangle|^2 = |\langle a_i|a_i\rangle|^2 = 1.$$

The expectation value of $\hat{A}$ in one of its eigenstates is trivially

$$\langle A\rangle_i = \langle a_i|\hat{A}|a_i\rangle = a_i.$$

Furthermore since§
$$\hat{A}^2|a_i\rangle = \hat{A}\hat{A}|a_i\rangle = a_i^2|a_i\rangle$$

we find the uncertainty of the measurement to be

$$\langle A^2\rangle_i - \langle A\rangle^2 = a_i^2 - a_i^2 = 0.$$

When a system is in an eigenstate of an operator observable, then that canonical observable is said to be *well-defined* or *precise*.

The uncertainties associated with simultaneous measurements of two observables $\hat{A}$ and $\hat{B}$ are related to the commutator of the operators. If $\hat{A}$ and $\hat{B}$ commute, it follows from Theorem VI that they have at least one set of common eigenstates. In any of these states both $\hat{A}$ and $\hat{B}$ can be measured with perfect precision, that is, with zero uncertainty (Postulate 5). On the other hand, if $\hat{A}$ and $\hat{B}$ do *not* commute, no common eigenstates exist (Corollary, Theorem IV). Consequently, no physical states exist in which A and B can be

---

‡ This is not a new postulate, but essentially follows from Postulate 4. Because of its importance, we list it separately.

§ The relationship $\hat{A}^2|a_i\rangle = a_i^2|a_i\rangle$ may be generalized to $\hat{A}^n|a_i\rangle = a_i^n|a_i\rangle$; using the definition of a transcendental operator function, we find $\hat{F}(\hat{A})|a_i\rangle = F(a_i)|a_i\rangle$.

measured simultaneously with perfect precision. For this reason commuting operators are termed *compatible*. The discussion above is formulated mathematically by the following theorem.

Theorem VII (Heisenberg's Uncertainty Principle) *If three Hermitian operator observables satisfy the commutation relation*

$$[\hat{A}, \hat{B}] = i\hat{C} \tag{3-55}$$

then regardless of the state of the system, $|\beta\rangle$, the results of measurements of these observables must conform to Heisenberg's inequality

$$\Delta A_\beta \, \Delta B_\beta \geqslant \tfrac{1}{2}\langle C\rangle_\beta. \tag{3-56}$$

Proof We begin by using Schwarz's inequality:

$$\langle \gamma|\gamma\rangle\langle \delta|\delta\rangle \geqslant |\langle \gamma|\delta\rangle|^2 \qquad \text{for any vectors } |\gamma\rangle \text{ and } |\delta\rangle. \tag{3-57}$$

A proof of this inequality may be found in most standard texts on linear vector spaces (see Problem 3-6). In a real three-dimensional space, Schwarz's inequality takes the form

$$a^2b^2 \geqslant (\mathbf{a}\cdot\mathbf{b})^2 = a^2b^2\cos^2\theta.$$

The inequality is certainly valid in real space since $\cos^2\theta \leqslant 1$. Proceeding with the proof of the theorem, we set

$$|\gamma\rangle = (\hat{A} - \langle A\rangle_\beta)|\beta\rangle$$
$$|\delta\rangle = (\hat{B} - \langle B\rangle_\beta)|\beta\rangle$$

so that

$$\langle \gamma|\gamma\rangle = \langle \beta|(\hat{A} - \langle A\rangle_\beta)(\hat{A} - \langle A\rangle_\beta)|\beta\rangle = \langle \beta|(\hat{A} - \langle A\rangle_\beta)^2|\beta\rangle = (\Delta A_\beta)^2$$

and

$$\langle \delta|\delta\rangle = \langle \beta|(\hat{B} - \langle B\rangle_\beta)(\hat{B} - \langle B\rangle_\beta)|\beta\rangle = \langle \beta|(\hat{B} - \langle B\rangle_\beta)^2|\beta\rangle = (\Delta B_\beta)^2.$$

Using Schwarz's inequality we find

$$(\Delta A_\beta)^2(\Delta B_\beta)^2 = \langle \gamma|\gamma\rangle\langle \delta|\delta\rangle$$
$$\geqslant |\langle \gamma|\delta\rangle|^2 = |\langle \beta|(\hat{A} - \langle A\rangle_\beta)(\hat{B} - \langle B\rangle_\beta)|\beta\rangle|^2. \tag{3-58}$$

We decompose the non-Hermitian‡ operator $\hat{G} = (\hat{A} - \langle A\rangle_\beta)(\hat{B} - \langle B\rangle_\beta)$ into

$$\hat{G} = \hat{D} + \tfrac{1}{2}i\hat{C}$$

---

‡ The product of two Hermitian operators is non-Hermitian unless they commute since
$$(\hat{A}\hat{B})^\dagger = (\hat{B}^\dagger\hat{A}^\dagger) = (\hat{B}\hat{A}) \neq (\hat{A}\hat{B}).$$

where

$$\hat{D} = \frac{(\hat{A} - \langle A \rangle_\beta)(\hat{B} - \langle B \rangle_\beta) + (\hat{B} - \langle B \rangle_\beta)(\hat{A} - \langle A \rangle_\beta)}{2}$$

and

$$\hat{C} = \frac{(\hat{A} - \langle A \rangle_\beta)(\hat{B} - \langle B \rangle_\beta) - (\hat{B} - \langle B \rangle_\beta)(\hat{A} - \langle A \rangle_\beta)}{i} = \frac{[\hat{A}, \hat{B}]}{i}.$$

The operators $\hat{D}$ and $\hat{C}$ are both Hermitian. Since $\langle D \rangle_\beta$ and $\langle C \rangle_\beta$ are both real, (3-58) takes the form

$$(\Delta A_\beta)^2 (\Delta B_\beta)^2 \geqslant |\langle G \rangle_\beta|^2 = |\langle D \rangle_\beta + \tfrac{1}{2} i \langle C \rangle_\beta|^2$$
$$= \langle D \rangle_\beta{}^2 + \tfrac{1}{4} \langle C \rangle_\beta{}^2 \geqslant \tfrac{1}{4} \langle C \rangle_\beta{}^2$$

or, finally,

$$\Delta A_\beta \, \Delta B_\beta \geqslant \tfrac{1}{2} \langle C \rangle_\beta. \tag{3-59}$$

Heisenberg's principle does not postulate the value of the product $\Delta A \, \Delta B$ in a state; rather it gives the lower bound of this product to be $\tfrac{1}{2} \langle C \rangle$. If two operators commute ($\hat{C} = 0$), then the lower bound to the product of uncertainties associated with the measurements of $\hat{A}$ and $\hat{B}$ is zero. This is consistent with our discussion above.

From the fundamental commutation relations

$$[\hat{x}_i, \hat{p}_j] = i\hbar \, \delta_{ij} \hat{I}$$

and

$$[\hat{x}_i, \hat{x}_j] = [\hat{p}_i, \hat{p}_j] = 0$$

we obtain $\Delta x_i \, \Delta x_j \geqslant 0$, $\Delta p_i \, \Delta p_j \geqslant 0$, and $\Delta x_i \, \Delta p_j \geqslant \tfrac{1}{2} \hbar \, \delta_{ij}$. Thus for example the operators $\hat{x}$ and $\hat{p}_x$ are totally incompatible and no state exists in which the products of their uncertainties is less than $\tfrac{1}{2} \hbar$. This implies that if we know the momentum of a particle precisely, we will not be able to locate it with any precision.

If a system is known to be in a state characterized by $|\beta\rangle$, then its dynamical features are determined probabilistically. Imagine that we have M identical atomic systems all in the state $|\beta\rangle$. Suppose also that we have a detector which measures the observable $\hat{A}$. When the " A meter " is applied to the first system, the probability that it will read the value a_i is $|\langle a_i | \beta \rangle|^2$. However once the measurement is made and the meter *does* register a_i, then *after* the measurement, we know that the first atomic system now has the *definite* value a_i. Consequently, the effect of the measurement, after a reading a_i is obtained, is to force the system from its original state $|\beta\rangle$ to the new state $|a_i\rangle$. Equivalently the effect of an actual measurement of $\hat{A}$ is to force the system into an eigenstate of $\hat{A}$ corresponding to the observed reading. In this way a measuring device acts as a *filter* and projects the state vector onto the eigenvector associated with the observable being measured. The student may be troubled by the fact that the state of a system is altered when a measurement is made. Fortunately this problem is only academic since we are assuming that the atomic state is reproducible, that is, we have $M - 1$ more atomic systems just like it. While a certain number of systems may have their states altered in the measurement process, the knowledge obtained is applicable to those systems yet unmeasured.‡

VIII Quantum Dynamics

At any instant the direction of the state vector determines the statistical nature of the physical system. As this vector rotates in the complex space, the physical characteristics of the system change in time. To find the equation of motion for the state vector, which we now write as $|\beta, t\rangle$, we must first establish a dynamical correspondence with classical mechanics. We shall therefore require that there be a correspondence between the expectation value of a quantum observable and the corresponding classical function, namely,

$$\langle \beta, t | \hat{A} | \beta, t \rangle = \langle A(t) \rangle_\beta \leftrightarrow A_{\text{class}}(t). \tag{3-60}$$

Equivalently we expect that the equation for the motion of $\langle A(t) \rangle_\beta$ produced by the rotation of $|\beta, t\rangle$ be analogous to that of the classical function $A_{\text{class}}(t)$. This requirement is known as Ehrenfest's theorem. We therefore postulate, in analogy with the classical form in (2-76), the equation of motion for the expectation value of an operator to be

$$\frac{d}{dt} \langle A(t) \rangle_\beta = \frac{d}{dt} \langle \beta, t | \hat{A} | \beta, t \rangle$$

$$= \langle \beta, t | \frac{[\hat{A}, \mathscr{H}]}{i\hbar} | \beta, t \rangle + \langle \beta, t | \frac{\partial \hat{A}}{\partial t} | \beta, t \rangle. \tag{3-61}$$

The time dependence in $\langle A(t) \rangle_\beta$ originates from the implicit dependence, that is, the variation of $|\beta, t\rangle$, and also possibly from the explicit dependence in

---

‡ For an excellent introductory treatment of the measurement process, see R. P. Feynman, R. B. Leighton, and M. Sands, "The Feynman Lectures on Physics," Vol. 3, Chapters 1 and 2. Addison-Wesley, Reading, Massachusetts, 1964.

the operator itself. We shall only on rare occasions consider operators with an explicit time dependence. In order to ensure (3-61), we offer the following postulate.

Postulate 6   The motion of the state vector for a physical system is determined by the *Schroedinger equation of motion* which is, in ket form,

$$\mathscr{H} \, |\beta, t\rangle = i\hbar \frac{d}{dt} |\beta, t\rangle \qquad (3\text{-}62)$$

or in bra form

$$\langle \beta, t| \, \mathscr{H} = -i\hbar \frac{d}{dt} \langle \beta, t| \qquad (\mathscr{H} = \mathscr{H}^{\dagger})$$

where $\mathscr{H}$ is the Hamiltonian operator for the system.

To verify that the Schroedinger equation of motion does indeed lead to (3-61), we differentiate and obtain

$$\frac{d}{dt} \langle \beta, t|\hat{A}|\beta, t\rangle = \langle \beta, t|\hat{A}| \left\{ \frac{d}{dt} |\beta, t\rangle \right\} + \left\{ \frac{d}{dt} \langle \beta, t| \right\} \hat{A}|\beta, t\rangle + \langle \beta, t| \frac{\partial A}{\partial t} |\beta, t\rangle.$$

The last term originates with the explicit time dependence in the operator $\hat{A}$. Using the bra and ket forms of the Schroedinger equation, we find

$$\frac{d}{dt} \langle \beta, t|\hat{A}|\beta, t\rangle = \langle \beta, t|\hat{A} \left\{ \frac{1}{i\hbar} \mathscr{H} |\beta, t\rangle \right\}$$

$$+ \left\{ -\frac{1}{i\hbar} \langle \beta, t|\mathscr{H} \right\} \hat{A}|\beta, t\rangle + \langle \beta, t| \frac{\partial \hat{A}}{\partial t} |\beta, t\rangle$$

or

$$\frac{d}{dt} \langle A \rangle_{\beta} = \left\langle \frac{[\hat{A}, \mathscr{H}]}{i\hbar} \right\rangle_{\beta} + \left\langle \frac{\partial \hat{A}}{\partial t} \right\rangle_{\beta} \qquad (3\text{-}63)$$

as required.

It should be emphasized that the Schroedinger "*picture*" in which the state vector rotates is only one of many ways in which to develop (3-63). It is also possible to have the state vector remain stationary and to allow the eigenvectors of $\hat{A}$ (as well as $\hat{A}$ itself) to change in such a way that the expectation values vary according to (3-63). This representation of quantum motion is known as the *Heisenberg picture*. In still a third picture, the *interaction picture*, both the state vectors and the eigenvectors of $\hat{A}$ move in such a way as to preserve (3-63) (see Appendix B). We shall work entirely within the Schroedinger picture.

As a consequence of (3-63) we note that if an operator $\hat{A}$ has no explicit time dependence (that is, $\partial \hat{A}/\partial t = 0$), then its expectation value is a constant of the motion if it commutes with the Hamiltonian, that is, if $[\hat{A}, \mathcal{H}] = 0$.

The solution to the Schroedinger equation (3-62) can be formally written as

$$|\beta, t\rangle = \left[\exp\left(-\frac{i}{\hbar}\mathcal{H}(t - t_0) \right) \right] |\beta, t_0\rangle \tag{3-64}$$

provided $\mathcal{H}$ is *independent* of time. The ket $|\beta, t_0\rangle$ represents the initial state of the system at $t = t_0$. To verify this solution, we differentiate both sides with respect to time and find

$$\frac{d}{dt}|\beta, t\rangle = \left[-\frac{i}{\hbar}\mathcal{H} \exp\left(-\frac{i}{\hbar}\mathcal{H}(t - t_0) \right) \right] |\beta, t_0\rangle.$$

Using (3-64), the above equation becomes

$$\frac{d}{dt}|\beta, t\rangle = -\frac{i}{\hbar}\mathcal{H}|\beta, t\rangle$$

which is the required equation of motion.

The time development in (3-64) can be formalized by writing

$$|\beta, t\rangle = \hat{U}(t, t_0)|\beta, t_0\rangle$$

where the *evolution operator* is

$$\hat{U}(t, t_0) = \exp\left(-\frac{i}{\hbar}\mathcal{H}(t - t_0) \right). \tag{3-65}$$

In the case where $\mathcal{H}$ depends on time, the evolution operator cannot be written as in (3-65). The evolution operator can be shown to satisfy the following integral equation:

$$\hat{U}(t, t_0) = \hat{1} + \frac{1}{i\hbar}\int_{t_0}^{t}\mathcal{H}(\tau_1)\hat{U}(\tau_1, t_0)\,d\tau_1. \tag{3-66}$$

It can be verified directly that when $\mathcal{H}$ is time independent, (3-65) satisfies (3-66).

We shall illustrate the postulates and theorems above with a simple but hypothetical example. Imagine that two observables for a particle are respectively the operator $\hat{S}$ and the operator $\hat{I}$. Let us assume that the characteristic values of $\hat{S}$ are s_1 and s_2. Also let the two characteristic states of $\hat{I}$ be I_1 and I_2. We next assume that the operators in the eigenbasis of $\hat{S}$ take the form

$$\hat{S} = \begin{pmatrix} 0 & 0 \\ 0 & 1 \end{pmatrix} \quad \text{and} \quad \hat{I} = \begin{pmatrix} \frac{1}{2} & -\frac{1}{2} \\ -\frac{1}{2} & \frac{1}{2} \end{pmatrix}.$$

Since we are working in the $\hat{S}$ eigenbasis ($\hat{S}$ is diagonal), its eigenvectors and eigenvalues are

$$|s_1\rangle = \begin{pmatrix} 1 \\ 0 \end{pmatrix} \qquad \text{for} \quad s_1 = 0$$

and

$$|s_2\rangle = \begin{pmatrix} 0 \\ 1 \end{pmatrix} \qquad \text{for} \quad s_2 = 1.$$

The eigenvectors of $\hat{I}$ may be found directly as

$$|I_1\rangle = \frac{1}{\sqrt{2}}\begin{pmatrix} 1 \\ 1 \end{pmatrix} \qquad \text{for} \quad I_1 = 0$$

and

$$|I_2\rangle = \frac{1}{\sqrt{2}}\begin{pmatrix} -1 \\ 1 \end{pmatrix} \qquad \text{for} \quad I_1 = 1.$$

Note that the eigenbases of these two nondegenerate Hermitian operators are distinct (Figure 3-2). This could have been foreseen from the fact that the matrices for $\hat{S}$ and $\hat{I}$ do not commute. The two operators are therefore incompatible and no state exists in which both are simultaneously defined with perfect precision.

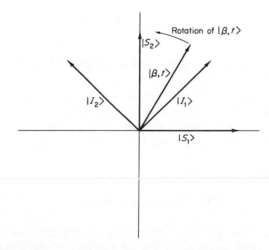

Figure 3-2 The geometrical relationship between the eigenvectors of $\hat{S}$ and $\hat{I}$ and the motion of the state vector $|\beta, t\rangle$.

Suppose that at some time the state vector $|\beta, t\rangle$ is given by

$$|\beta, t\rangle = \begin{pmatrix} \frac{1}{2} \\ \frac{1}{2}\sqrt{3} \end{pmatrix}.$$

The probabilities of observing the system in the various states are

$$\mathscr{P}_\beta(s_1) = |\langle s_1|\beta, t\rangle|^2 = \left[(1, 0)\begin{pmatrix} \frac{1}{2} \\ \frac{1}{2}\sqrt{3} \end{pmatrix}\right]^2 = (\tfrac{1}{2})^2 = \tfrac{1}{4}$$

$$\mathscr{P}_\beta(s_2) = |\langle s_2|\beta, t\rangle|^2 = \left[(0, 1)\begin{pmatrix} \frac{1}{2} \\ \frac{1}{2}\sqrt{3} \end{pmatrix}\right]^2 = (\tfrac{1}{2}\sqrt{3})^2 = \tfrac{3}{4}$$

$$\mathscr{P}(I_1) = |\langle I_1|\beta, t\rangle|^2 = \left[\left(\frac{1}{\sqrt{2}}, \frac{1}{\sqrt{2}}\right)\begin{pmatrix} \frac{1}{2} \\ \frac{1}{2}\sqrt{3} \end{pmatrix}\right]^2 = \tfrac{1}{8}(1 + \sqrt{3})^2$$

$$= \tfrac{1}{8}(4 + 2\sqrt{3}) = \tfrac{1}{2} + \tfrac{1}{4}\sqrt{3} = 0.933$$

$$\mathscr{P}(I_2) = |\langle I_2|\beta, t\rangle|^2 = \left[\left(-\frac{1}{\sqrt{2}}, \frac{1}{\sqrt{2}}\right)\begin{pmatrix} \frac{1}{2} \\ \frac{1}{2}\sqrt{3} \end{pmatrix}\right]^2 = \tfrac{1}{8}(-1 + \sqrt{3})^2$$

$$= \tfrac{1}{8}(4 - 2\sqrt{3}) = \tfrac{1}{2} - \tfrac{1}{4}\sqrt{3} = 0.067.$$

The expectation value of each observable in this state is

$$\langle \beta, t|\hat{S}|\beta, t\rangle = (\tfrac{1}{2}, \tfrac{1}{2}\sqrt{3})\begin{pmatrix} 0 & 0 \\ 0 & 1 \end{pmatrix}\begin{pmatrix} \frac{1}{2} \\ \frac{1}{2}\sqrt{3} \end{pmatrix} = \tfrac{3}{4}$$

and

$$\langle \beta, t|\hat{I}|\beta, t\rangle = (\tfrac{1}{2}, \tfrac{1}{2}\sqrt{3})\begin{pmatrix} \frac{1}{2} & -\frac{1}{2} \\ -\frac{1}{2} & \frac{1}{2} \end{pmatrix}\begin{pmatrix} \frac{1}{2} \\ \frac{1}{2}\sqrt{3} \end{pmatrix} = 0.067.$$

These expectation values can be interpreted as representing some intermediate values of $\hat{S}$ and $\hat{I}$.

As the state vector evolves (rotates) in time, the probabilities and the expectation values associated with $\hat{S}$ and $\hat{I}$ vary and give the effect of quantum motion. If $\hat{S}$ and $\hat{I}$ were truly canonical variables, we could construct the Hamiltonian and evolution operators for the system and predict the evolution of the state.

IX Stationary States

If at $t = t_0 = 0$ the system is characterized by a state vector $|\beta, t_0\rangle = |\beta, 0\rangle$, then its future evolution is obtained using the operation

$$|\beta, t\rangle = \hat{U}(t, 0)|\beta, 0\rangle \tag{3-67}$$

where $\hat{U}$ is related to the Hamiltonian of the system by

$$U(t, 0) = e^{-i\mathcal{H}t/\hbar}.$$

Let us suppose that the initial state was in fact an eigenstate of the Hamiltonian, which we denote by $|\beta, 0\rangle = |\varepsilon_i\rangle$. In such a state the energy would be precisely equal to the corresponding energy eigenvalue ε_i. Using (3-67) we find the state at later times to be

$$|\beta, t\rangle = e^{-i\mathcal{H}t/\hbar}|\varepsilon_i\rangle.$$

However if $|\varepsilon_i\rangle$ is an eigenvector of $\mathcal{H}$ with eigenvalue ε_i it must also be an eigenvector of $e^{-i\mathcal{H}t/\hbar}$ with eigenvalue $e^{-i\varepsilon_i t/\hbar}$. The state's evolution becomes

$$|\beta, t\rangle = e^{-i\omega_i t}|\varepsilon_i\rangle \qquad \text{where} \quad \omega_i = \varepsilon_i/\hbar.$$

The oscillatory factor is a complex scalar of unit magnitude and does not affect the directionality of the original state vector. Note also that

$$\langle\beta, t|\beta, t\rangle = \langle\varepsilon_i|e^{i\varepsilon_i t/\hbar}e^{-i\varepsilon_i t/\hbar}|\varepsilon_i\rangle$$

$$= \langle\varepsilon_i|\varepsilon_i\rangle = 1.$$

Thus the norm of the state also remains equal to unity. The state $|\beta, t\rangle$ is physically equivalent to $|\varepsilon_i\rangle$ and no evolution occurs at all! We are thus led to the following rule:

Rule   If a system is initially in an (energy) eigenstate of the Hamiltonian, it will remain in that state indefinitely and all expectation values of observables will be stationary.

This rule explains why the hydrogenic electron in one of the Bohr orbits remains stationary indefinitely unless acted upon by an external agent (for example, an electromagnetic field). The importance of the Hamiltonian and its energy eigenstates cannot be emphasized enough. The Hamiltonian determines the evolution of all states and is known as the generator of displacements in time. Its eigenstates are stationary as well as being states of characteristic energy.

X   The Dimensionality of "Quantum Space"

The theorems and postulates above were established without reference to N, the dimensionality of the space. We now offer three reasons to suggest that this dimensionality must be *infinite*.

First, consider the fundamental commutation relation:

$$[\hat{x}, \hat{p}_x] = i\hbar\hat{1}. \tag{3-68}$$

It is simple to verify that in a finite-dimensional space, the trace of a product of *any* two operators is independent of their order in the product. It follows then that for any two operators we must have

$$\text{Tr } \hat{A}\hat{B} - \text{Tr } \hat{B}\hat{A} = \text{Tr}(\hat{A}\hat{B} - \hat{B}\hat{A}) = \text{Tr}[\hat{A}, \hat{B}] = 0.$$

However the trace of the identity operator $\hat{1}$ is necessarily

$$\text{Tr } \hat{1} = N \neq 0.$$

Taking the trace of both sides of (3-68), we are led to an inconsistency which can be resolved only in an infinite-dimensional space.‡

Second, the number of characteristic values of an observable must equal the number of distinct eigenvalues, which in turn can be no greater than the dimensionality of the space in which the operator is represented. But the number of observed values of a classical observable is generally infinite. For example, the characteristic energies conjectured by Planck for a one-dimensional oscillator is

$$\varepsilon_n = nh\nu$$

where n is an integer ranging from zero to *infinity*. The representation of the oscillator Hamiltonian therefore requires an infinite-dimensional space.

There is yet a third and perhaps more important reason for requiring an infinite-dimensional space. Consider the characteristic states of *position* determined by the equation

$$\hat{\mathbf{r}}|\mathbf{r}_i\rangle = \mathbf{r}_i|\mathbf{r}_i\rangle. \tag{3-69}$$

Each ket $|\mathbf{r}_i\rangle$ represents a state in which the particle is characteristically at the point $\mathbf{r}_i$, that is, (x_i, y_i, z_i), in space. The uncertainty regarding the position is zero when the system is in one of these eigenstates. From the uncertainty principle it follows that the momentum characteristics are totally uncertain in these states. However it is an observed fact that *all* points in space are characteristic, that is, a precise measurement of the position of the particle may be performed anywhere. Equivalently, position is not quantized. This does not mean that (3-69) is not valid but that the spectrum of $\hat{\mathbf{r}}$ is *continuous* rather than discrete (denumerable).§ To indicate the continuous spectrum in (3-69), we remove the subscript and write

$$\hat{\mathbf{r}}|\mathbf{r}\rangle = \mathbf{r}|\mathbf{r}\rangle. \tag{3-70}$$

---

‡ The infinite-dimensional space with the properties described above is known mathematically as a vector *Hilbert* space.

§ Eigenvectors belonging to the continuous spectrum do not have a finite norm and are therefore not normalizable by conventional means. Strictly speaking, these vectors are not part of a Hilbert space.

Since $\hat{\mathbf{r}}$ and $\hat{\mathbf{p}}$ appear symmetrically in the fundamental commutation relations, it is not surprising that the spectrum of linear momentum is also continuous and we may write

$$\hat{\mathbf{p}}|\mathbf{p}\rangle = \mathbf{p}|\mathbf{p}\rangle. \qquad (3\text{-}71)$$

In any eigenstate of $\hat{\mathbf{p}}$ the particle is precisely moving with a velocity $\mathbf{v} = \mathbf{p}/m$, but of course its position is vague. *A finite-dimensional space cannot be used to represent operators having a continuous spectrum.*

The student at this point should be able to visualize conceptually discrete vectors in an infinite-dimensional space by permitting $N \to \infty$. In a discrete eigenbasis the operators become matrices with an infinite number of rows and columns. The ket vectors become infinite column matrices, that is, they have an infinite number of components. However, representation of a vector or an operator in a continuous eigenbasis cannot be displayed as a conventional matrix. We must therefore rely on an abstract mathematical framework.

The orthonormality of two continuous eigenvectors of the $\hat{\mathbf{r}}$ operator will be written

$$\langle \mathbf{r}_i|\mathbf{r}_j\rangle = \delta_{ij} \to \langle \mathbf{r}'|\mathbf{r}\rangle = \delta(\mathbf{r} - \mathbf{r}'). \qquad (3\text{-}72)$$

The *Dirac delta function* (3-72) is a direct generalization of the Kronecker delta and has the properties:

(a) $\delta(\mathbf{r} - \mathbf{r}') = \infty$ for $\mathbf{r} = \mathbf{r}'$
(b) $\delta(\mathbf{r} - \mathbf{r}') = 0$ for $\mathbf{r} \neq \mathbf{r}'$
(c) $\int \delta(\mathbf{r} - \mathbf{r}')\, d\mathbf{r}' = 1 = $ area under function
(d) $\int f(\mathbf{r}')\, \delta(\mathbf{r} - \mathbf{r}')\, d\mathbf{r}' = f(\mathbf{r})$.

The limits of the integrals cover all space. The last property listed is known as the "sifting" property and shall be used frequently. For an exercise involving the delta function, see Problem 3-10.

XI The Coordinate Representation‡

The expansion of a vector in the continuous $\hat{\mathbf{r}}$ eigenbasis is accomplished using

$$|\beta, t\rangle = \sum_i \langle \mathbf{r}_i|\beta, t\rangle|\mathbf{r}_i\rangle \to \int d\mathbf{r}(\langle \mathbf{r}|\beta, t\rangle)|\mathbf{r}\rangle. \qquad (3\text{-}73)$$

---

‡ In the *momentum representation*, vectors are expanded in the continuous eigenbasis of the $\hat{\mathbf{p}}$ operator. See Problem 5-1.

This is a direct generalization of (3-45) using an integral replacement for the sum. Since the $\hat{\mathbf{r}}$ eigenvalue is a continuous variable, the coefficient in (3-73)

$$\Psi_\beta(\mathbf{r}, t) = \langle \mathbf{r} | \beta, t \rangle$$

may be regarded as a continuous function. Generalizing Postulate 4, the quantity

$$|\langle \mathbf{r} | \beta, t \rangle|^2 \, d\mathbf{r} = |\Psi_\beta|^2 \, d\mathbf{r} \qquad (3\text{-}74)$$

represents the probability that the system will be found in a differential volume of space $d\mathbf{r}$ about the point $\mathbf{r}$. The quantity $\mathscr{P}_\beta = |\Psi_\beta|^2$ represents the *probability density* of finding the particle at the point $\mathbf{r}$ and time t when the system is characterized by $|\beta, t\rangle$. As we shall see below $\Psi_\beta(\mathbf{r}, t)$ is called the *state function* of the system and contains all the dynamical information regarding the system in the $\hat{\mathbf{r}}$ eigenbasis.

The completeness relation for the eigenvectors of $\hat{\mathbf{r}}$ is written

$$\sum_i |\mathbf{r}_i\rangle\langle\mathbf{r}_i| = \hat{1} \to \int d\mathbf{r} |\mathbf{r}\rangle\langle\mathbf{r}| = \hat{1}. \qquad (3\text{-}75)$$

We shall always assume that $\hat{\mathbf{r}}$ is nondegenerate so that the $\hat{\mathbf{r}}$ eigenbasis is orthonormal, and unique.‡

There is a distinct advantage in working in the continuous eigenbasis of the operator $\hat{\mathbf{r}}$ (coordinate representation). In a discrete basis $|a_i\rangle$ an arbitrary state vector $|\beta, t\rangle$ is specified by giving an infinite set of components, that is,

$$\beta^{(a_i)} = \langle a_i | \beta, t \rangle .$$

On the other hand, in the coordinate representation a state vector is completely characterized by a single continuous state function

$$\Psi_\beta(\mathbf{r}, t) = \langle \mathbf{r} | \beta, t \rangle .$$

The coordinate representation is probably most useful when a single particle is involved. However, it must be stressed that this representation is only possible when $\hat{\mathbf{r}}$ is an observable.

Any vector may be represented in the coordinate representation using

$$|\beta\rangle = \int \Psi_\beta(\mathbf{r}) |\mathbf{r}\rangle \, d\mathbf{r} \qquad (3\text{-}76a)$$

or in bra form

$$\langle\beta| = \int \Psi_\beta{}^*(\mathbf{r})\langle\mathbf{r}| \, d\mathbf{r}. \qquad (3\text{-}76b)$$

---

‡ *Completeness* of the $\hat{\mathbf{r}}$ eigenbasis is also assumed for the purpose of expanding arbitrary states involving the particle's spatial characteristics.

The inner product of two vectors in the coordinate representation becomes

$$\langle \alpha | \beta \rangle = \left\{ \int \langle \mathbf{r}' | \Psi_\alpha{}^*(\mathbf{r}') \, d\mathbf{r}' \right\} \left\{ \int \Psi_\beta(\mathbf{r}) | \mathbf{r} \rangle \, d\mathbf{r} \right\}$$

$$= \iint d\mathbf{r} \, d\mathbf{r}' \Psi_\alpha{}^*(\mathbf{r}') \Psi_\beta(\mathbf{r}) \langle \mathbf{r}' | \mathbf{r} \rangle.$$

Using the orthonormality of the $\hat{\mathbf{r}}$ eigenbasis (that is, $\langle \mathbf{r}' | \mathbf{r} \rangle = \delta(\mathbf{r} - \mathbf{r}')$) and the sifting property of the delta function, we find

$$\langle \alpha | \beta \rangle = \int \Psi^*{}_\alpha(\mathbf{r}) \Psi_\beta(\mathbf{r}) \, d\mathbf{r}. \tag{3-77}$$

Rule   The inner product of two state vectors may be evaluated by integrating the product of their respective state functions in the coordinate representation.

The function associated with an eigenvector $|a_i\rangle$ of an operator $\hat{A}$ is called an *eigenfunction* of $\hat{A}$ and is written

$$\psi_{a_i}(\mathbf{r}) = \langle \mathbf{r} | a_i \rangle. \tag{3-78}$$

This eigenfunction characterizes an eigenstate of $\hat{A}$. Thus instead of finding the eigenvectors of observables, it is sufficient to calculate the eigenfunctions associated with the corresponding characteristic states.

The eigenfunctions of an operator are simpler to obtain than the abstract eigenvectors themselves because while the latter satisfy an infinite dimensional vector eigenvalue equation, the former, as we shall show, satisfy a differential equation. The representation of quantum mechanics in the $\hat{\mathbf{r}}$ eigenbasis is also known as *wave mechanics* and was first developed by E. Schroedinger. We shall next develop the mathematical formulation of quantum theory in the differential language of wave mechanics. All physical quantities, that is, inner products, probabilities, expectation values, uncertainties, etc., remain the same regardless of the representation. We shall rely on wave mechanics when possible. In problems where $\hat{\mathbf{r}}$ is not a canonical observable, for example, in representing intrinsic spin and in quantizing the electromagnetic field, we must use the more general Dirac formalism rather than Schroedinger's wave mechanics.

XII   The Transition to Wave Mechanics

We shall restrict our present discussion to a single particle whose relevant observables are functions of $\hat{\mathbf{p}}$ and $\hat{\mathbf{r}}$. We wish to establish the eigenvalues and eigenfunctions of a particular operator function $\hat{A}(\hat{\mathbf{p}}, \hat{\mathbf{r}})$. The vector eigenvalue equation reads

$$\hat{A} | a_i \rangle = a_i | a_i \rangle. \tag{3-79}$$

Taking the inner product of $\langle \mathbf{r} |$ with both sides of this equation, we obtain the scalar equation

$$\langle \mathbf{r} | \hat{A} | a_i \rangle = a_i \langle \mathbf{r} | a_i \rangle = a_i \psi_{a_i}(\mathbf{r}). \tag{3-80}$$

Although the right side contains the eigenfunction, it is not yet clear what the left side represents. To evaluate the matrix element $\langle \mathbf{r} | \hat{A} | a_i \rangle$ we shall assume the operator to be of the form

$$\hat{A}(\hat{\mathbf{p}}, \hat{\mathbf{r}}) = \hat{G}(\hat{\mathbf{p}}) + \hat{K}(\hat{\mathbf{r}}) \tag{3-81}$$

(for example, $\hat{A}$ could be the Hamiltonian operator $\mathscr{H}$ which contains a kinetic energy term and a potential energy term). Although this form is somewhat special, the results remain quite general. Using (3-81) in (3-80), we find

$$\langle \mathbf{r} | \hat{G}(\hat{\mathbf{p}}) | a_i \rangle + \langle \mathbf{r} | \hat{K}(\hat{\mathbf{r}}) | a_i \rangle = a_i \psi_{a_i}(\mathbf{r}). \tag{3-82}$$

The second term on the left is simple to evaluate. Since $\langle \mathbf{r} |$ is an eigenbra of $\hat{\mathbf{r}}$ it must also be an eigenbra of $\hat{K}(\hat{\mathbf{r}})$ with the property

$$\langle \mathbf{r} | \hat{K}(\mathbf{r}) = K(\mathbf{r}) \langle \mathbf{r} |.$$

Equation (3-82) becomes

$$\langle \mathbf{r} | \hat{G}(\hat{\mathbf{p}}) | a_i \rangle + K(\mathbf{r}) \psi_{a_i}(\mathbf{r}) = a_i \psi_{a_i}(\mathbf{r}). \tag{3-83}$$

To complete the analysis we must evaluate the matrix element $\langle \mathbf{r} | \hat{G}(\hat{\mathbf{p}}) | a_i \rangle$, which can be done once we know the effect of $\hat{\mathbf{p}}$ on an eigenvector $\langle \mathbf{r} |$. To accomplish this we shall require the following theorem:

Theorem VIII   *The operator $\hat{p}_x$ is the generator of infinitesimal translations along the x axis.*‡ *Mathematically we have*

$$\left\{ \hat{1} + \frac{\hat{p}_x \, dx}{i\hbar} \right\} | x, y, z \rangle = | x + dx, y, z \rangle \tag{3-84}$$

that is, the effect of the operator in brackets on a state, $| \mathbf{r} \rangle = | x, y, z \rangle$, is to produce a translation dx of the particle along the x axis.

Proof   We wish to prove that the term on the left in (3-84) is indeed an eigenvector of $\hat{x}$ with the eigenvalue $x + dx$, that is,

$$\hat{x} \left[\left\{ \hat{1} + \frac{\hat{p}_x \, dx}{i\hbar} \right\} | x, y, z \rangle \right] = (x + dx) \left[\left\{ \hat{1} + \frac{\hat{p}_x \, dx}{i\hbar} \right\} | x, y, z \rangle \right]. \tag{3-85}$$

---

‡ See Chapter 2, Eq. (2-81). The operator in brackets is actually the first two terms of a Taylor expansion of

$$\hat{U}_{p_x}(x) = \exp(-i\hat{p}_x \, x/\hbar)$$

which produces a finite translation x along the x axis.

The left side may be rewritten, using the commutation relation $\hat{x}\hat{p}_x = i\hbar\hat{1} + \hat{p}_x\hat{x}$, as

$$\left\{\hat{x} + dx\,\hat{1} + \frac{\hat{p}_x\hat{x}\,dx}{i\hbar}\right\}|x, y, z\rangle = \left\{x + dx + \frac{x\,dx\,\hat{p}_x}{i\hbar}\right\}|x, y, z\rangle.$$

However this is exactly equal to the right side of (3-85) as can be seen by multiplying out and dropping terms of order dx^2, that is,

$$\left\{x + \frac{x\,dx}{i\hbar}\hat{p}_x + dx + \frac{dx^2\,\hat{p}_x}{i\hbar}\right\}|x, y, z\rangle \simeq \left\{x + dx + \frac{x\,dx\,\hat{p}_x}{i\hbar}\right\}|x, y, z\rangle.$$

This establishes (3-85) and proves Theorem VIII.

Equation (3-84) can be put in a more useful form by writing it as

$$\hat{p}_x|x, y, z\rangle = \frac{i\hbar(|x + dx, y, z\rangle - |x, y, z\rangle)}{dx}$$

or in bra form as

$$\langle x, y, z|\hat{p}_x = \frac{-i\hbar(\langle x + dx, y, z| - \langle x, y, z|)}{dx}.$$

Finally, the matrix element takes the form

$$\langle x, y, z|\hat{p}_x|a_i\rangle = -\frac{i\hbar(\langle x + dx, y, z|a_i\rangle - \langle x, y, z|a_i\rangle)}{dx}$$

or

$$\langle \mathbf{r}|\hat{p}_x|a_i\rangle = -\frac{i\hbar(\psi_{a_i}(x + dx, y, z) - \psi_{a_i}(x, y, z))}{dx} = \left(\frac{\hbar}{i}\right)\left(\frac{\partial\psi_{a_i}}{\partial x}\right).$$

More generally for the operator $\hat{\mathbf{p}}$ we find

$$\langle \mathbf{r}|\hat{\mathbf{p}}|a_i\rangle = \frac{\hbar}{i}\nabla\psi_{a_i}$$

or

$$\langle \mathbf{r}|\hat{G}(\hat{\mathbf{p}})|a_i\rangle = \hat{G}\left(\frac{\hbar}{i}\nabla\right)\psi_{a_i} \tag{3-86}$$

where $\hat{G}(\hbar/i)\nabla)$ is the corresponding differential operator function. Finally (3-80) becomes the differential eigenvalue equation

$$\langle \mathbf{r}|\hat{A}|a_i\rangle = \hat{A}\left(\frac{\hbar}{i}\nabla, \mathbf{r}\right)\psi_{a_i}$$

$$= \left\{\hat{G}\left(\frac{\hbar}{i}\nabla\right) + K(\mathbf{r})\right\}\psi_{a_i}(\mathbf{r}) = a_i\psi_{a_i}(\mathbf{r}). \tag{3-87}$$

Rule To obtain the wave-mechanical (differential) operator for a canonical observable, make the replacements

$$\hat{\mathbf{p}} \to \frac{\hbar}{i} \nabla, \qquad \hat{\mathbf{r}} \to \mathbf{r}, \qquad \text{and} \qquad \hat{A}(\hat{\mathbf{p}}, \hat{\mathbf{r}}) \to \hat{A}\left(\frac{\hbar}{i} \nabla, \mathbf{r}\right). \qquad (3\text{-}88)$$

For example, suppose we wish to find the energy eigenfunctions (stationary states) of the nonrelativistic Hamiltonian $\mathscr{H} = (\hat{p}^2/2m) + \hat{V}(\hat{\mathbf{r}})$. First we construct the wave-mechanical operator

$$\mathscr{H} = \frac{\frac{\hbar}{i} \nabla \cdot \frac{\hbar}{i} \nabla}{2m} + V(\mathbf{r}) = -\frac{\hbar^2}{2m} \nabla^2 + V(\mathbf{r}).$$

We then solve the differential eigenvalue equation

$$\mathscr{H} \psi_{\varepsilon_i} = \varepsilon_i \psi_{\varepsilon_i}$$

$$\left\{ -\frac{\hbar^2}{2m} \nabla^2 + V(\mathbf{r}) \right\} \psi_{\varepsilon_i}(\mathbf{r}) = \varepsilon_i \psi_{\varepsilon_i}(\mathbf{r})$$

for the energy eigenfunctions and eigenvalues.

Matrix elements of the form $\langle \alpha | \hat{A} | \beta \rangle$ may be evaluated quite simply in wave mechanics. Using the completeness relation, $\int |\mathbf{r}\rangle\langle\mathbf{r}| \, d\mathbf{r} = \hat{1}$, and (3-87) we write

$$\langle \alpha | \hat{A} | \beta \rangle = \int \langle \alpha | \mathbf{r} \rangle \langle \mathbf{r} | \hat{A} | \beta \rangle \, d\mathbf{r} = \int \Psi_\alpha^{\ *}(\mathbf{r}) \hat{A}\left(\frac{\hbar}{i} \nabla, \mathbf{r}\right) \Psi_\beta(\mathbf{r}) \, d\mathbf{r}. \quad (3\text{-}89)$$

Rule The matrix element $\langle \alpha | \hat{A} | \beta \rangle$ is evaluated by operating with the wave-mechanical operator $\hat{A}\left(\dfrac{\hbar}{i} \nabla, \mathbf{r}\right)$ on $\Psi_\beta(\mathbf{r})$, multiplying by $\Psi_\alpha^*(\mathbf{r})$, and integrating over all space.

For example, the element $\langle \alpha | \hat{\mathbf{p}} | \beta \rangle$ is evaluated in wave mechanics as

$$\langle \alpha | \hat{\mathbf{p}} | \beta \rangle = \int \Psi_\alpha^*(\mathbf{r}) \frac{\hbar}{i} \nabla \Psi_\beta(\mathbf{r}) \, d\mathbf{r}.$$

XIII   The Schroedinger Wave Equation

In wave mechanics, the state of a system is represented by the *state function*

$$\Psi_\beta(\mathbf{r}, t) = \langle \mathbf{r} | \beta, t \rangle.$$

As the state vector rotates, the state function evolves in time from its initial value $\Psi_\beta(\mathbf{r}, t_0) = \langle \mathbf{r} | \beta, t_0 \rangle$. We can deduce an equation of motion for the state function using the Schroedinger picture in which the state vector satisfies

$$\mathcal{H} | \beta, t \rangle = i\hbar \frac{d | \beta, t \rangle}{dt}.$$

Taking the inner product of $\langle \mathbf{r} |$ with both sides of this equation we find

$$\langle \mathbf{r} | \mathcal{H} | \beta, t \rangle = i\hbar \frac{\partial}{\partial t} \langle \mathbf{r} | \beta, t \rangle$$

or

$$\mathcal{H} \left(\frac{\hbar}{i} \nabla, \mathbf{r} \right) \Psi_\beta(\mathbf{r}, t) = i\hbar \frac{\partial}{\partial t} \Psi_\beta(\mathbf{r}, t).$$

Using a Hamiltonian of the general form $\mathcal{H} = (\hat{p}^2/2m) + V(\hat{\mathbf{r}})$, we obtain the *Schroedinger wave equation* for the state (or wave) function $\Psi_\beta(\mathbf{r}, t)$,

$$-\frac{\hbar^2}{2m} \nabla^2 \Psi_\beta + V(\mathbf{r})\Psi_\beta = i\hbar \frac{\partial \Psi_\beta}{\partial t}. \tag{3-90}$$

While Ψ_β is itself complex and unmeasurable, its absolute value squared is related to the probability of finding the particle in various regions of space. The fact that Ψ_β satisfies a wave equation explains the wavelike qualities inherent in matter discussed in Chapter 1. We shall see presently that the wave function of a free particle is essentially what de Broglie was alluding to in his more primitive theory.

Since the wave equation is first order in time, only the initial value $\Psi_\beta(\mathbf{r}, t_0)$ is required to determine the evolution of the state function. The solution to the wave equation is easily obtained by taking the inner product of $\langle \mathbf{r} |$ with both sides of (3-65) which is

$$| \beta, t \rangle = \hat{U}(t, t_0) | \beta, t_0 \rangle$$

giving

$$\Psi_\beta(\mathbf{r}, t) = \langle \mathbf{r} | \beta, t \rangle = \langle \mathbf{r} | \hat{U}(t, t_0) | \beta, t_0 \rangle.$$

This may be re-expressed using the closure relation, $\int d\mathbf{r}' |\mathbf{r}'\rangle\langle\mathbf{r}'| = \hat{1}$, as

$$\Psi_\beta(\mathbf{r}, t) = \int d\mathbf{r}' \langle \mathbf{r} | \hat{U}(t, t_0) | \mathbf{r}' \rangle \langle \mathbf{r}' | \beta, t_0 \rangle. \tag{3-91}$$

Defining the propagation kernel or *propagator* by

$$G(\mathbf{r}, \mathbf{r}', t, t') = \langle \mathbf{r} | \hat{U}(t, t') | \mathbf{r}' \rangle, \tag{3-92}$$

(3-91) can be expressed as

$$\Psi_\beta(\mathbf{r}, t) = \int d\mathbf{r}' G(\mathbf{r}, \mathbf{r}', t, t_0)\Psi_\beta(\mathbf{r}', t_0). \tag{3-93}$$

Using $\hat{U} = \exp(-i\mathcal{H}(t - t_0)/\hbar)$, the dependence of G on the Hamiltonian becomes quite clear. Once the Hamiltonian is known and the initial state function specified, the wave function may be calculated at later times using (3-93).

Evaluating the propagator G in closed form from the defining relation (3-92) is usually difficult. A series expression is possible if we employ the completeness of the $\mathcal{H}$ eigenbasis, that is, $\sum_i |\varepsilon_i\rangle\langle\varepsilon_i| = \hat{1}$. Setting

$$\hat{U} = \exp[-i\mathcal{H}(t - t_0)/\hbar]$$

(3-92) becomes

$$
\begin{aligned}
G(\mathbf{r}, \mathbf{r}', t, t') &= \sum_i \langle\mathbf{r}|\exp[-i\mathcal{H}(t - t_0)/\hbar]|\varepsilon_i\rangle\langle\varepsilon_i|\mathbf{r}'\rangle \\
&= \sum_i \exp[-i\varepsilon_i(t - t_0)/\hbar]\langle\mathbf{r}|\varepsilon_i\rangle\langle\varepsilon_i|\mathbf{r}'\rangle \\
&= \sum_i \exp[-i\varepsilon_i(t - t_0)/\hbar]\psi_{\varepsilon_i}(\mathbf{r})\psi_{\varepsilon_i}^*(\mathbf{r}').
\end{aligned} \tag{3-94}
$$

Once the eigenvalues and eigenfunctions of the Hamiltonian have been found, the propagator can be constructed using (3-94).

XIV   The Schroedinger Wave Equation and Probability Flow

As the state function evolves in time, the probability density associated with the system changes throughout space. Since the total probability is always unity, the density flows from one region to another very much like a fluid.

The connection between probability flow and the Schroedinger equation is established using the latter,

$$\left\{-\frac{\hbar^2}{2m}\nabla^2 + V(\mathbf{r})\right\}\Psi_\beta = i\hbar\frac{\partial\Psi_\beta}{\partial t}$$

and its conjugate form

$$\left\{-\frac{\hbar^2}{2m}\nabla^2 + V(\mathbf{r})\right\}\Psi_\beta^* = -i\hbar\frac{\partial\Psi_\beta^*}{\partial t}.$$

Multiplying the first equation by $\Psi_\beta{}^*$, the second by Ψ_β, and subtracting, we obtain

$$\frac{-\hbar^2}{2m}(\Psi_\beta{}^* \nabla^2 \Psi_\beta - \Psi_\beta \Delta^2 \Psi_\beta{}^*) = i\hbar\left(\Psi_\beta{}^* \frac{\partial \Psi_\beta}{\partial t} + \Psi_\beta \frac{\partial \Psi_\beta{}^*}{\partial t}\right)$$

or equivalently

$$\nabla \cdot \frac{i\hbar}{2m}(\Psi_\beta{}^* \nabla\Psi_\beta - \Psi_\beta \nabla\Psi_\beta{}^*) = \frac{\partial}{\partial t}|\Psi_\beta|^2. \qquad (3\text{-}95)$$

Defining the *probability flux* or probability current vector by

$$\mathbf{J}_\beta(\mathbf{r}, t) = \frac{-i\hbar}{2m}(\Psi_\beta{}^* \nabla\Psi_\beta - \Psi_\beta \nabla\Psi_\beta{}^*) \qquad (3\text{-}96)$$

and using the definition of probability density, $\mathscr{P}_\beta(\mathbf{r}, t) = |\Psi_\beta|^2$, (3-95) becomes an equation of continuity, that is,

$$\nabla \cdot \mathbf{J}_\beta + \frac{\partial \mathscr{P}_\beta}{\partial t} = 0. \qquad (3\text{-}97)$$

Equation (3-97) implies that probability is locally conserved, that is, it is neither created nor destroyed. Equivalently, any net decrease (or increase) of probability within a region in space is accompanied by a net outflow (or inflow) of probability. Also, if the system is in a stationary state, then $\partial \mathscr{P}_\beta/\partial t = 0$ and the probability flow is solenoidal, that is, $\nabla \cdot \mathbf{J}_\beta = 0$. If a state function is real, then from (3-96) it follows that $\mathbf{J} = \mathbf{0}$.

Note that the derivation of the equation of continuity for probability flow depends on the potential energy function being real. While complex potentials are somewhat unphysical because they lead to complex energy eigenvalues, they have nevertheless been used extensively in modeling nuclear scattering interactions. It is simple to verify (Problem 3-14) that the imaginary part of the potential leads directly to a "sink" term in (3-97), resulting in a loss of probability. This destruction of probability is to be interpreted as absorption of particles by nuclei.

There are certain restrictions to be imposed on the state function for a system since its square represents a probability density. These are contained in the following postulate.

Postulate 7   A state function and its derivative representing a real physical system must be everywhere finite, continuous,‡ and single valued.

---

‡ In the unphysical situation where the potential energy function suffers an infinite discontinuity, the continuity requirement on the derivative of Ψ_β must be relaxed.

Table 3-1

The Relationship between the Dirac (vector) and Schroedinger (function) Formalisms

	Quantum mechanics in a vector space	Quantum mechanics in the coordinate representation (wave mechanics)
Operator	$\hat{A}(\hat{\mathbf{p}}, \hat{\mathbf{f}})$	$\hat{A}\left(\dfrac{\hbar}{i}\nabla, \mathbf{r}\right)$
Characteristic state	$\lvert a_i\rangle,\ \langle a_i\rvert$	$\psi_{a_i}(\mathbf{r}) = \langle \mathbf{r}\lvert a_i\rangle,\ \psi_{a_i}^*(\mathbf{r}) = \langle a_i\lvert \mathbf{r}\rangle$ (eigenfunctions)
Eigenvalue	a_i	a_i
Eigenvalue equation	$\hat{A}\lvert a_i\rangle = a_i\lvert a_i\rangle$	$\hat{A}\left(\dfrac{\hbar}{i}\nabla, \mathbf{r}\right)\psi_{a_i}(\mathbf{r}) = a_i\,\psi_{a_i}(\mathbf{r})$
Arbitrary state	$\lvert \beta, t\rangle,\ \langle \beta, t\rvert$	$\Psi_\beta(\mathbf{r}, t) = \langle \mathbf{r}\lvert \beta, t\rangle,\ \Psi_\beta^*(\mathbf{r}, t) = \langle \beta, t\lvert \mathbf{r}\rangle$
Inner product	$\langle \alpha\lvert \beta\rangle$	$\displaystyle\int \psi_\alpha^*(\mathbf{r})\psi_\beta(\mathbf{r})\, d\mathbf{r}$
Matrix element	$\langle \alpha\lvert \hat{A}\rvert \beta\rangle$	$\displaystyle\int d\mathbf{r}\,\psi_\alpha^*(\mathbf{r})\hat{A}\left(\dfrac{\hbar}{i}\nabla, \mathbf{r}\right)\psi_\beta(\mathbf{r})$
Hermitian property	$\langle \alpha\lvert \hat{A}\rvert \beta\rangle = (\langle \beta\lvert \hat{A}\rvert \alpha\rangle)^*$	$\displaystyle\int d\mathbf{r}\,\psi_\alpha^*(\mathbf{r})\hat{A}\left(\dfrac{\hbar}{i}\nabla, \mathbf{r}\right)\psi_\beta(\mathbf{r}) = \left[\int d\mathbf{r}\,\psi_\beta^*(\mathbf{r})\hat{A}\left(\dfrac{\hbar}{i}\nabla, \mathbf{r}\right)\psi_\alpha\right]^*$
Schroedinger equation of motion	$\hat{\mathscr{H}}(\hat{\mathbf{p}}, \hat{\mathbf{f}})\lvert \beta, t\rangle = i\hbar\dfrac{d\lvert \beta, t\rangle}{dt}$	$\mathscr{H}\left(\dfrac{\hbar}{i}\nabla, \mathbf{r}\right)\Psi_\beta(\mathbf{r}, t) = i\hbar\dfrac{\partial\Psi_\beta(\mathbf{r}, t)}{\partial t}$
Time development of an energy eigenstate	$\lvert \varepsilon_i, t\rangle = \exp\left[-\dfrac{i\varepsilon_i(t-t_0)}{\hbar}\right]\lvert \varepsilon_i, t_0\rangle$	$\Psi_{\varepsilon_i}(\mathbf{r}, t) = \exp\left[-\dfrac{i\varepsilon_i(t-t_0)}{\hbar}\right]\Psi_{\varepsilon_i}(\mathbf{r}, t_0)$
Time development of an arbitrary state	$\lvert \beta, t\rangle = \exp\left[-\dfrac{i\hat{\mathscr{H}}(t-t_0)}{\hbar}\right]\lvert \beta, t_0\rangle$	$\Psi_\beta(\mathbf{r}, t) = \displaystyle\int d\mathbf{r}'\,G(\mathbf{r}, \mathbf{r}', t-t_0)\Psi_\beta(\mathbf{r}', t_0)$
Expansion	$\lvert \beta\rangle = \sum_i \beta^{(a_i)}\lvert a_i\rangle$ where $\beta^{(a_i)} = \langle a_i\lvert \beta\rangle$	$\psi_\beta(\mathbf{r}) = \sum_i \beta^{(a_i)}\psi_{a_i}(\mathbf{r})$ where $\beta^{(a_i)} = \displaystyle\int \psi_{a_i}^*(\mathbf{r})\psi_\beta(\mathbf{r})\, d\mathbf{r}$
Completeness relation	$\sum_i \lvert a_i\rangle\langle a_i\rvert = \hat{1}$	$\sum_i \psi_{a_i}^*(\mathbf{r}')\psi_{a_i}(\mathbf{r}) = \delta(\mathbf{r} - \mathbf{r}')$

In Chapter 4 we turn our attention to the stationary (energy) eigenfunctions and eigenvalues of various one-dimensional Hamiltonians. Since we shall be dealing with one-particle systems, the wave-mechanical representation will be used. Table 3-1 summarizes the relationship between Dirac's formalism and Schroedinger's wave mechanics. The last property (completeness) is established using

$$\langle \mathbf{r} | \sum_i | a_i \rangle \langle a_i | \, | \mathbf{r}' \rangle = \langle \mathbf{r} | \hat{1} | \mathbf{r}' \rangle$$

$$\sum_i \psi_{a_i}^*(\mathbf{r}') \psi_{a_i}(\mathbf{r}) = \delta(\mathbf{r} - \mathbf{r}').$$

Suggested Reading

Borowitz, S., "Fundamentals of Quantum Mechanics." Benjamin, New York, 1967.
Dirac, P. A. M., "The Principles of Quantum Mechanics," 4th ed., 1958. Oxford Univ. Press, London and New York, 1958.
  (This text is one of the great classic works on modern quantum theory. While it is mathematically sophisticated it is nevertheless unique in its approach, style, and notation, and much of the material in Chapter 3 can be found in Dirac's book.)
Feynman R. P., Leighton, R. B., and Sands, M., "The Feynman Lectures on Physics," Vol. 3. Addison-Wesley, Reading, Massachusetts, 1964.
Gottfried, K., "Quantum Mechanics," Vol. I. Benjamin, New York, 1966.
Merzbacher, E. "Quantum Mechanics," 2nd ed. Wiley, New York, 1970.
Messiah, A., "Quantum Mechanics," Vol. I. Wiley, New York, 1958.
Stehle, P. "Quantum Mechanics." Holden-Day, San Francisco, 1966.

Problems

3-1. (a) Find the eigenvalues and *one* set of orthonormal eigenvectors of the Hermitian matrix

$$\hat{A} = \begin{pmatrix} 2 & 0 & 0 \\ 0 & 1 & 1 \\ 0 & 1 & 1 \end{pmatrix}.$$

(b) Verify that the eigenvectors are orthogonal.

3-2. Consider the matrix:

$$\hat{B} = \begin{pmatrix} 0 & 1 & 1 \\ 1 & 0 & 1 \\ 1 & 1 & 0 \end{pmatrix}.$$

(a) Show that this matrix commutes with the matrix $\hat{A}$ given in Problem 3-1.

(b) Find at least one orthonormal set of eigenvectors common to $\hat{A}$ and $\hat{B}$.

(c) Display the matrices in this common eigenbasis.

3-3. Using the rules of bra-ket algebra, verify that in general

$$(\hat{A}\hat{B})^{\dagger} = \hat{B}^{\dagger}\hat{A}^{\dagger} \quad \text{and} \quad (\hat{A}\hat{B})^{-1} = \hat{B}^{-1}\hat{A}^{-1}.$$

3-4. Consider a set of orthonormal vectors in a given basis:

$$|\beta_1\rangle = \begin{pmatrix} 1 \\ 0 \\ 0 \end{pmatrix}; \quad |\beta_2\rangle = \frac{1}{\sqrt{2}}\begin{pmatrix} 0 \\ 1 \\ 1 \end{pmatrix}; \quad |\beta_3\rangle = \frac{1}{\sqrt{2}}\begin{pmatrix} 0 \\ 1 \\ -1 \end{pmatrix}.$$

(a) Find the corresponding projection matrices $\hat{P}_1$, $\hat{P}_2$, and $\hat{P}_3$.

(b) Verify that $\hat{P}_1 + \hat{P}_2 + \hat{P}_3 = \hat{1}$.

(c) Let

$$|a\rangle = \begin{pmatrix} 1 \\ 1 \\ 1 \end{pmatrix}.$$

Verify that $\hat{P}_2$ effects a projection on $|\beta_2\rangle$, that is, $\hat{P}_2|a\rangle = (\langle\beta_2|a\rangle)|\beta_2\rangle$.

3-5. Explain why the relation $e^{\hat{A}+\hat{B}} = e^{\hat{A}}e^{\hat{B}}$ is *not* valid for arbitrary operators. Under what circumstances does the relation hold? Express the matrix $e^{\lambda\hat{A}}$ to order λ^2 using

$$\hat{A} = \begin{pmatrix} 1 & 2 \\ 2 & 1 \end{pmatrix}.$$

3-6. Verify the vector identity:

$$\langle\gamma|\gamma\rangle\langle\delta|\delta\rangle - |\langle\gamma|\delta\rangle|^2 \equiv \langle\beta|\beta\rangle\langle\delta|\delta\rangle$$

where

$$|\beta\rangle = |\gamma\rangle - \left[\frac{\langle\delta|\gamma\rangle}{\langle\delta|\delta\rangle}\right]|\delta\rangle.$$

Use this identity to prove Schwarz's inequality, (3-57).

3-7. Using the commutation relation $[\hat{x}, \hat{p}_x] = i\hbar\hat{1}$, verify the relation

$$\hat{p}_x = \frac{m}{i\hbar}[\hat{x}, \mathscr{H}]$$

where $\mathscr{H}$ is any Hamiltonian of the form

$$\mathscr{H} = \frac{\hat{p}_x^2}{2m} + \hat{V}(\hat{x}).$$

3-8. Show that for any bound stationary state we have $\langle p_x \rangle_n = 0$. (Hint: Take the expectation value of both sides of the relation derived in Problem 3-7 with respect to an eigenstate of $\mathscr{H}$, for example, $|\varepsilon_n\rangle$.)

3-9. Use the rules of matrix multiplication to show that in a three-dimensional space

$$\mathrm{Tr}\ \hat{A}\hat{B} = \mathrm{Tr}\ \hat{B}\hat{A}.$$

3-10. Consider the function defined by (Figure 3-3)

$$\begin{aligned} f(x, x') &= 1/a & \text{for } -\tfrac{1}{2}a + x' \leqslant x \leqslant \tfrac{1}{2}a + x' \\ &= 0 & \text{otherwise.} \end{aligned}$$

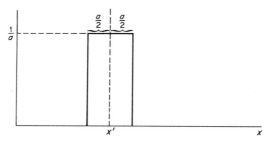

Figure 3-3

(a) Show that as $a \to 0$, this function becomes a Dirac delta function, that is,

$$\lim_{a \to 0} f(x, x') \to \delta(x - x').$$

(b) Verify that this delta function has the "sifting" property

$$g(x') = \int_{-\infty}^{\infty} g(x)\ \delta(x - x')\ dx$$

using the function $g(x) = bx^2$. (Hint: Perform the integration for finite a and then let $a \to 0$.)

3-11. Show that if we keep the state ket fixed at its initial value, that is, $|\beta\rangle \equiv |\beta, 0\rangle$, we can preserve the classical form of the equation of motion for $\langle A(t)\rangle$ in Eq. (3-61) provided we let the operators vary according to

$$\hat{A}_H(t) = e^{i\mathscr{H}t/\hbar}\hat{A}e^{-i\mathscr{H}t/\hbar}.$$

Here $\hat{A}_H(t)$ and $\hat{A}$ are respectively the operators in Heisenberg and Schroedinger pictures (see Appendix B).

3-12. Show that

$$\left[x, \frac{\hbar}{i} \frac{\partial}{\partial x} \right] = i\hbar \hat{1}$$

holds in wave mechanics, by operating with both sides on an arbitrary function ψ.

3-13. Consider the operator $1/\hat{p}_x$. Discuss a general form for this operator in wave mechanics. (Hint: Find the operator which satisfies

$$\frac{1}{\hat{p}_x} \psi(x) = \varphi(x)$$

subject to the condition that $\varphi(-\infty) = 0$. Do so by operating with $\hat{p}_x$ on both sides.)

3-14. Show that if the potential is a complex function then a sink term appears in (3-97) of the form

$$\text{probability absorption} = -2 \operatorname{Im} V |\Psi|^2 / \hbar.$$

4 | Wave Mechanics in One Dimension

Every physical system is characterized by its Hamiltonian, which in turn is usually determined by the potential energy function $V(\mathbf{r})$. To illustrate the ideas developed in Chapter 3, we shall investigate the properties of one-dimensional, one-particle systems whose Hamiltonians are independent of time, with particular emphasis on the stationary eigenstate solutions. The stationary eigenstates are of primary interest for the following reasons:

(a) Each state is of characteristic energy ε_i.

(b) If the system is initially in such a state, it remains in that state indefinitely.

(c) Once the energy eigenvalues and eigenfunctions are obtained, the propagator $G(\mathbf{r}, \mathbf{r}', t, t_0)$ may be constructed using (3-93), and the evolution of arbitrary states determined.

I Classification of Stationary States in Wave Mechanics

The eigenfunctions and eigenvalues of stationary states are determined by the solutions of the Schroedinger (energy-eigenvalue) equation‡

$$\mathscr{H}\psi_{\varepsilon_i}(\mathbf{r}) = \left\{ -\frac{\hbar^2 \, \nabla^2}{2m} + V(\mathbf{r}) \right\} \psi_{\varepsilon_i}(\mathbf{r}) = \varepsilon_i \, \psi_{\varepsilon_i}(\mathbf{r}). \tag{4-1}$$

We classify a state as bound or unbound according to whether the corresponding function (or probability density) vanishes at infinity. If the state function associated with $|\beta\rangle$ has the property

$$\lim_{r \to \infty} \langle \mathbf{r}|\beta\rangle = \lim_{r \to \infty} \Psi_\beta(\mathbf{r}) \to 0 \tag{4-2}$$

then the state is bound; otherwise it is unbound. We shall see shortly that the following rule holds:

Rule An energy eigenstate will generally be part of a *discrete* spectrum if it is bound and part of a *continuous* spectrum if it is unbound.

It is also important to have a rule which determines the type of spectrum associated with a given Hamiltonian.

Rule A Hamiltonian operator will have bound (discrete) or unbound (continuous) eigenstates according to whether or not its classical counterpart supports bound orbits.

For example, classically a free particle governed by a Hamiltonian with $V = 0$ moves in a straight line at constant speed and eventually tends toward infinity. The corresponding quantum-mechanical Hamiltonian has only a continuous spectrum. On the other hand, the classical oscillator Hamiltonian ($V = \frac{1}{2}kr^2$) allows only bound motions regardless of the energy. For all finite energies, there exists a maximum displacement from the origin referred to as the amplitude of the oscillation. Consequently, the quantum oscillator supports only bound states and has a discrete spectrum for all energies. The Kepler or Coulomb Hamiltonian ($V = -k/r$) supports bound motion (ellipses) for $\varepsilon < 0$ and unbound motion (parabolas and hyperbolas) for $\varepsilon \geq 0$ (Figure 4-1). We may therefore expect a discrete spectrum for $\varepsilon < 0$ and a continuous one for $\varepsilon \geq 0$.§

---

‡ This time-independent Schroedinger equation should not be confused with the (time-dependent) Schroedinger wave equation (3-90).

§ Here ε refers to the energy and is not to be confused with the eccentricity introduced in Chapter 2.

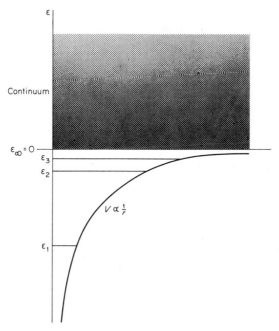

Figure 4-1   Bound (discrete) spectrum for $\varepsilon < 0$ and the unbound (continuous spectrum) for $\varepsilon > 0$ for the energies associated with a $1/r$ potential.

The theories associated with bound and unbound stationary eigenstates are fundamentally different. For bound states, the vanishing of the eigenfunctions at infinity leads directly to a discrete spectrum. The problem then is to find the eigenfunctions and corresponding energy eigenvalues and to evaluate the expectation values of various operator observables. On the other hand, for unbound states, the energy eigenvalues are continuous, that is they are not quantized. Furthermore, the form of the eigenfunctions for unbound states commonly leads to normalization procedures that leave the expectation values of some operators undefined. More important is the *probability current* since this current is related to the steady state scattering process. In fact, as we shall see in Chapter 8, the ratio between the fluxes traveling in the different directions and regions of space determines the scattering cross section of a collision process. For this reason unbound or continuous states are also called *scattering states*.

While it is not possible to require that the eigenfunction for an unbound state vanish at infinity, it is nevertheless necessary to impose certain asymptotic conditions suggested by the particular scattering experiment. These conditions dictate the general nature of the eigenfunction as we approach infinity.

We shall illustrate the energy-eigenvalue problem for one-dimensional systems. While one-dimensional problems usually represent only approximations to physical situations, they are discussed here primarily because they illustrate the mathematical techniques of quantum theory. Often they can easily be generalized to three-dimensional problems. Our discussion begins with continuous states of Hamiltonians in one dimension and examines their relationship to scattering from potential barriers.

II   The Free Particle in One Dimension

We begin by considering a free particle, that is, one for which $V = 0$. The Hamiltonian for this system is $\mathscr{H} = \hat{p}^2/2m$ and leads to the Schroedinger energy eigenvalue equation

$$-\frac{\hbar^2}{2m}\frac{d^2}{dx^2}\psi = \varepsilon\psi$$

or

$$\psi'' + k^2\psi = 0 \qquad \text{where} \quad k = \left(\frac{2m\varepsilon}{\hbar^2}\right)^{1/2}.$$

This equation has the two linearly independent solutions

$$\psi_k^+ = e^{ikx} \qquad \text{and} \quad \psi_k^- = e^{-ikx}.$$

Each solution is a simultaneous eigenfunction of the $\hat{p}$ operator. Operating with $\hat{p}$ on $\psi_k^{\pm}$ we find

$$\hat{p}\psi_k^{\pm} = \frac{\hbar}{i}\frac{d}{dx}\psi_k^{\pm} = \pm\frac{\hbar}{i}ik\,\psi_k^{\pm} = \pm\hbar k\psi_k^{\pm}$$

so that the corresponding momentum eigenvalues are $p^{\pm} = \pm\hbar k$. Thus the state $\psi_k^+ = e^{ikx}$ represents a particle with momentum $\hbar k$ moving along the positive x axis. Similarly ψ_k^- represents a free particle moving toward the left with momentum $-\hbar k$. Both states are degenerate in energy since both have

$$\varepsilon = \frac{\hbar^2 k^2}{2m}.$$

The probability flux associated with $\psi_k^{\pm}$ is, from (3-96),

$$J^{\pm} = -\frac{i\hbar}{2m}\left[\psi_k^{\pm*}\frac{d}{dx}\psi_k^{\pm} - \psi_k^{\pm}\frac{d}{dx}\psi_k^{\pm*}\right] = \pm\frac{\hbar k}{m} = \pm\frac{p}{m} = \pm v.$$

Thus the flux is equal to the velocity of the particle and flows to the right for $\psi_k{}^+$ and to the left for $\psi_k{}^-$. The general solution to our free-particle Schroedinger energy-eigenvalue equation is

$$\psi_{gen} = A\psi_k{}^+ + B\psi_k{}^-.$$

III Scattering from One-Dimensional Barriers

In a one-dimensional scattering problem, a beam of particles is incident from a given direction and is partially reflected from a potential barrier $V(x)$; the remainder of the beam is transmitted (Figure 4-2).

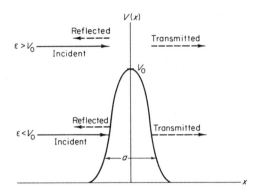

Figure 4-2 Scattering from a one-dimensional barrier.

The barrier may be repulsive $(V > 0)$ or attractive $(V < 0)$; however, all that we shall assume is that $V \to 0$ as $x \to \pm\infty$. Thus the particles of the beam are essentially free at large distances from the origin. While the shape of the barrier is not of particular importance, it may be qualitatively characterized by a height V_0 and a width a as in Figure 4-2. The particles of the beam may be of greater or lower energy than V_0.

The energy eigenfunctions associated with this potential are found by solving the one-dimensional Schroedinger equation

$$\mathscr{H}\left(\frac{\hbar}{i}\frac{d}{dx}, x\right)\psi_\varepsilon(x) = \left\{-\frac{\hbar^2}{2m}\frac{d^2}{dx^2} + V(x)\right\}\psi_\varepsilon(x) = \varepsilon\psi_\varepsilon(x). \qquad (4\text{-}3)$$

For scattering states the energy eigenvalues are continuous and we shall solve (4-3) for any choice of $\varepsilon > 0$. As $|x| \to \infty$ the potential vanishes and it may be verified that e^{ikx} and e^{-ikx} $(k = (2m\varepsilon/\hbar^2)^{1/2})$ are two linearly independent solutions. As shown previously these solutions represent particles traveling to the right and left respectively. We shall assume that the beam is incident from the left. Far to the left of the barrier we would then expect an incident beam ($\sim e^{ikx}$) and a reflected beam ($\sim e^{-ikx}$). However, far to the right we must only have a transmitted beam ($\sim e^{ikx}$). We therefore impose the following *asymptotic* conditions on the acceptable solutions to (4-3):

$$\psi_\varepsilon(x) \underset{x \to -\infty}{\sim} A e^{ikx} + B e^{-ikx} \qquad \text{(incidence and reflection)}$$

$$\psi_\varepsilon(x) \underset{x \to +\infty}{\sim} E e^{ikx} \qquad \text{(transmission)} \tag{4-4}$$

$$\left(k = \left(\frac{2m\varepsilon}{\hbar^2} \right)^{1/2} \right).$$

The scattering problem reduces to finding those (exact) solutions (for fixed ε) to the Schroedinger equation (4-3) which have the required asymptotic form. When the exact solution has been obtained and the asymptotic form established, the coefficients A, B, and E may be used to predict the extent to which the beam is reflected and transmitted.

The incident flux is related to the asymptotic eigenfunction $\psi_{\text{inc}} = A e^{ikx}$, using (3-96), which gives

$$J_{\text{inc}} = \frac{-i\hbar}{2m} \left(\psi_{\text{inc}}^* \frac{d}{dx} \psi_{\text{inc}} - \psi_{\text{inc}} \frac{d}{dx} \psi_{\text{inc}}^* \right)$$

$$= \frac{\hbar k}{m} |A|^2.$$

An application to the reflected and transmitted functions gives

$$J_{\text{refl}} = \frac{\hbar k}{m} |B|^2 \qquad \text{and} \qquad J_{\text{trans}} = \frac{\hbar k}{m} |E|^2.$$

It is convenient to define the reflection and transmission coefficients as

$$\mathscr{R} = \frac{J_{\text{refl}}}{J_{\text{inc}}} = \left| \frac{B}{A} \right|^2 \tag{4-5}$$

and

$$\mathscr{T} = \frac{J_{\text{trans}}}{J_{\text{inc}}} = \left| \frac{E}{A} \right|^2. \tag{4-6}$$

Strictly speaking, $\mathscr{R}$ and $\mathscr{T}$ refer to the probability flow; however, they also represent the fractions of the incident beam reflected and transmitted. Since probability is conserved, we expect that $\mathscr{R} + \mathscr{T} = 1$. Note that $\mathscr{R}$ and $\mathscr{T}$ depend on ε as well as on the characteristics of the barrier, V_0 and a.

IV   The Rectangular Barrier

We illustrate the essential features of scattering in one dimension using a rectangular barrier potential, $V(x) = V_0$ for $0 \leqslant x \leqslant a$ and $V(x) = 0$ otherwise (Figure 4-3).

The discontinuity at $x = 0$ and $x = a$ makes this barrier somewhat unphysical. The Schroedinger equation must be solved in the following three regions:

$$-\frac{\hbar^2}{2m}\frac{d^2}{dx^2}\,\psi_\varepsilon^{I}(x) = \varepsilon\psi_\varepsilon^{I}(x) \qquad (x < 0)$$

$$\left\{-\frac{\hbar^2}{2m}\frac{d^2}{dx^2} + V_0\right\}\psi_\varepsilon^{II}(x) = \varepsilon\psi_\varepsilon^{II}(x) \qquad (0 \leqslant x \leqslant a)$$

$$-\frac{\hbar^2}{2m}\frac{d^2}{dx^2}\,\psi_\varepsilon^{III}(x) = \varepsilon\psi_\varepsilon^{III}(x) \qquad (a < x). \tag{4-7}$$

The "exterior" solutions are easily shown to be of the general form

$$\psi_\varepsilon^{I}(x) = A\,e^{ikx} + B\,e^{-ikx}$$

$$\left(k = \left(\frac{2m\varepsilon}{\hbar^2}\right)^{1/2}\right). \tag{4-8}$$

$$\psi_\varepsilon^{III}(x) = E\,e^{ikx} + F\,e^{-ikx}.$$

The "interior" solution is either

$$\psi_\varepsilon^{II}(x) = C e^{ik'x} + D e^{-ik'x} \qquad \left(k' = \left(\frac{2m(\varepsilon - V_0)}{\hbar^2}\right)^{1/2}\right)$$

when $\varepsilon > V_0$ or $\hspace{4cm}$ (4-9)

$$\psi_\varepsilon^{II}(x) = C e^{Kx} + D e^{-Kx} \qquad \left(K = \left(\frac{2m(V_0 - \varepsilon)}{\hbar^2}\right)^{1/2}\right)$$

when $\varepsilon < V_0$.

We observe that the asymptotic requirement $\psi_\varepsilon \underset{x\to\infty}{\sim} e^{ikx}$ can only be satisfied if $F = 0$ in region III. The remaining coefficients can be evaluated using the continuity requirements (Postulate 7) on ψ_ε which require the continuity of ψ_ε and its derivative at $x = 0$ and $x = a$. We are thus led to four equations for the remaining five coefficients (Table 4-1). These equations uniquely determine the ratios B/A and E/A whose squares respectively give the reflection and transmission coefficients.

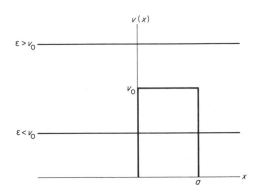

Figure 4-3   The rectangular barrier and two possible scattering energies.

The details are left as an exercise but the results are listed below. For $\varepsilon > V_0$, we find

$$\mathscr{R} = \left|\frac{B}{A}\right|^2 = \frac{\sin^2 k'a}{(4\varepsilon/V_0)(\varepsilon/V_0 - 1)}\left[1 + \frac{\sin^2 k'a}{(4\varepsilon/V_0)(\varepsilon/V_0 - 1)}\right]^{-1}$$

$$\mathscr{T} = \left|\frac{E}{A}\right|^2 = \left[1 + \frac{\sin^2 k'a}{(4\varepsilon/V_0)(\varepsilon/V_0 - 1)}\right]^{-1}$$

(4-10a)

while for $\varepsilon < V_0$

$$\mathscr{R} = \left|\frac{B}{A}\right|^2 = \frac{\sinh^2 Ka}{(4\varepsilon/V_0)(1 - \varepsilon/V_0)}\left[1 + \frac{\sinh^2 Ka}{(4\varepsilon/V_0)(1 - \varepsilon/V_0)}\right]^{-1}$$

$$\mathscr{T} = \left|\frac{E}{A}\right|^2 = \left[1 + \frac{\sinh^2 Ka}{(4\varepsilon/V_0)(1 - \varepsilon/V_0)}\right]^{-1}.$$

(4-10b)

Note that $\mathscr{R} + \mathscr{T} = 1$ is satisfied in both cases.

Table 4-1

Generating Four Equations for A, B, C, D, and E using the Continuity Requirements on ψ_ε (for $\varepsilon > V_0$)

$x = 0$	$x = a$
$\psi_\varepsilon^{\mathrm{I}}(0) = \psi_\varepsilon^{\mathrm{II}}(0)$ $A + B = C + D$	$\psi_\varepsilon^{\mathrm{II}}(a) = \psi_\varepsilon^{\mathrm{III}}(a)$ $C\,e^{ik'a} + D\,e^{-ik'a} = E\,e^{ika}$
$\psi_\varepsilon'^{\mathrm{I}}(0) = \psi_\varepsilon'^{\mathrm{II}}(0)$ $ik(A - B) = ik'(C - D)$	$\psi_\varepsilon'^{\mathrm{II}}(a) = \psi_\varepsilon'^{\mathrm{III}}(a)$ $ik'(C\,e^{ik'a} - D\,e^{-ik'a}) = ikE\,e^{ika}$

Our solutions to the barrier problem have the following interesting properties:

(a)   Reflection may occur even when $\varepsilon > V_0$. This is entirely a quantum-mechanical effect and must be attributed to the wavelike behavior of the particle. The classical limit is obtained at large incident energies‡ ($\varepsilon/V_0 \gg 1$). In that case, (4-10) correctly reduces to $\mathscr{R} \to 0$ and $\mathscr{T} \to 1$ as expected.

(b)   Transmission may occur even when $\varepsilon < V_0$, the reasoning being similar to that described in (a). The classical limit‡ occurs in (4-10) for $V_0/\varepsilon \gg 1$ and gives $\mathscr{T} \to 0$ and $\mathscr{R} \to 1$ as expected. The ability to penetrate a classically forbidden region ($p^2 < 0$) is the quantum-mechanical phenomenon of *tunneling* and is observed in a variety of experiments. Note that the probability of finding the particle in the classically forbidden region is not zero, although in the classical limit this probability is vanishingly small. The scattering state described by (4-8) and (4-9) represents a situation in which reflection and transmission is measured far from the barrier. Any actual measurement of the particle's characteristics within the classically forbidden region localizes the particle, modifies the barrier, and changes the entire nature of the problem.

(c)   For $\varepsilon > V_0$, there exists a set of resonant incident energies for which $\mathscr{T} = 1$ ($\mathscr{R} = 0$) and for which the barrier becomes "invisible" (Figure 4-4). From (4-10), it follows that these energies are the roots of the equation

$$\frac{\sin^2 k_n'a}{4\dfrac{\varepsilon_n}{V_0}\left(\dfrac{\varepsilon_n}{V_0} - 1\right)} = 0 \qquad \left(k_n' = \frac{[2m(\varepsilon_n - V_0)]^{1/2}}{\hbar}\right)$$

for in that case $\mathscr{T}$ is clearly unity. The left-hand side is zero when

$$\frac{2\pi}{\lambda'}a = k_n'a = n\pi \qquad (n = 1, 2, 3, \ldots)$$

or equivalently when

$$a = \tfrac{1}{2}n\lambda'$$

that is, when the barrier width is a multiple of one half of the de Broglie wavelength within the barrier. This is an interference effect of quantum mechanics. We shall see later that a similar effect occurs in three dimensions

---

‡ Strictly speaking, the classical limit should be reached when $h \to 0$. In this limit we would expect $\mathscr{R} = 1$, $\mathscr{T} = 0$ when $\varepsilon < V_0$, and $\mathscr{R} = 0$, $\mathscr{T} = 1$ when $\varepsilon > V_0$. In the case where $\varepsilon < V_0$, (4–10b), we find $\sinh Ka \to \infty$ as $h \to 0$ and the classical result is, in fact, recovered. However for $\varepsilon > V_0$, (4–10a), some difficulties are encountered in taking the limit. These difficulties are related to the unphysical discontinuity of the potential at $x = 0$ and $x = a$ and will generally be absent with a "smooth" barrier.

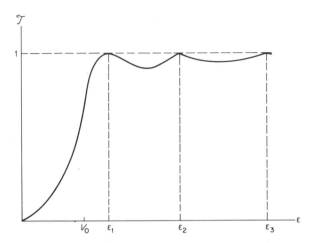

Figure 4-4 The resonances ε_n in the transmission coefficient. Note that as $\varepsilon \to \infty$, $\mathcal{T} \to 1$ (classical limit).

with electron scattering from noble gas atoms, where the transparency of the atomic scatterers at certain incident energies is known as the *Ramsauer* effect.

The barrier considered above was repulsive in character ($V_0 > 0$) and supported only scattering states. The attractive barrier or well ($V_0 < 0$) can also support scattering states for $\varepsilon > 0$. The transmission coefficients are found in a similar manner to that used for the repulsive barrier. For negative energies ($\varepsilon < 0$), we shall show in Section IX that only bound states exist in the well. For these states we must require that ψ vanish at $x = \pm\infty$.

V   Bound Stationary States in One Dimension

In bound state problems, we are concerned with finding the allowable eigenfunctions and energies of a Hamiltonian. We seek solutions of the Schroedinger equation

$$\left\{ -\frac{\hbar^2}{2m}\frac{d^2}{dx^2} + V(x) \right\} \psi_i(x) = \varepsilon_i \psi_i(x) \tag{4-11}$$

subject to the boundary condition $\psi_i \xrightarrow[x \to \pm\infty]{} 0$. We will see that if we choose ε_i arbitrarily, the solution to (4-11) will not tend to zero as we approach infinity. Only for certain choices of $\varepsilon = \varepsilon_i$ will acceptable solutions exist. This is in fact how the bound nature of the eigenfunction leads to a discrete spectrum. Once the set of acceptable ε_i and ψ_i are found, expectation values of various operator observables may be calculated.

We shall demonstrate on numerous occasions that the general nature of the energy eigenstates may be predicted by analyzing the inherent characteristics of the Hamiltonian (for example, its symmetry properties). We begin by proving the following theorem regarding degeneracies in one-dimensional Hamiltonians.

Theorem IX *The bound states of a one-dimensional Hamiltonian are always nondegenerate.*‡

Proof (Contradiction) Let ψ_i and ψ_j be two degenerate linearly independent normalized eigensolutions to the Schroedinger equation, that is,

$$\left\{-\frac{\hbar^2}{2m}\frac{d^2}{dx^2} + V(x)\right\}\psi_i(x) = \varepsilon_i\,\psi_i(x)$$

and

$$\left\{-\frac{\hbar^2}{2m}\frac{d^2}{dx^2} + V(x)\right\}\psi_j(x) = \varepsilon_j\psi_j(x)$$

with $\varepsilon_j = \varepsilon_i = \varepsilon$. Multiplying the first by ψ_j and the second by ψ_i and subtracting, we obtain

$$-\frac{\hbar^2}{2m}(\psi_i\psi_j'' - \psi_j\psi_i'') = 0$$

or

$$\frac{-\hbar^2}{2m}\frac{d}{dx}(\psi_i\psi_j' - \psi_j\psi_i') = 0.$$

It follows that§

$$(\psi_i\psi_j' - \psi_j\psi_i') = \text{const.} \tag{4-12}$$

Since $\psi_j(\infty) = \psi_i(\infty) = 0$, as is required for bound states, it follows that the constant above is zero and we have for all x

$$\psi_i\psi_j' - \psi_j\psi_i' = 0. \tag{4-13}$$

---

‡ This theorem cannot be generalized to three dimensions. Furthermore there are some *unphysically* singular one-dimensional potentials (for example, $V = -1/|x|$) which support bound states and have a degeneracy. These potentials will not be treated here.

§ The quantity

$$W = \psi_i\psi_j' - \psi_j\psi_i' = \begin{vmatrix} \psi_i & \psi_j \\ \psi_i' & \psi_j' \end{vmatrix}$$

is called the *Wronskian* of the functions ψ_i and ψ_j.

Equation (4-13) may be solved as

$$\frac{\psi_j'}{\psi_j} = \frac{\psi_i'}{\psi_i}, \qquad \frac{d\psi_j}{\psi_j} = \frac{d\psi_i}{\psi_i},$$

$$\ln \psi_j = \ln \psi_i + \ln C, \qquad \psi_j = C\psi_i. \tag{4-14}$$

However since ψ_i and ψ_j are both assumed normalized it follows that we may set $C = 1$ and that ψ_i and ψ_j are in fact identical, proving the theorem by contradiction.

In fact, we could replace ψ_j by ψ_i^* in the proof of Theorem IX, obtaining the result $\psi_i^* = \psi_i$ and proving that the bound eigenfunctions of a one-dimensional Hamiltonian are also essentially real.

As a consequence of Theorem II in Chapter 3 it follows that the bound eigenfunctions are unique and mutually orthogonal, that is,

$$\int_{-\infty}^{\infty} \psi_j^*(x)\psi_i(x)\, dx = \delta_{ji}.$$

VI The Infinite Well

An infinite well is described by a potential $V(x)$, where $V(x) = 0$ for $0 \leqslant x \leqslant a$ and $V(x) = \infty$ otherwise (Figure 4-5). The Hamiltonian for a particle in this well is $\mathscr{H} = (p^2/2m) + V(x)$. Classically, the motion is confined to remain in the well, being maintained by repeated reflections from the walls of the well. The corresponding Hamiltonian operator supports only bound states and has a discrete spectrum. Since V is infinite for $x > a$ and $x < 0$, we shall require that ψ_n vanish everywhere on the exterior of the well. We must therefore only find the interior solution for ψ_n which satisfies

$$-\frac{\hbar^2}{2m}\frac{d^2}{dx^2}\psi_n = \varepsilon_n \psi_n$$

and which vanishes at the boundary, $\psi_n(0) = \psi_n(a) = 0$. The general solution is

$$\psi_n = A_n \sin k_n x + B_n \cos k_n x \qquad \left(k_n = \left(\frac{2m\varepsilon_n}{\hbar^2}\right)^{1/2}\right).$$

Imposing the continuity condition at the origin, that is, $\psi_n(0) = 0$, we find $B_n = 0$. At $x = a$ we require $\psi_n(a) = 0$ and obtain

$$\psi_n(a) = A_n \sin k_n a = 0. \tag{4-15}$$

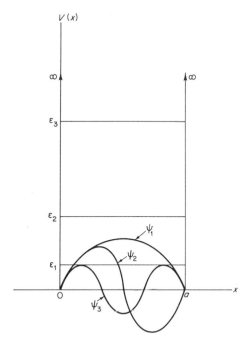

Figure 4-5 The infinite well and a few of its eigenvalues and eigenfunctions.

Equation (4-15) can be solved (nontrivially, $A_n \neq 0$) only if the argument of the sine function is restricted to the values

$$k_n a = n\pi \qquad (n = 1, 2, 3, \ldots)$$

or

$$k_n = \frac{n\pi}{a}. \tag{4-16}$$

This generates eigenfunctions of the form $\psi_n = A_n \sin (n\pi/a)x$. The coefficients are determined by the normalization requirement‡

$$\int \psi_n^* \psi_n \, dx = |A_n|^2 \int_0^a dx \sin^2\left(\frac{n\pi}{a} x\right) = |A_n|^2 \frac{1}{2} a = 1$$

or $A_n = (2/a)^{1/2}$. The eigenfunctions for the well now take the form§

$$\psi_n = \left(\frac{2}{a}\right)^{1/2} \sin \frac{n\pi}{a} x \qquad (0 \leqslant x \leqslant a)$$

$$\psi_n = 0 \qquad\qquad\qquad \text{(otherwise)} \tag{4-17}$$

---

‡ The normalization constant will always be taken to be real and positive.

§ Note that the derivative of ψ_n is not continuous at $x = 0$ and $x = a$; this may be attributed to the *infinite* discontinuity of $V(x)$ at these points. See footnote at the bottom of page 81.

with the energy eigenvalues determined by (4-16) as

$$\varepsilon_n = \frac{\hbar^2 k_n^{\,2}}{2m} = \left(\frac{\hbar^2 \pi^2}{2ma^2}\right)n^2. \tag{4-18}$$

The first few eigenfunctions and eigenvalues have been plotted in Figure 4-5. Note that these functions are real and nondegenerate as expected. A simple calculation would verify that they are indeed mutually orthogonal. It is interesting that the ground state ($n = 1$) energy is not zero. We shall see next that this phenomenon is a direct consequence of the uncertainty principle.

A simple calculation shows the expectation value of $\hat{p}$ in the ground state to be

$$\langle p \rangle_1 = \langle \varepsilon_1 | \hat{p} | \varepsilon_1 \rangle = \int_0^a \psi_1^*(x) \frac{\hbar}{i} \frac{d}{dx} \psi_1(x)\, dx$$

$$= \frac{\hbar}{i} \frac{2}{a} \int_0^a \sin\left(\frac{\pi}{a}x\right) \cos\left(\frac{\pi}{a}x\right) dx = 0.$$

However since $\hat{p}^2 = 2m\mathscr{H}$ (inside the well), we have

$$\langle p^2 \rangle_1 = 2m \langle \mathscr{H} \rangle_1 = 2m\varepsilon_1$$

so that the uncertainty in the ground state becomes

$$\Delta p_1 = [\langle p^2 \rangle_1 - \langle p \rangle_1^2]^{1/2} = (\langle p^2 \rangle_1)^{1/2} = (2m\varepsilon_1)^{1/2} = \frac{\hbar \pi}{a}. \tag{4-19}$$

Note that as the width of the well, which localizes the particle, decreases, the uncertainty of $\hat{p}$ increases as required by the uncertainty principle. However, since Δp_1 cannot vanish (for finite a), (4-19) implies that ε_1 cannot be zero. In fact, the last equality in (4-19) can be used to estimate the ground-state energy of a particle confined to a well of width a, regardless of its specific shape.

Using the eigenfunctions and energies as given by (4-17) and (4-18), it becomes possible to construct the propagator $G(x, x', t - t_0)$. The motion of an arbitrary state within the well may be determined using (3-93).

VII The Infinite Symmetric Well

The symmetry of a Hamiltonian is often reflected in its eigenstates. We shall illustrate this by translating the well of the previous section a distance $a/2$ along the negative x axis. Physical intuition would suggest that the energy eigenvalues remain unaffected by this translation. The new potential

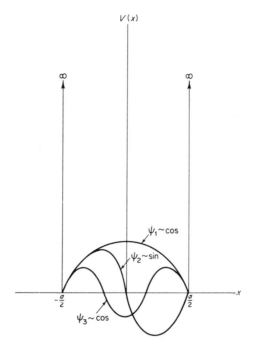

Figure 4-6   The infinite symmetric well and the symmetry and antisymmetry of its eigenfunctions.

$V(x)$ and the eigenfunctions are obtained using the shift $x \to x + a/2$. The potential (which is now symmetric) and the eigenfunctions become (Figure 4-6)

$$V(x) = 0 \qquad \text{for} \quad -a/2 \leqslant x \leqslant a/2$$
$$V(x) = \infty \qquad \text{otherwise}$$

and

$$\psi_n(x) = \left(\frac{2}{a}\right)^{1/2} \sin \frac{n\pi}{a}\left(x + \frac{a}{2}\right) = \left(\frac{2}{a}\right)^{1/2} \sin\left(\frac{n\pi x}{a} + \frac{n\pi}{2}\right).$$

Now, however, we generate two sets of solutions, that is,

$$\psi_n(x) = \left(\frac{2}{a}\right)^{1/2} \cos \frac{n\pi}{a} x \qquad (n = 1, 3, 5, \ldots)$$

$$(-a/2 \leqslant x \leqslant a/2).$$

$$\psi_n(x) = \left(\frac{2}{a}\right)^{1/2} \sin \frac{n\pi}{a} x \qquad (n = 2, 4, 6, \ldots)$$

The cosine eigenfunctions have the property $\psi_n(-x) = \psi_n(x)$, that is, they are symmetric with respect to reflections through the origin. The sine eigenfunctions have the property $\psi_n(-x) = -\psi_n(x)$, and they are antisymmetric functions. We shall see in the next section that symmetric and antisymmetric functions are said to be of even and odd parity respectively. In both even and odd states, the density $|\psi_n|^2$ is an even function. This reflection symmetry is no mere accident but follows from the symmetry property of the Hamiltonian as we next show.

VIII Parity

We introduce the (Hermitian) parity operator‡ $\hat{P}$ and define it by the commutation property

$$\hat{P}\hat{\mathbf{r}} = -\hat{\mathbf{r}}\hat{P} \tag{4-20}$$

or

$$[\mathbf{r}, \hat{P}]_+ = 0$$

where the *anticommutator* is defined by

$$[\hat{A}, \hat{B}]_+ = \hat{A}\hat{B} + \hat{B}\hat{A}.$$

Equation (4-20) also holds (see Problem 4-8) when we replace $\hat{\mathbf{r}}$ by $\hat{\mathbf{p}}$. From (4-20), it follows that

$$\hat{P}\hat{F}(\hat{\mathbf{r}}, \hat{\mathbf{p}}) = \hat{F}(-\hat{\mathbf{r}}, -\hat{\mathbf{p}})\hat{P}$$

where $\hat{F}$ is an arbitrary function of $\hat{\mathbf{r}}$ and $\hat{\mathbf{p}}$. The last relation will become obvious if it is recalled that $\hat{F}$ is *defined* by a Taylor series. Now, if $\hat{F}$ is an even operator function of $\hat{\mathbf{r}}$ and $\hat{\mathbf{p}}$ then $\hat{F}(-\hat{\mathbf{r}}, -\hat{\mathbf{p}}) = \hat{F}(\hat{\mathbf{r}}, \hat{\mathbf{p}})$. *Therefore $\hat{P}$ commutes with any even function of $\hat{\mathbf{p}}$ and $\hat{\mathbf{r}}$, that is,*

$$[\hat{P}, \hat{F}_{\text{even}}(\hat{\mathbf{p}}, \hat{\mathbf{r}})] = 0. \tag{4-21}$$

Using the commutation properties of the parity operator, it is possible to determine the effect of $\hat{P}$ on the eigenket $|\mathbf{r}\rangle$. Operating with $\hat{P}$ on the eigenvalue equation $\hat{\mathbf{r}}|\mathbf{r}\rangle = \mathbf{r}|\mathbf{r}\rangle$ and using (4-20) we find

$$\hat{P}\hat{\mathbf{r}}|\mathbf{r}\rangle = \mathbf{r}\hat{P}|\mathbf{r}\rangle$$
$$-\hat{\mathbf{r}}\{\hat{P}|\mathbf{r}\rangle\} = \mathbf{r}\{\hat{P}|\mathbf{r}\rangle\}$$

or

$$\hat{\mathbf{r}}\{\hat{P}|\mathbf{r}\rangle\} = -\mathbf{r}\{\hat{P}|\mathbf{r}\rangle\}. \tag{4-22}$$

---

‡ The parity operator is not a canonical function of $\hat{\mathbf{p}}$ and $\hat{\mathbf{r}}$, and thus has no classical analog. It is an example of a symmetry operator.

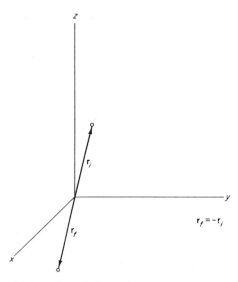

Figure 4-7   The reflection effect of the parity operator on a particle initially situated at $\mathbf{r}_i$.

The vector in brackets is an eigenket of $\hat{\mathbf{r}}$ with the eigenvalue $-\mathbf{r}$; thus we may make the identification

$$\hat{P}|\mathbf{r}\rangle = |-\mathbf{r}\rangle. \qquad (4\text{-}23)$$

The effect of the parity operator therefore is to produce a state in which the particle has been reflected through the origin (Figure 4-7).

Operating twice with $\hat{P}$ on $|\mathbf{r}\rangle$, and expecting a double reflection to be equivalent to an identity operation, we may write

$$\hat{P}\hat{P}|\mathbf{r}\rangle = \hat{P}^2|\mathbf{r}\rangle = |\mathbf{r}\rangle$$

so that‡

$$\hat{P}^2 = \hat{P}\hat{P} = \hat{1}. \qquad (4\text{-}24)$$

From (4-24) we may deduce (see Problem 4-14) that the eigenvalues of $\hat{P}$ are ± 1. Those eigenstates of $\hat{P}$ with the $+1$ eigenvalue are states of even parity. The -1 eigenvalue corresponds to states of odd parity. Mathematically we may write

$$\hat{P}|\beta(\text{even})\rangle = +1|\beta(\text{even})\rangle$$

and

$$\hat{P}|\beta(\text{odd})\rangle = -1|\beta(\text{odd})\rangle. \qquad (4\text{-}25)$$

---

‡ Note that $\hat{P}$ is both Hermitian *and* unitary since $\hat{P} = \hat{P}^\dagger$ and $\hat{P}\hat{P} = \hat{P}^\dagger\hat{P} = \hat{1}$.

Since a reflection-invariant Hamiltonian‡ $(\mathcal{H}(-\hat{\mathbf{r}}) = \mathcal{H}(\hat{\mathbf{r}}))$ commutes with the parity operator, $\hat{P}$, it follows that the two operators must have at least one set of common eigenvectors. We may conclude that every symmetric potential has at least one set of eigenstates that are of definite parity, that is, odd or even.

The eigenfunction for an even parity state has the property

$$\begin{aligned} \psi_{\beta_{\text{even}}}(\mathbf{r}) = \langle \mathbf{r} | \beta \text{ (even)} \rangle &= \langle \mathbf{r} | \{\hat{P} | \beta \text{ (even)} \rangle \} \\ &= \{\langle \mathbf{r} | \hat{P} \} | \beta \text{ (even)} \rangle \\ &= \langle -\mathbf{r} | \beta \text{ (even)} \rangle = \psi_{\beta_{\text{even}}}(-\mathbf{r}), \end{aligned} \qquad (4\text{-}26)$$

that is, it is invariant upon reflection through the origin. It may similarly be shown that for odd parity states

$$\psi_{\beta_{\text{odd}}}(\mathbf{r}) = -\psi_{\beta_{\text{odd}}}(-\mathbf{r}). \qquad (4\text{-}27)$$

Since one-dimensional bound eigenstates are nondegenerate and therefore unique, it follows that those belonging to a symmetric Hamiltonian are necessarily of definite parity, even or odd.

IX   The Finite Symmetric Well

To show how symmetry may be applied to simplify a problem, we shall consider a particle bound in a *finite* symmetric well, $V(x) = 0$ for $-a/2 \leqslant x \leqslant a/2$ and $V(x) = V_0$ otherwise (Figure 4-8).

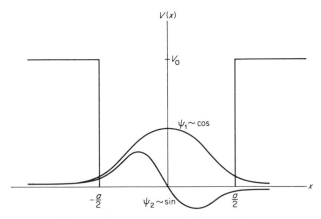

Figure 4-8   The finite symmetric well and a few of its eigenfunctions.

---

‡ Since we are assuming the Hamiltonian to be of the form $\mathcal{H} = (\hat{p}^2/2m) + \hat{V}(\hat{\mathbf{r}})$, $\mathcal{H}$ is automatically reflection-invariant with respect to a replacement of $\hat{p}$ by $-\hat{p}$. Equivalently the kinetic energy always commutes with the parity operator.

We concern ourselves with bound states for which $\varepsilon < V_0$. The Schroedinger equation has the general solutions

$$\psi_n^{I}(x) = C_n e^{K_n x} + E_n e^{-K_n x} \qquad (x < -a/2)$$

$$\psi_n^{II}(x) = A_n \sin k_n x + B \cos k_n x \qquad (-a/2 \leqslant x \leqslant a/2)$$

$$\psi_n^{III}(x) = D_n e^{-K_n x} + F_n e^{K_n x} \qquad (a/2 < x) \qquad (4\text{-}28)$$

where

$$k_n = \left(\frac{2m\varepsilon_n}{\hbar^2}\right)^{1/2} \quad \text{and} \quad K_n = \left(\frac{2m(V_0 - \varepsilon_n)}{\hbar^2}\right)^{1/2}. \qquad (4\text{-}29)$$

Since bound states require $\psi_n(\pm\infty) = 0$, we must set $E_n = F_n = 0$ in the exterior solutions. Next, we recall that a symmetric one-dimensional Hamiltonian's bound eigenfunctions must be of either odd (antisymmetric) or even (symmetric) parity. We may generate the even functions by noting that $\psi_n(-x) = \psi_n(x)$ can only be satisfied if $A_n = 0$ and $C_n = D_n$. It is convenient to assign odd quantum numbers to these states; the reasoning for this will become clear shortly. The even eigenfunctions therefore have the form

$$\psi_n^{I} = D_n e^{K_n x}$$

$$\psi_n^{II} = B_n \cos k_n x \qquad (n = 1, 3, 5, \ldots; \quad \text{even parity}). \qquad (4\text{-}30a)$$

$$\psi_n^{III} = D_n e^{-K_n x}$$

Assigning even integers to the odd functions and noting that $\psi_n(-x) = -\psi_n(x)$ requires $B_n = 0$ and $C_n = -D_n$, we obtain

$$\psi_n^{I} = C_n e^{K_n x}$$

$$\psi_n^{II} = A_n \sin k_n x \qquad (n = 2, 4, 6, \ldots; \quad \text{odd parity}). \qquad (4\text{-}30b)$$

$$\psi_n^{III} = -C_n e^{-K_n x}$$

For either parity state, we require continuity of the eigenfunction and its derivative at $x = \pm a/2$. Due to the symmetry of ψ_n, continuity at $x = +a/2$ automatically ensures the same at $x = -a/2$. The continuity requirements at $x = a/2$ have been summarized in Table 4-2. Dividing (b) by (a) in Table 4-2, we obtain the relations

$$k_n \tan \tfrac{1}{2}k_n a = K_n \qquad \text{(even parity)} \qquad (4\text{-}31a)$$

$$k_n \cot \tfrac{1}{2}k_n a = -K_n \qquad \text{(odd parity)}. \qquad (4\text{-}31b)$$

Furthermore, we may use (a) to express C_n or D_n in terms of B_n or A_n. The eigenfunctions now take the form

$$\psi_n^{I} = B_n(e^{K_n a/2} \cos \tfrac{1}{2}k_n a)e^{K_n x}$$

$$\psi_n^{II} = B_n \cos k_n x \qquad (n = 1, 3, 5, \ldots; \quad \text{even parity}) \quad (4\text{-}32a)$$

$$\psi_n^{III} = B_n(e^{K_n a/2} \cos \tfrac{1}{2}k_n a)e^{-K_n x}$$

Table 4-2

The Continuity Requirements for Odd and Even Functions of a Rectangular Well

	Even parity		Odd parity
a.	$\psi_n^{II}(a/2) = \psi_n^{III}(a/2)$ $B_n \cos \frac{1}{2}k_n a = C_n e^{-K_n a/2}$	a.	$\psi_n^{II}(a/2) = \psi_n^{III}(a/2)$ $A_n \sin \frac{1}{2}k_n a = D_n e^{-K_n a/2}$
b.	$\psi_n'^{II}(a/2) = \psi_n'^{III}(a/2)$ $-k_n B_n \sin \frac{1}{2}k_n a = -K_n C_n e^{-K_n a/2}$	b.	$\psi_n'^{II}(a/2) = \psi_n'^{III}(a/2)$ $k_n A_n \cos \frac{1}{2}k_n a = -K_n D_n e^{-K_n a/2}$

and

$$\psi_n^{I} = -A_n(e^{K_n a/2} \sin \tfrac{1}{2}k_n a)e^{K_n x}$$

$$\psi_n^{II} = A_n \sin k_n x \qquad\qquad (n = 2, 4, 6, \ldots; \quad \text{odd parity}). \quad (4\text{-}32b)$$

$$\psi_n^{III} = A_n(e^{K_n a/2} \sin \tfrac{1}{2}k_n a)e^{-K_n x}$$

The coefficients A_n and B_n are determined by normalization requirements. Once the ε_n (k_n and K_n) are found, the eigenfunctions are uniquely determined. These energy eigenvalues are obtained from (4-31a) and (4-31b). As these are transcendental equations, they are best solved graphically. We set

$$R = \left(\frac{mV_0 a^2}{2\hbar^2}\right)^{1/2} \quad \text{and} \quad \xi_n = \left(\frac{m\varepsilon_n a^2}{2\hbar^2}\right)^{1/2} \qquad (4\text{-}33)$$

in terms of which (4-31a) and (4-31b) become

$$\xi_n \tan \xi_n = (R^2 - \xi_n^2)^{1/2} \quad \text{and} \quad -\xi_n \cot \xi_n = (R^2 - \xi_n^2)^{1/2}.$$

The roots ξ_n give the energy eigenvalues ε_n in (4-33) and are themselves determined by the intersections of the circle function $(R^2 - \xi^2)^{1/2}$ with the trigonometric functions $\xi \tan \xi$ and $-\xi \cot \xi$ (Figure 4-9).

Thus we can serially order the eigenfunctions by alternating even and odd parity states, that is, $n = 1, 3, 5, \ldots$ for even parity and $n = 2, 4, 6, \ldots$ for odd parity. Observe that the first root ξ_1 corresponds to the (even parity) ground state, while the second root ξ_2 belongs to the (odd parity) first excited state, the third root ξ_3 to even parity again, etc.

The deeper the well, that is, the larger V_0, the larger the radius R of the circle function; thus there will be more intersections with the trigonometric functions and consequently more bound states. From Figure 4-9 it can be seen that there is always at least one intersection so that *even the shallowest of wells can support at least one bound state*, this being of even parity.

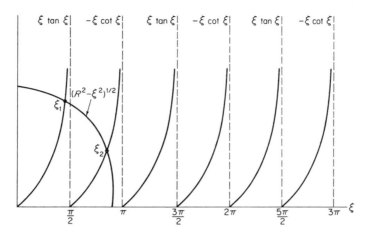

Figure 4-9 Graphical solution for the eigenvalues of a finite symmetric well.

Finally we verify the correspondence with the infinite well by considering the limit $V_0 \to \infty$. From (4-29), it follows that $K_n \to \infty$ and the exterior solutions in (4-30) do indeed vanish for both even and odd states. Furthermore, the radius of the circle function (Figure 4-9) tends to infinity, intersecting the $\xi \tan \xi$ and $-\xi \cot \xi$ functions at the asymptotes $\xi_n = \frac{1}{2}n\pi$. The eigenvalues follow, from (4-33), as

$$\frac{m\varepsilon_n a^2}{2\hbar^2} = \xi_n{}^2 = \frac{1}{4}n^2\pi^2$$

or

$$\varepsilon_n = \frac{\hbar^2\pi^2}{2ma^2}n^2.$$

This agrees with the previous results for the energy eigenvalues of an infinite well.

X   The Harmonic Oscillator

The quantum mechanics of a particle bound to the origin by an elastic potential $\frac{1}{2}kx^2$ is of both practical and academic importance. In physical problems involving electromagnetic radiation, lattice vibrations in solids, and molecular vibrations, the motions may be decomposed into one-dimensional simple harmonic oscillations. We shall first obtain the eigenvalues and eigenfunctions in Schroedinger's wave mechanics and then demonstrate that these solutions could have been obtained using the operator formalism.

The Schroedinger equation for the oscillator is

$$\left\{-\frac{\hbar^2}{2m}\frac{d^2}{dx^2} + \frac{1}{2}m\omega^2 x^2\right\}\psi_n = \varepsilon_n \psi_n \tag{4-34}$$

where $\omega^2 = k/m$. As we have already pointed out, this symmetric Hamiltonian supports only bound states and therefore has a discrete spectrum. The eigenfunctions are also real, nondegenerate, unique, mutually orthogonal, and of definite parity.

Changing to the dimensionless variable $\xi = \sqrt{\alpha}\,x$, where $\alpha = m\omega/\hbar$, (4-34) becomes

$$\frac{d^2\psi_n(\xi)}{d\xi^2} + \left(\frac{\lambda_n}{\alpha} - \xi^2\right)\psi_n(\xi) = 0 \tag{4-35}$$

with $\lambda_n = 2m\varepsilon_n/\hbar^2$. Only those solutions which satisfy the requirement $\psi_n(\pm\infty) = 0$ are acceptable. We shall see that if we choose ε arbitrarily, then the solution to (4-35) will generally diverge at infinity. Therefore, the energy eigenvalues for our problem are those for which (4-35) has solutions which vanish at infinity.

The method of solution for (4-35) is based on the standard theory of differential equations familiar to students of mathematics. We begin by assuming that a solution exists of the form

$$\psi = G(\xi)H(\xi) \tag{4-36}$$

where $G(\xi)$ is a function describing the general behavior of ψ for large $|\xi|$, and $H(\xi)$ is an infinite series. The function $G(\xi)$ is called the asymptotic solution and may be obtained by investigating the behavior of the differential equation far from the origin. For asymptotic ξ $(|\xi| \to \infty)$ we may set $\xi^2 \gg \lambda_n/\alpha$ and write (4-35) as

$$\frac{d^2\psi}{d\xi^2} - \xi^2\psi \simeq 0$$

with the two linearly independent asymptotic solutions $e^{\xi^2/2}$ and $e^{-\xi^2/2}$. The first solution diverges at infinity and leads to eigenfunctions which are unacceptable for bound states. Equation (4-36) suggests that we use a trial solution of the form $\psi = e^{-\xi^2/2}H(\xi)$, in which case (4-35) becomes

$$\frac{d^2H(\xi)}{d\xi^2} - 2\xi\frac{dH(\xi)}{d\xi} + \left(\frac{\lambda_n}{\alpha} - 1\right)H(\xi) = 0. \tag{4-37}$$

Equation (4-37) is Hermite's differential equation. According to our assumption, $H(\xi)$ should be representable by an infinite series of the form

$$H = \sum_{j=0}^{\infty} a_j \xi^j. \tag{4-38}$$

Substituting (4-38) into Hermite's equation and differentiating each term of the series, we obtain

$$\sum_{j=2}^{\infty} j(j-1)a_j \xi^{j-2} - 2\xi \sum_{j=1}^{\infty} j a_j \xi^{j-1} + \left(\frac{\lambda_n}{\alpha} - 1\right) \sum_{j=0}^{\infty} a_j \xi^j = 0$$

or collecting like powers of ξ^j,

$$\sum_{j=0}^{\infty} \left\{ (j+1)(j+2)a_{j+2} + \left[\frac{\lambda_n}{\alpha} - (1+2j)\right] a_j \right\} \xi^j = 0. \qquad (4\text{-}39)$$

Since (4-39) is to be valid for all ξ, it follows that each coefficient must separately vanish and we obtain the *two-term recursion* formula

$$a_{j+2} = - \left[\frac{(\lambda_n/\alpha) - (1+2j)}{(j+1)(j+2)} \right] a_j . \qquad (4\text{-}40)$$

By choosing a_0 arbitrarily, we find one solution to (4-37). Equation (4-40) may be used to express all successive even coefficients in terms of a_0. For example, for $j = 0$ and $j = 2$, we would have respectively

$$a_2 = - \left[\frac{(\lambda_n/\alpha) - 1}{2} \right] a_0$$

and

$$a_4 = \frac{-[(\lambda_n/\alpha) - 5]}{12} a_2 = \frac{[(\lambda_n/\alpha) - 5] [(\lambda_n/\alpha) - 1]}{12} a_0 .$$

In this way we generate an even series of the form

$$H^{\text{even}}(\xi) = a_0 \left[1 - \frac{((\lambda_n/\alpha) - 1)}{2} \xi^2 + \frac{((\lambda_n/\alpha) - 5)(\lambda_n/\alpha) - 1)}{(12)(2)} \xi^4 - \cdots \right]. \qquad (4\text{-}41)$$

If we choose a_1 arbitrarily, we obtain the odd series solution

$$H^{\text{odd}}(\xi) = a_1 \left[\xi - \frac{((\lambda_n/\alpha) - 3)}{6} \xi^3 + \frac{((\lambda_n/\alpha) - 3)((\lambda_n/\alpha) - 7)}{(6)(20)} \xi^5 - \cdots \right]. \qquad (4\text{-}42)$$

Since a_0 and a_1 are arbitrary and the two series linearly independent, the general solution to Hermite's equation is of the form $H_{\text{general}}(\xi) = H^{\text{even}}(\xi) + H^{\text{odd}}(\xi)$. The solution to the Schroedinger equation (4-35) is now

$$\psi_{\text{general}}(\xi) = e^{-\xi^2/2} [H_{\text{even}}(\xi) + H_{\text{odd}}(\xi)] \qquad (4\text{-}43)$$

and is valid for any choice of λ_n/α.

Unfortunately, no quantization has thus far evolved. *This is because (4-43) does not yet satisfy the required boundary conditions at infinity.* In fact, a comparison test indicates that for $|\xi| \gg 1$, both the even and odd series tend to infinity at least as rapidly as the function e^{ξ^2}. Thus for large ξ we have

$$\psi = e^{-\xi^2/2}H(\xi) \sim e^{-\xi^2/2}e^{\xi^2} \xrightarrow[|\xi| \to \infty]{} \infty$$

and it follows that ψ diverges at infinity and is therefore unacceptable.

We shall now examine the solution to (4-35) when

$$\frac{\lambda_n}{\alpha} = 2n + 1 \qquad (n = 0, 1, 2, \ldots).$$

If n is even, then the recursion formula (4-40) indicates that a_{n+2} and all higher coefficients vanish. Now, the solution for even n takes the form

$$\psi = e^{\xi^2/2}(H_n(\text{even polynomial of degree } n) + H^{\text{odd}}(\text{series})).$$

Since we have seen that the (odd) series gives unacceptable solutions, we retain only the (even) polynomial (that is, set $a_1 = 0$). Similarly if n is odd, we retain only the odd terms which lead to an (odd) polynomial of degree n. It is well known that the exponential $e^{-\xi^2/2}$ dominates over any polynomial for large $|\xi|$; hence, we will obtain acceptable eigenfunctions only when $\lambda_n/\alpha = 2n + 1$. Using the definitions of λ_n and α we find that bound eigenfunctions exist for the energies

$$\frac{\lambda_n}{\alpha} = \frac{2\varepsilon_n}{\hbar\omega} = 2n + 1$$

or

$$\varepsilon_n = \hbar\omega(n + \tfrac{1}{2}) \qquad (n = 0, 1, 2, \ldots). \tag{4-44}$$

The eigenfunctions take the form

$$\psi_n(\xi) = e^{-\xi^2/2}H_n(\xi) \tag{4-45}$$

where $H_n(\xi)$ is an even or odd polynomial of degree n. Since $e^{-\xi^2/2}$ is an even function, the parity is even or odd according to the value of n. The coefficients are determined from (4-40) once a_0 and a_1 have been specified. It is conventional to choose

$$a_0^{(n)} = \frac{(-1)^{n/2}n!}{[\tfrac{1}{2}n]!} \qquad (n \text{ even})$$

and

$$a_1^{(n)} = \frac{(-1)^{(n-1)/2}2(n!)}{[\tfrac{1}{2}(n-1)]!} \qquad (n \text{ odd})$$

in which case the $H_n(\xi)$ become the so-called Hermite polynomials. The first few are

$$\begin{aligned}
H_0 &= 1 & H_1 &= 2\xi \\
H_2 &= -2 + 4\xi^2 & H_3 &= -12\xi + 8\xi^3 \\
H_4 &= 12 - 48\xi^2 + 16\xi^4 & H_5 &= 120\xi - 160\xi^3 + 32\xi^5.
\end{aligned} \tag{4-46}$$

In terms of these Hermite polynomials, the eigenfunctions take the form

$$\psi_n(\xi) = N_n e^{-\xi^2/2} H_n(\xi)$$

or

$$\psi_n(x) = N_n e^{-\alpha x^2/2} H_n(\sqrt{\alpha}\, x), \tag{4-47a}$$

where the normalization constant is

$$N_n = \left(\frac{\alpha}{\pi}\right)^{1/4} \left(\frac{1}{2^n n!}\right)^{1/2}. \tag{4-47b}$$

The first few normalized eigenfunctions and eigenvalues for the oscillator are

$$\psi_0 = \left(\frac{\alpha}{\pi}\right)^{1/4} e^{-\alpha x^2/2}[1] \qquad\qquad (\varepsilon_0 = \tfrac{1}{2}\hbar\omega)$$

$$\psi_1 = \left(\frac{\alpha}{\pi}\right)^{1/4} \sqrt{\frac{1}{2}}\, e^{-\alpha x^2/2}[2\sqrt{\alpha}\, x] \qquad (\varepsilon_1 = \tfrac{3}{2}\hbar\omega) \qquad (4\text{-}48)$$

$$\psi_2 = \left(\frac{\alpha}{\pi}\right)^{1/4} \sqrt{\frac{1}{8}}\; e^{-\alpha x^2/2}[4\alpha x^2 - 2] \quad (\varepsilon_2 = \tfrac{5}{2}\hbar\omega).$$

Note that the eigenvalues are equally spaced by an energy $\hbar\omega$. The general nature of the spectrum therefore agrees with Planck's hypothesis discussed in Chapter 1. We shall verify below that the existence of a zero-point energy, $\varepsilon_0 = \tfrac{1}{2}\hbar\omega$, absent in the Planck theory, appears here as a direct consequence of the uncertainty principle as it did for a potential well.

XI Properties of Oscillator Eigenfunctions

The calculations of matrix elements are facilitated using the following properties of Hermite polynomials:

(1) $$e^{\xi^2 - (z-\xi)^2} = \sum_{n=0}^{\infty} H_n(\xi) \frac{z^n}{n!} \qquad \begin{array}{l}\text{(generating}\\ \text{function)}\end{array}$$

(2) $$H_n(\xi) = (-1)^n e^{\xi^2} \frac{d^n}{d\xi^n}(e^{-\xi^2}) \qquad \begin{array}{l}\text{(Rodrigues'}\\ \text{formula)}\end{array}$$

(3) $$H_{n+1}(\xi) - 2\xi H_n(\xi) + 2n H_{n-1}(\xi) = 0 \tag{4-49}$$

(4) $$H_n'(\xi) = 2n\, H_{n-1}(\xi)$$

(5) $$H_n''(\xi) = 4n(n-1)H_{n-2}(\xi).$$

For example, the matrix elements of $\hat{x}$ may be expressed as

$$\langle \varepsilon_m | \hat{x} | \varepsilon_n \rangle = \int \psi_m^*(x) x \psi_n(x)\, dx$$

$$= \frac{1}{\alpha} \int \psi_m^*(\xi) \xi \psi_n(\xi)\, d\xi. \qquad (4\text{-}50)$$

Multiplying both sides of property (3) above by $e^{-\xi^2/2}$, we obtain the relation

$$\xi e^{-\xi^2/2} H_n = n e^{-\xi^2/2} H_{n-1} + \frac{1}{2} e^{-\xi^2/2} H_{n+1}$$

$$\xi \frac{\psi_n}{N_n} = n \frac{\psi_{n-1}}{N_{n-1}} + \frac{1}{2} \frac{\psi_{n+1}}{N_{n+1}}$$

or

$$\xi \psi_n = n \frac{N_n}{N_{n-1}} \psi_{n-1} + \frac{1}{2} \frac{N_n}{N_{n+1}} \psi_{n+1}. \qquad (4\text{-}51)$$

Substituting (4-51) into (4-50), the matrix element becomes

$$\langle \varepsilon_m | \hat{x} | \varepsilon_n \rangle = \frac{N_n}{N_{n-1}} \frac{n}{\alpha} \int \psi_m^* \psi_{n-1}\, d\xi + \frac{N_n}{2N_{n+1}\alpha} \int \psi_m^* \psi_{n+1}\, d\xi$$

$$= \frac{N_n}{N_{n-1}} \frac{n}{\sqrt{\alpha}} \int \psi_m^* \psi_{n-1}\, dx + \frac{N_n}{2N_{n+1}\sqrt{\alpha}} \int \psi_m^* \psi_{n+1}\, dx$$

$$= \frac{N_n}{N_{n-1}} \frac{n}{\sqrt{\alpha}} \delta_{m,\,n-1} + \frac{N_n}{2N_{n+1}} \frac{1}{\sqrt{\alpha}} \delta_{m,\,n+1}. \qquad (4\text{-}52)$$

Thus the only nonvanishing matrix elements are

$$\langle \varepsilon_{n-1} | \hat{x} | \varepsilon_n \rangle = \frac{N_n n}{N_{n-1}\sqrt{\alpha}} = \left(\frac{n}{2\alpha} \right)^{1/2} = \left(\frac{n\hbar}{2m\omega} \right)^{1/2} \qquad (4\text{-}53a)$$

and

$$\langle \varepsilon_{n+1} | \hat{x} | \varepsilon_n \rangle = \frac{N_n}{N_{n+1}} \frac{1}{2\sqrt{\alpha}} = \left(\frac{n+1}{2\alpha} \right)^{1/2} = \left[\frac{(n+1)\hbar}{2m\omega} \right]^{1/2}. \qquad (4\text{-}53b)$$

The $\hat{x}$-operator may be displayed in the $\mathcal{H}$ eigenbasis as

$$\hat{x} \rightarrow \left(\frac{\hbar}{2m\omega} \right)^{1/2} \begin{pmatrix} 0 & \sqrt{1} & 0 & 0 & \cdots \\ \sqrt{1} & 0 & \sqrt{2} & 0 & \cdots \\ 0 & \sqrt{2} & 0 & \sqrt{3} & \cdots \\ 0 & 0 & \sqrt{3} & 0 & \\ \vdots & \vdots & \vdots & \vdots & \ddots \end{pmatrix} \mathcal{H} \text{ eigenbasis}$$

From (4-52), the diagonal elements vanish and we find that $\langle \varepsilon_n | \hat{x} | \varepsilon_n \rangle = 0$; therefore the average position of a particle in an oscillator eigenstate is at the origin. A calculation involving property (4) above indicates that the average momentum in any state is also zero (see Problem 3-8), that is, $\langle p \rangle_n = 0$. The expectation values $\langle x^2 \rangle_n$ and $\langle p^2 \rangle_n$ are not so simple to evaluate. We shall use the virial theorem‡ of classical mechanics which specifies that the time averages of $\bar{T}$ and $\bar{V}$ for an oscillator are equal. Generalizing this to quantum expectation values, we may set $\langle T \rangle_n = \langle V \rangle_n$ for any oscillator state. However, since $\langle T + V \rangle_n = \langle \mathcal{H} \rangle_n = \varepsilon_n$ it follows that $\langle T \rangle_n = \langle V \rangle_n = \frac{1}{2}\varepsilon_n$, from which we deduce

$$\langle T \rangle_n = \frac{1}{2m} \langle p^2 \rangle_n = \frac{1}{2}\varepsilon_n = \frac{1}{2}\hbar\omega(n + \frac{1}{2})$$

and

$$\langle V \rangle_n = \frac{1}{2}m\omega^2 \langle x^2 \rangle_n = \frac{1}{2}\varepsilon_n = \frac{1}{2}\hbar\omega(n + \frac{1}{2}).$$

Using these results the uncertainties of the canonical variables may be expressed as

$$\Delta p_n = [\langle p^2 \rangle_n - \langle p \rangle^2{}_n]^{1/2} = [(n + \frac{1}{2})\hbar m\omega]^{1/2}$$

and

$$\Delta x_n = [\langle x^2 \rangle_n - \langle x \rangle^2{}_n]^{1/2} = [(n + \frac{1}{2})\hbar/m\omega]^{1/2}.$$

The product of the uncertainties in the nth oscillator state is therefore

$$\Delta x_n \, \Delta p_n = (n + \frac{1}{2})\hbar. \tag{4-54}$$

Note that for any state this product is never smaller than $\frac{1}{2}\hbar$ as required by the uncertainty principle. The lower bound is actually reached in the ground state ($n = 0$). Were it not for the zero-point energy $\frac{1}{2}\hbar\omega$, (4-54) would have read $\Delta p_n \, \Delta x_n = n\hbar$, thereby violating the uncertainty principle in the ground state.

XII Oscillations in Nonstationary States—Classical Correspondence

If a particle is under the influence of an elastic restoring force and its initial state is characterized by a (stationary) eigenstate of $\mathcal{H}$, then the expectation values of $\hat{p}$ and $\hat{x}$ are stationary and $\langle p \rangle = 0$ and $\langle x \rangle = 0$ for all times. If

‡ For a discussion of the "virial theorem" see, for example, J. Marion, "Classical Dynamics of Particles and Systems," 2nd ed., p. 233. Academic Press, New York, 1970 or see Problems 2-3 and 4-10.

however the particle's initial state (at $t = t_0 = 0$) is an arbitrary state $\Psi(x) \neq \psi_n$, then its evolution is determined by

$$\Psi(x, t) = \int G(x, x', t)\Psi(x') \, dx' \tag{4-55}$$

where the propagator‡ is, from (3-94), (4-44), and (4-47),

$$G(x, x', t) = \sum_{n=0}^{\infty} \psi_n^*(x')\psi_n(x) \exp \frac{-i\varepsilon_n t}{\hbar}$$

$$= \left(\frac{\alpha}{\pi}\right)^{1/2} \sum_{n=0}^{\infty} \left(\frac{1}{2^n n!}\right) \exp\left[-\frac{1}{2}\alpha(x^2 + x'^2)\right] H_n(\sqrt{\alpha}\, x)$$

$$\cdot H_n(\sqrt{\alpha}\, x') \exp[-i(n + \tfrac{1}{2})\omega t].$$

Due to the half-integral factor $(n + \tfrac{1}{2})$ in the exponent of the oscillatory factor, it follows that the propagator has a period equal to twice the classical period $\tau_{\text{class}} = 2\pi/\omega$. In fact after a time τ_{class} has elapsed, the propagator and consequently the state function ψ in (4-55) develop an extra factor $\exp[i(n + \tfrac{1}{2})2\pi] = -1$; hence they change sign after one classical period. However, since expectation values involve integrals of $|\psi|^2$ rather than ψ, observables do in fact oscillate with the natural frequency ω.

We may establish the frequencies of the oscillation of $\langle x \rangle$ and $\langle p \rangle$ using the equations of motion (3-61) which are

$$\frac{d\langle x \rangle}{dt} = -\frac{\langle [\mathcal{H}, \hat{x}] \rangle}{i\hbar} \quad \text{and} \quad \frac{d\langle p \rangle}{dt} = -\frac{\langle [\mathcal{H}, \hat{p}] \rangle}{i\hbar}$$

where $\mathcal{H} = (\hat{p}^2/2m) + \tfrac{1}{2}m\omega^2\hat{x}^2$. Since $\hat{x}$ commutes with the potential energy, and $\hat{p}$ with the kinetic energy, in $\mathcal{H}$, the above equations become

$$\frac{d\langle x \rangle}{dt} = -\frac{1}{2mi\hbar} \langle [\hat{p}^2, \hat{x}] \rangle = -\frac{1}{2mi\hbar} \langle\langle (\hat{p}\hat{p}\hat{x} - \hat{x}\hat{p}\hat{p}) \rangle\rangle \tag{4-56}$$

and

$$\frac{d\langle p \rangle}{dt} = -\frac{m\omega^2}{2i\hbar} \langle [\hat{x}^2, \hat{p}] \rangle = -\frac{m\omega^2}{2i\hbar} \langle\langle (\hat{x}\hat{x}\hat{p} - \hat{p}\hat{x}\hat{x}) \rangle\rangle. \tag{4-57}$$

Using the commutation relation $\hat{x}\hat{p} = i\hbar\hat{1} + \hat{p}\hat{x}$ twice, we obtain

$$\hat{x}\hat{p}\hat{p} = (\hat{x}\hat{p})\hat{p} = (i\hbar\hat{1} + \hat{p}\hat{x})\hat{p}$$

$$= i\hbar\hat{p} + \hat{p}(\hat{x}\hat{p})$$

$$= i\hbar\hat{p} + \hat{p}(i\hbar\hat{1} + \hat{p}\hat{x}) = 2i\hbar\hat{p} + \hat{p}\hat{p}\hat{x}.$$

---

‡ This propagator may also be expressed in closed form as

$$G(x, x', t) = \left[\frac{m\omega}{2\pi\hbar i \sin \omega t}\right]^{1/2} \exp\left\{\frac{im\omega}{2\hbar \sin \omega t} [(x^2 + x'^2) \cos \omega t - 2xx']\right\}.$$

Similarly, it may be verified that

$$\hat{p}\hat{x}\hat{x} = 2ih\hat{x} + \hat{x}\hat{x}\hat{p}$$

in which case (4-56) and (4-57) become

$$\frac{d\langle x\rangle}{dt} = \frac{\langle p\rangle}{m} \quad \text{and} \quad \frac{d\langle p\rangle}{dt} = -m\omega^2\langle x\rangle.$$

Differentiating both equations with respect to time and decoupling them, we find the two second-order equations

$$\frac{d^2}{dt^2}\langle p\rangle = -\omega^2\langle p\rangle \quad \text{and} \quad \frac{d^2}{dt^2}\langle x\rangle = -\omega^2\langle x\rangle$$

with solutions

$$\langle p\rangle = \langle p\rangle_0 \cos(\omega t + \phi_p) \tag{4-58}$$

$$\langle x\rangle = \langle x\rangle_0 \cos(\omega t + \phi_x). \tag{4-59}$$

Thus, confirming Ehrenfest's theorem, the expectation values do indeed exhibit simple harmonic motion with the classical frequency, but only if the initial state is not an energy eigenstate. In such a state, we have shown that $\langle p\rangle_0 = \langle x\rangle_0 = 0$, and (4-58) and (4-59) verify that $\langle x\rangle$ and $\langle p\rangle$ remain identically zero.

It is important to recognize that if we associate $\langle p\rangle$ and $\langle x\rangle$ with p_{class} and x_{class}, no vibrations occur in stationary oscillator states. If the oscillator involves a charged particle, then no radiation is emitted while the system remains in an energy eigenstate. As we shall show in Chapter 7, vibrations accompanied by radiative emission occur when the particle makes transitions between stationary states.

XIII The Oscillator Problem in Dirac Notation—
The Ladder Method

In this final section, we wish to demonstrate that wave mechanics is only one of many representations of quantum mechanics and that the spectrum of a Hamiltonian is fundamentally determined by its dependence on $\hat{p}$ and $\hat{r}$ and on the commutation relations between these variables. We now obtain the eigenvalues of the oscillator Hamiltonian and the matrix elements of $\hat{x}$ in the $\mathcal{H}$ eigenbasis without resorting to the coordinate representation.

We wish to obtain the eigenvalues of the equation

$$\mathcal{H}|n\rangle = \left\{\frac{\hat{p}^2}{2m} + \frac{1}{2}m\omega^2\hat{x}^2\right\}|n\rangle = \varepsilon_n|n\rangle \tag{4-60}$$

where the canonical variables satisfy $[\hat{x}, \hat{p}] = i\hbar\hat{1}$. We introduce the (dimensionless) "reduced" coordinate and momentum operators

$$\hat{x}' = \left(\frac{m\omega}{2\hbar}\right)^{1/2}\hat{x} \quad \text{and} \quad \hat{p}' = \left(\frac{1}{2m\hbar\omega}\right)^{1/2}\hat{p}$$

in terms of which (4-60) becomes

$$\mathscr{H}|n\rangle = \hbar\omega(\hat{p}'^2 + \hat{x}'^2)|n\rangle = \varepsilon_n|n\rangle.$$

Note that the Hamiltonian may be factored into

$$\mathscr{H} = \hbar\omega\{(\hat{x}' - i\hat{p}')(\hat{x}' + i\hat{p}') - i[\hat{x}', \hat{p}']\}. \tag{4-61}$$

Since $\hat{x}'$ and $\hat{p}'$ do not commute, the last term is required to cancel the cross terms of the product. The commutator is evaluated as

$$[\hat{x}', \hat{p}'] = \left(\frac{m\omega}{2\hbar}\right)^{1/2}\left(\frac{1}{2m\hbar\omega}\right)^{1/2}[\hat{x}, \hat{p}] = \frac{1}{2}i\hat{1}. \tag{4-62}$$

We next introduce the operator

$$\hat{\eta} = \hat{x}' + i\hat{p}' \tag{4-63a}$$

and its adjoint

$$\hat{\eta}^\dagger = \hat{x}' - i\hat{p}' \tag{4-63b}$$

in terms of which the Hamiltonian, (4-61), becomes

$$\mathscr{H} = \hbar\omega(\hat{N} + \tfrac{1}{2}) \tag{4-64}$$

where $\hat{N} = \hat{\eta}^\dagger\hat{\eta}$. We may regard the Hamiltonian to be a function of the new coordinate $\hat{\eta}$ and its canonically conjugate momentum $\hat{\eta}^\dagger$. From (4-62) it may be shown that the commutation relation for these variables is $[\hat{\eta}, \hat{\eta}^\dagger] = \hat{1}$. Note that these new variables are not Hermitian ($\hat{\eta}^\dagger \neq \hat{\eta}$) and are therefore not observables, but they nevertheless enable us to find ε_n.

In order to find the eigenvectors and eigenvalues of the Hamiltonian in (4-64) it suffices to solve the eigenvalue equation

$$\hat{N}|n\rangle = N_n|n\rangle. \tag{4-65}$$

The kets $|n\rangle$ represent common eigenvectors of $\mathscr{H}$ and $\hat{N}$ with the eigenvalues related by

$$\varepsilon_n = \hbar\omega(N_n + \tfrac{1}{2}). \tag{4-66}$$

It remains to demonstrate that the eigenvalues of $\hat{N}$ are nonnegative integers, that is, $N_n = n$ where $n = 0, 1, 2, \ldots$. Operating with $\hat{\eta}$ on (4-65) we obtain

$$\hat{\eta}\hat{N}|\eta\rangle = \hat{\eta}\hat{\eta}^\dagger\hat{\eta}|n\rangle = N_n\hat{\eta}|n\rangle.$$

Using the commutation relation $\hat{\eta}\hat{\eta}^\dagger = \hat{1} + \hat{\eta}^\dagger\hat{\eta}$, this becomes

$$\{\hat{1} + \hat{\eta}^\dagger\hat{\eta}\}\hat{\eta}|n\rangle = N_n\hat{\eta}|\eta\rangle$$

or, transposing,

$$\hat{N}\{\hat{\eta}|n\rangle\} = (N_n - 1)\{\hat{\eta}|n\rangle\}. \tag{4-67}$$

From the last relation (4-67) it follows that if N_n is an eigenvalue then so is $N_n - 1$, indicating that the eigenvalues are integrally spaced. Furthermore the term in brackets must be proportional to the eigenvector corresponding to the $N_n - 1$ eigenvalue, that is,

$$\hat{\eta}|n\rangle = C_n^-|n - 1\rangle$$

where C_n^- is a proportionality constant. Similar reasoning for the $\hat{\eta}^\dagger$ operator leads to the result‡

$$\hat{\eta}^\dagger|n\rangle = C_n^+|n + 1\rangle.$$

Multiplying both relations by their bra forms we find

$$\langle n|\hat{\eta}^\dagger\hat{\eta}|n\rangle = |C_n^-|^2\langle n - 1|n - 1\rangle$$
$$\langle n|\hat{N}|n\rangle = N_n = |C_n^-|^2 \tag{4-68a}$$

and

$$\langle n|\hat{\eta}\hat{\eta}^\dagger|n\rangle = |C_n^+|^2\langle n + 1|n + 1\rangle$$
$$\langle n|1 + \hat{N}|n\rangle = 1 + N_n = |C_n^+|^2. \tag{4-68b}$$

From (4-68a) it follows that $N_n \geq 0$; thus there must exist a nonnegative minimum eigenvalue, $N_{n_{\min}}$. A lowering operation on the lowest eigenstate must not produce any new state, that is,

$$\hat{\eta}|n_{\min}\rangle = C_{n_{\min}}^-|n - 1\rangle = 0$$

or using C_n^- from (4-68a),

$$\hat{\eta}|n_{\min}\rangle = (N_{n_{\min}})^{1/2}|n - 1\rangle = 0.$$

The last relation can be satisfied if $N_{n_{\min}} = 0$. Successive raising operations indicate that the spectrum of $\hat{N}$ must be composed of a set of positive integers and zero. Setting $N_n = n$, it follows from (4-66) that the spectrum of $\mathcal{H}$ is

$$\varepsilon_n = \hbar\omega(n + \tfrac{1}{2}) \qquad (n = 0, 1, 2, \ldots)$$

in agreement with our wave-mechanical treatment.

---

‡ The operators $\hat{\eta}$ and $\hat{\eta}^\dagger$ are respectively termed lowering (or destruction) and raising (or creation) operators.

The coefficients C_n may now be evaluated from (4-68a) and (4-68b) as $C_n^- = \sqrt{n}$ and $C_n^+ = (n+1)^{1/2}$ and the lowering and raising operations written

$$|n-1\rangle = \frac{1}{\sqrt{n}} \hat{\eta} |n\rangle \qquad\qquad (4\text{-}69)$$

and

$$|n+1\rangle = \frac{1}{(n+1)^{1/2}} \hat{\eta}^\dagger |n\rangle. \qquad\qquad (4\text{-}70)$$

The spectrum of $\mathcal{H}$ has therefore been determined; once any one eigenvector is obtained all others may be generated using raising and lowering operations. In particular, the first excited state may be obtained from the ground state by the raising operation $|1\rangle = (1/\sqrt{1})\hat{\eta}^\dagger |0\rangle$. In fact, performing n successive raising operations on $|0\rangle$ we generate the nth eigenvector, that is,

$$|n\rangle = \frac{1}{(n!)^{1/2}} (\hat{\eta}^\dagger)^n |0\rangle. \qquad\qquad (4\text{-}71)$$

To complete our analysis we shall evaluate the matrix elements $\langle m|\hat{x}|n\rangle$. Using (4-63), the $\hat{x}$-operator may be expressed in terms of the raising and lowering operators as

$$\hat{x} = (2\hbar/m\omega)^{1/2}\hat{x}' = (2\hbar/m\omega)^{1/2}(\hat{\eta}^\dagger + \hat{\eta})/2$$

and the matrix elements become

$$\begin{aligned}
\langle m|\hat{x}|n\rangle &= \tfrac{1}{2}(2\hbar/m\omega)^{1/2}[\langle m|\hat{\eta}^\dagger |n\rangle + \langle m|\hat{\eta}|n\rangle] \\
&= \tfrac{1}{2}(2\hbar/m\omega)^{1/2}[(n+1)^{1/2}\langle m|n+1\rangle + \sqrt{n}\langle m|n-1\rangle] \\
&= \tfrac{1}{2}(2\hbar/m\omega)^{1/2}[(n+1)^{1/2}\delta_{m,n+1} + \sqrt{n}\delta_{m,n-1}].
\end{aligned}$$

We may therefore write the nonvanishing elements as

$$\langle n+1|\hat{x}|n\rangle = \left[\frac{(n+1)\hbar}{2m\omega}\right]^{1/2}$$

and

$$\langle n-1|\hat{x}|n\rangle = \left[\frac{n\hbar}{2m\omega}\right]^{1/2}$$

both of which agree with our previous results, (4-53).

We again emphasize that these matrix elements and eigenvalues have been obtained without resorting to any representation, using only the properties of the relevant operators. In one-body problems more complicated than the one encountered here, it is usually simpler to rely on the coordinate representation provided the latter exists.

Finally, we shall demonstrate that the oscillator *eigenfunctions* can also be generated using the ladder method. In the coordinate representation we set $|n\rangle \to \psi_n$, $\hat{x} \to x$, and $\hat{p} \to (\hbar/i)\, d/dx$. The lowering and raising operations in (4-69) and (4-70) take the form

$$\hat{\eta}\psi_n = \left\{ \left(\frac{m\omega}{2\hbar}\right)^{1/2} x + i\left(\frac{1}{2m\hbar\omega}\right)^{1/2} \frac{\hbar}{i} \frac{d}{dx}\right\}\psi_n = \sqrt{n}\,\psi_{n-1} \qquad (4\text{-}72)$$

$$\hat{\eta}^{\dagger}\psi_n = \left\{ \left(\frac{m\omega}{2\hbar}\right)^{1/2} x - i\left(\frac{1}{2m\hbar\omega}\right)^{1/2} \frac{\hbar}{i} \frac{d}{dx}\right\}\psi_n = (n+1)^{1/2}\psi_{n+1}. \qquad (4\text{-}73)$$

Setting $n = 0$ in the first equation, we obtain the differential equation for the ground state eigenfunction, that is,

$$\frac{1}{(2\alpha)^{1/2}} \left(\frac{d}{dx} + \alpha x\right)\psi_0 = 0 \qquad \left(\alpha = \frac{m\omega}{\hbar}\right)$$

whose solution is

$$\psi_0 = \left(\frac{\alpha}{\pi}\right)^{1/4} e^{-\alpha x^2/2}$$

where the coefficient has been chosen to ensure normalization. Comparison with (4-48) shows our present result to be identical with the wave-mechanical result. The first excited state may be generated setting $n = 0$ in the raising equation (4-73). This gives

$$\frac{1}{(2\alpha)^{1/2}} \left(\frac{d}{dx} - \alpha x\right)\psi_0 = \sqrt{1}\,\psi_1$$

$$\frac{1}{(2\alpha)^{1/2}} \left(\frac{d}{dx} - \alpha x\right)\left(\frac{\alpha}{\pi}\right)^{1/4} e^{-\alpha x^2/2} = \psi_1$$

or

$$\psi_1 = \left(\frac{\alpha}{\pi}\right)^{1/4} \frac{1}{\sqrt{2}} e^{-\alpha x^2/2}[2\sqrt{\alpha}\, x]$$

and also agrees with previous results. The entire set of normalized eigenfunctions may thus be generated from successive raising operations.

The one-dimensional examples discussed in this chapter, while interesting enough, are particularly important when the results can be applied to the three-dimensional world. In Chapter 5, we shall turn our attention to eigenstates of Hamiltonians in three dimensions. Since bound and unbound states require basically different techniques, we shall defer a discussion of continuous states to Chapter 8 (Scattering).

Suggested Reading

Arfken, G., "Mathematical Methods for Physicists," 2nd ed., Chapter 9. Academic Press, New York, 1970.

Bohm, D., "Quantum Theory," Chapter 11. Prentice-Hall, Englewood Cliffs, New Jersey, 1951.

Borowitz, S., "Fundamentals of Quantum Mechanics," Chapter 9. Benjamin, New York, 1967.

Eisberg, R. M., "Fundamentals of Modern Physics," Chapter 8. Wiley, New York, 1961.

Merzbacher, E., "Quantum Mechanics," 2nd ed., Chapter 6. Wiley, New York, 1970.

Messiah, A., "Quantum Mechanics," Vol. I, Chapter 3. Wiley, New York, 1961.

Pauling, L., and Wilson, E. B., "Introduction to Quantum Mechanics," Chapter 3. McGraw-Hill, New York, 1935.

Saxon, D. S., "Elementary Quantum Mechanics," Chapter 6. Holden-Day, San Francisco, 1964.

Problems

4-1. Find the reflection coefficient for a beam of particles incident on the step potential in Figure 4-10.

4-2. Verify that the reflection and transmission coefficients for a repulsive barrier are correctly given by Eq. (4-34).

4-3. (a) Show for $\varepsilon \ll V_0$ and $a \gg 1$ (that is, $Ka \gg 1$) that the transmission coefficient in Eq. (4-10) has the asymptotic form $\mathcal{T} \sim e^{-2Ka}$.

(b) Assuming that the approximate transmission coefficient for a set of adjacent barriers (Figure 4-11) is equal to the product of the individual coefficients, show that for N such barriers

$$\mathscr{T} \sim \exp\left(-2\sum_{i=1}^{N} K_i a_i\right) \quad \text{where} \quad K_i = \frac{[2m(V_{0_i} - \varepsilon)]^{1/2}}{\hbar}.$$

(c) Use the result above to show that, for an arbitrary but slowly varying repulsive barrier, the approximate transmission coefficient is given by

$$\mathscr{T} \sim \exp\left\{-2\int_{x_1}^{x_2} \frac{[2m(V(x) - \varepsilon)]^{1/2}}{\hbar}\, dx\right\}$$

where x_1 and x_2 are the classical turning points of the motion (Figure 4-12). What assumptions have been made in deriving this expression? (See Appendix A.)

4-4. Use the "semiclassical" result given in Problem 4-3c to find the approximate transmission coefficient for the triangle potential (Figure 4-13) when $\varepsilon = \frac{1}{4}V_0$.

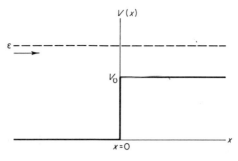

Figure 4-10

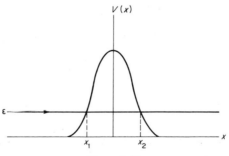

Figure 4-11

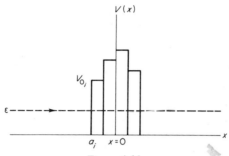

Figure 4-12

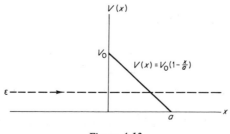

Figure 4-13

4-5. Calculate the reflection coefficient for the "double" barrier in Figure 4-14 when $\varepsilon < V_0$.

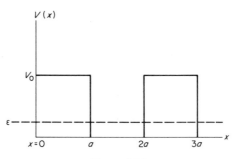

Figure 4-14

4-6. Find the transmission coefficient for a beam of particles incident on a delta-function potential $V = A\delta(x)$. (Hint: Use Eq. (4-10b) and let $a \to 0$ and $V_0 \to \infty$, while keeping the product $V_0 a = A = \text{constant.}$)

4-7. (a) Verify the following by performing the necessary integrals for a particle in an infinite asymmetric well:

$$\langle x \rangle_n = \tfrac{1}{2}a \qquad\qquad \langle p \rangle_n = 0$$

$$\langle x^2 \rangle_n = \frac{a^2}{3} - \frac{a^2}{2\pi^2 n^2} \qquad \langle p^2 \rangle_n = \frac{\hbar^2 \pi^2}{a^2} n^2 \qquad (n = 1, 2, \ldots).$$

(b) Show that $\Delta p_n{}^2 \, \Delta x_n{}^2 \geqslant \tfrac{1}{4}\hbar^2$ holds for all n values. What is $\Delta p_n \, \Delta x_n$ in the ground state, $n = 1$?

4-8. Show, using the relation $[\hat{x}, \hat{p}_x] = i\hbar \hat{I}$, that if the parity operator anticommutes with $\hat{x}$ (that is, $\hat{P}\hat{x} = -\hat{x}\hat{P}$) then it also anticommutes with $\hat{p}_x$. (Hint: The identity operator commutes with all operators including parity.)

4-9. Using the commutation relation $[\hat{x}, \hat{p}_x] = i\hbar \hat{I}$, show that

$$[\hat{p}_x, \hat{x}^n] = -i\hbar\, n\hat{x}^{n-1} \qquad \text{and} \qquad [\hat{x}, \hat{p}_x{}^n] = i\hbar\, n\hat{p}_x{}^{n-1}.$$

(More generally, $[\hat{p}_x, \hat{F}(\hat{x})] = -i\hbar(d\hat{F}/d\hat{x})$ and $[\hat{x}, \hat{G}(\hat{p}_x)] = i\hbar(d\hat{G}/d\hat{p}_x)$.)

4-10. Use the results in Problem 4-9 to show that for any Hamiltonian of the form

$$\mathcal{H} = \frac{\hat{p}_x{}^2}{2m} + a\hat{x}^n,$$

the following operator relation holds:

$$[\hat{x}\hat{p}_x, \mathcal{H}] = i\hbar \left(\frac{\hat{p}_x{}^2}{m} - an\hat{x}^n \right) = i\hbar(2\hat{T} - n\,\hat{V}).$$

(Note that for a bound *stationary* state, Ehrenfest's theorem requires

$$\frac{d\langle xp_x \rangle}{dt} = \frac{1}{i\hbar} \langle [\hat{x}\hat{p}_x, \mathcal{H}] \rangle = 0.$$

Thus we have $2\langle T \rangle = n\langle V \rangle$. This result is known as the quantum virial theorem for a power-law potential. For an oscillator ($n = 2$), we find $\langle T \rangle = \langle V \rangle$.)

4-11. Find the energy eigenvalues and eigenfunctions for a particle in a potential $V = \frac{1}{2}m\omega^2 (x^2 - 2bx)$. (Hint: Complete the square in the potential and compare this Hamiltonian to the one for the oscillator potential, that is, $V = \frac{1}{2}m\omega^2 x^2$.)

4-12. Estimate the ground-state energy of an oscillator by minimizing

$$\varepsilon = \frac{\Delta p^2}{2m} + \frac{1}{2}m\omega^2 \Delta x^2,$$

subject to the "uncertainty" restriction

$$\Delta p \, \Delta x = \tfrac{1}{2}\hbar.$$

4-13. Calculate graphically the ground-state energy of an electron

$$(m = 9 \times 10^{-28} \text{ gm})$$

in a finite one-dimensional well of height $V_0 \sim 10$ eV $\sim 1.6 \times 10^{-11}$ erg and width $a = 10^{-8}$ cm.

4-14. Prove that if an operator $\hat{A}$ is Hermitian and satisfies $\hat{A}^2 = \hat{1}$, its eigenvalues are ± 1. An operator with the property $\hat{A}^2 = \hat{1}$ is termed *involutory*.

4-15. Consider a system governed by a free particle (one-dimensional) Hamiltonian, $\mathcal{H} = \hat{p}^2/2m$.
(a) Show using Eq. (3-94) that the free particle propagator may be expressed as

$$G(x, x', t, t_0) = \int_{-\infty}^{\infty} \frac{dp}{(2\pi\hbar)} e^{-ip^2\tau/2m\hbar} e^{ipX/\hbar}$$

where $X = x - x'$ and $\tau = t - t_0$.

(b) Perform the integration and show that

$$G(X, \tau) = \left[\frac{m}{2\pi i\hbar\tau}\right]^{1/2} e^{imX^2/2\hbar\tau}.$$

(Hint: Complete the square in the exponent and use

$$\int_{-\infty}^{\infty} e^{-u^2} \, du = \sqrt{\pi}.)$$

(c)   Assume that at $t_0 = 0$, a free particle is in a state represented by the wave packet

$$\Psi(x, 0) = \frac{1}{(\pi\delta^2)^{1/4}} e^{-x^2/2\delta^2} e^{i p_0 x/\hbar}.$$

Note that at $t_0 = 0$, the probability density is a Gaussian of mean width $= \delta$. Show, using Eq. (3-93), that as time progresses the wave packet spreads.

5 | Wave Mechanics in Three Dimensions

The problems to be considered in this chapter include a particle in a rectangular well (box), the three-dimensional oscillator, and the Coulomb problem (the hydrogen atom). Special attention will be paid to the last topic because of its importance in atomic theory.

As a rule, the number of indices (quantum numbers) required to label a state completely is equal to the number of degrees of freedom associated with the corresponding classical system. In the one-dimensional problem discussed in Chapter 4 a single index (n) was sufficient, but in three dimensions three quantum numbers will be required. We shall see below how these numbers come about.

Unlike bound states in one dimension, the eigenstates of a Hamiltonian in three dimensions are often highly degenerate.‡ The degeneracy invariably results from some special symmetry in the physical problem. The energy eigenfunctions are therefore no longer unique and no longer necessarily

---

‡ Theorem IX which excludes the possibility of a degeneracy in one-dimensional bound state problems is not valid in three dimensions.

mutually orthogonal. Any eigenfunction set must be tested for orthonormality. Furthermore any linear combination of degenerate eigenfunctions
is a new and distinct eigenfunction of the same energy. We can therefore
generate hosts of new degenerate sets once any one set has been found. The
physical significance of any one set and the reason that particular set has been
generated will become clear in the discussion to follow.

I   The Eigenvalue Problem in Three Dimensions

The Schroedinger energy-eigenvalue equation in three dimensions takes
the form

$$\mathcal{H}|\varepsilon_i\rangle = \varepsilon_i|\varepsilon_i\rangle \rightarrow \left\{ -\frac{\hbar^2}{2m}\nabla^2 + V(\mathbf{r}) \right\}\psi_i(\mathbf{r}) = \varepsilon_i \psi_i(\mathbf{r}) \tag{5-1}$$

where the index i will be shown to be an abbreviation for a set of three
indices. Equation (5-1) is a partial differential equation to be solved for the
function $\psi_i(\mathbf{r})$. Since we are dealing with bound states here, we seek those
solutions which vanish at infinity, that is,

$$\psi_i(\mathbf{r}) = \langle \mathbf{r}|\varepsilon_i\rangle \xrightarrow[r\to\infty]{} 0.$$

A standard technique of solution is the method of separation of variables.
The separation may be carried out in Cartesian coordinates, or it might become
necessary to transform to a different coordinate system (for example, spherical
or cylindrical polar coordinates). We shall see that in the case of a degeneracy
the coordinate system used in the separation of variables will determine the
particular set of eigenfunctions obtained. It then becomes our task to identify
characteristics of that set. We begin by considering some problems which are
soluble in Cartesian coordinates.

II   The Free Particle (Cartesian Coordinates)

The free particle Hamiltonian, $V = 0$, supports only unbound states and
rightfully belongs in Chapter 8 on scattering. However, we shall briefly discuss
it here to illustrate how a degeneracy arises in quantum theory, and its
significance.

The Schroedinger equation takes the form

$$\mathcal{H}\psi_\varepsilon = -\frac{\hbar^2}{2m}\nabla^2\psi_\varepsilon = \varepsilon\psi_\varepsilon. \tag{5-2}$$

The method of separation of variables suggests that we try a solution of the form

$$\psi_\varepsilon(\mathbf{r}) = X(x)Y(y)Z(z). \tag{5-3}$$

If this trial solution ultimately leads to three ordinary differential equations for x, y, and z, then we say that (5-2) is separable in Cartesian coordinates. Substitution of (5-3) into (5-2) gives

$$-\frac{\hbar^2}{2m}\left\{YZ\frac{d^2X}{dx^2} + XZ\frac{d^2Y}{dy^2} + XY\frac{d^2Z}{dz^2}\right\} = \varepsilon XYZ$$

or

$$\left\{\frac{1}{X}\frac{d^2X}{dx^2} + \frac{1}{Y}\frac{d^2Y}{dy^2} + \frac{1}{Z}\frac{d^2Z}{dz^2}\right\} = -k^2$$

where $k^2 = 2m\varepsilon/\hbar^2$.

Each term on the left of this equation is a function of a single variable; the sum can be a constant only if each term is independently constant, that is,

$$\frac{1}{X}X'' = -k_x{}^2, \qquad \frac{1}{Y}Y'' = -k_y{}^2, \qquad \frac{1}{Z}Z'' = -k_z{}^2$$

where $k_x{}^2 + k_y{}^2 + k_z{}^2 = k^2 = 2m\varepsilon/\hbar^2$.

It is convenient to abbreviate the three (continuous) quantum numbers k_x, k_y, and k_z by a single *vector* quantum number $\mathbf{k} = k_x\mathbf{i} + k_y\mathbf{j} + k_z\mathbf{k}$. The solutions to the three equations above are $X \sim e^{ik_x x}$, $Y \sim e^{ik_y y}$, and $Z \sim e^{ik_z z}$ so that the total free particle eigenfunctions and eigenvalues may be written‡

$$\psi_{\mathbf{k}}(\mathbf{r}) = N_{\mathbf{k}}\,e^{ik_x x}e^{ik_y y}e^{ik_z z} = N_{\mathbf{k}}e^{i\mathbf{k}\cdot\mathbf{r}} \tag{5-4a}$$

with

$$\varepsilon_{\mathbf{k}} = \frac{\hbar^2|\mathbf{k}|^2}{2m}. \tag{5-4b}$$

The quantum vector $\mathbf{k}(k_x, k_y, k_z)$ labels the Cartesian eigenfunction and its eigenvalue.

The existence of a degeneracy should be made clear. Note that the energy

---

‡ Setting $N_k = (1/2\pi\hbar)^{3/2}$ normalizes these functions according to the conventional rule

$$\langle\mathbf{k}|\mathbf{k}'\rangle = \int \psi_{\mathbf{k}}{}^*(\mathbf{r})\psi_{\mathbf{k}'}(\mathbf{r})\,d\mathbf{r} = \delta(\mathbf{k} - \mathbf{k}')/\hbar^3.$$

However, an alternative procedure (box normalization) is also used where the particle is regarded as being in a large box of volume $\mathscr{V}$. Here we set $N_k = 1/\sqrt{\mathscr{V}}$ and when all calculations are complete we take the limit as $\mathscr{V} \to \infty$.

depends only on the magnitude of **k**. Changing the direction of **k** (varying k_x, k_y, k_z) subject to the restriction

$$|\mathbf{k}| = (k_x^2 + k_y^2 + k_z^2)^{1/2} = k_0$$

results in generating new eigenfunctions in (5-4) without affecting the eigenvalues. Thus a linear combination of these degenerate eigenfunctions is also an energy eigenfunction. Because of the continuous nature of the spectrum this combination takes the form of an integral

$$\psi_{k_0}(\mathbf{r}) = \int_{|\mathbf{k}|=k_0} a(\mathbf{k}) e^{i\mathbf{k}\cdot\mathbf{r}} \, d\Omega_\mathbf{k} \tag{5-5}$$

where the integral is performed over a sphere of radius k_0 in **k** space to ensure that only degenerate eigenfunctions are mixed. Thus for any choice of the mixing coefficients $a(\mathbf{k})$ we generate a new eigenfunction $\psi_{k_0}(\mathbf{r})$ with the same energy $\varepsilon_{k_0} = \hbar^2 k_0^2 / 2m$.

Now that we see that the solutions to (5-2) are not unique, we should ask the question: What is the physical significance of the eigenfunction set $e^{i\mathbf{k}\cdot\mathbf{r}}$, and why did it evolve rather than some other set? We observe that for a free particle ($\mathscr{H} = \hat{p}^2/2m$) we have $[\hat{\mathbf{p}}, \mathscr{H}] = 0$, from which it follows that there exists a set of eigenfunctions common to $\mathscr{H}$ and $\hat{\mathbf{p}}$. Naturally, we shall inquire as to whether $\psi_\mathbf{k}$ is that set. Operating on $\psi_\mathbf{k}$ with $\hat{\mathbf{p}} = (\hbar/i)\,\nabla$, we find

$$\frac{\hbar}{i}\nabla\psi_\mathbf{k} = \frac{\hbar}{i}\nabla e^{i\mathbf{k}\cdot\mathbf{r}} = \hbar\mathbf{k}e^{i\mathbf{k}\cdot\mathbf{r}} = \hbar\mathbf{k}\psi_\mathbf{k}. \tag{5-6}$$

Comparing the right- and left-hand sides we observe that $\psi_\mathbf{k}$ is indeed a simultaneous eigenfunction of $\hat{\mathbf{p}}$ with the eigenvalue $\hat{\mathbf{p}} = \hbar\mathbf{k}$. Thus our Cartesian solution to the free particle quite naturally leads us to a *momentum* eigenfunction.

The time evolution of this momentum eigenfunction is [see (3-67)]

$$\psi_\mathbf{k}(\mathbf{r}, t) = e^{-i\varepsilon_k t/\hbar}\psi_\mathbf{k}(\mathbf{r}) = e^{-i\varepsilon_k t/\hbar}e^{i\mathbf{k}\cdot\mathbf{r}} = e^{i(\mathbf{k}\cdot\mathbf{r}-\omega_k t)} \tag{5-7}$$

and is equivalent to the propagation of a plane wave (along **k**) with

$$\omega_k = \frac{\varepsilon_k}{\hbar} = \frac{\hbar^2 k^2}{\hbar 2m} = \frac{\hbar k^2}{2m}$$

and $\mathbf{k} = \mathbf{p}/\hbar$. From the last relation it follows that the wavelength, $\lambda = 2\pi/k = h/p$, is consistent with de Broglie's hypothesis and we may identify this momentum free-particle eigenfunction with a de Broglie wave.

However, the unphysical nature of the wave becomes apparent when we calculate its phase velocity

$$v_{\text{phase}} = \frac{\omega_k}{k} = \frac{\hbar k^2}{2mk} = \frac{\hbar k}{2m} = \frac{p}{2m} = \frac{1}{2}v_{\text{particle}}.$$

Contrary to what might have been expected, the phase velocity is not equal to the particle's velocity. It should be recalled that only the probabilistic aspects of ψ_k have direct physical meaning. Applying the current operation (3-96) to ψ_k we obtain

$$-\frac{i\hbar}{2m}(\psi_k^*\nabla\psi_k - \psi_k\nabla\psi_k^*) = \frac{\hbar k}{m}\psi_k^*\psi_k$$

$$\mathbf{J} = \frac{\mathbf{p}}{m}\psi_k^*\psi_k = \mathbf{v}_{\text{particle}}\mathscr{P}_k . \qquad (5\text{-}8)$$

Therefore the velocity of the probability current is equal to that of the free particle.‡

In our analysis of the free particle equation (5-2) we were naturally led to the linear momentum eigenfunctions $\psi_k = N_k e^{i\mathbf{k}\cdot\mathbf{r}}$ because we chose to separate this equation in Cartesian coordinates. Had we separated the equation in spherical coordinates, we would have been led to functions of the form

$$\psi_{klm_l}(r,\theta,\phi) = j_l(kr)Y_{lm_l}(\theta,\phi)$$

where j_l is a *spherical Bessel function* and Y_{lm_l} is termed a *spherical harmonic*. These functions will be shown to be simultaneous eigenfunctions of energy and angular momentum (rather than linear momentum) and have the mathematical form associated with spherical (rather than plane) waves. The quantum numbers l and m_l refer to states of angular momentum while k refers to the energy according to $\varepsilon_k = \hbar^2 k^2/2m$. Note that the two eigenfunction sets $\psi_k(xyz)$ and $\psi_{klm_l}(r,\theta,\phi)$ are distinct since $\hat{\mathbf{p}}$ and $\hat{\mathbf{L}}$ are incompatible (that is, $[\hat{\mathbf{L}},\hat{\mathbf{p}}] \neq 0$). However, as we shall see in Section VIII, each linear momentum eigenfunction is at worst a linear combination of the degenerate angular momentum eigenfunctions.

III   The Particle in a Box

As a generalization of a one-dimensional infinite well, we consider the *bound states* for a particle in a rectangular box potential. The Schroedinger equation reads

$$\left\{-\frac{\hbar^2}{2m}\nabla^2 + V(\mathbf{r})\right\}(\psi_i\mathbf{r}) = \varepsilon_i\psi_i(\mathbf{r}) \qquad (5\text{-}9)$$

---

‡ This current velocity is also equal to the so-called "group" velocity of the de Broglie wave, defined by $v_{\text{group}} = \partial\omega/\partial k$.

where

$$V(\mathbf{r}) = 0 \qquad \text{for} \quad 0 \leqslant x \leqslant a,\, 0 \leqslant y \leqslant b, \text{ and } 0 \leqslant z \leqslant c$$
$$V = \infty \qquad \text{otherwise.}$$

As in the one-dimensional case in Chapter 7, we require that $\psi(\mathbf{r})$ vanish at the well boundaries. The rectangular nature of the potential boundary suggests that this problem should be separable in Cartesian coordinates. In fact, as in the previous case, substitution of $\psi_i = X(x)Y(y)Z(z)$ immediately leads to the same ordinary differential equations

$$X'' = -k_x^2 X, \qquad Y'' = -k_y^2 Y, \qquad Z'' = -k_z^2 Z$$

with $k_x^2 + k_y^2 + k_z^2 = k^2 = 2m\varepsilon/\hbar^2$. Since the eigenfunctions must vanish exterior to the well and at its boundaries (recall $V = \infty$ there), the acceptable solutions must now be of the form

$$X_{n_x} = \left(\frac{2}{a}\right)^{1/2} \sin k_x x, \qquad Y_{n_y} = \left(\frac{2}{b}\right)^{1/2} \sin k_y y, \qquad Z_{n_z} = \left(\frac{2}{c}\right)^{1/2} \sin k_z z$$

where

$$k_x = \frac{n_x \pi}{a}, \qquad k_y = \frac{n_y \pi}{b}, \qquad \text{and} \qquad k_z = \frac{n_z \pi}{c} \tag{5-10}$$

and n_x, n_y, $n_z = 1, 2, 3, \ldots$. The eigenfunctions are characterized by the three quantum numbers n_x, n_y, and n_z and take the form

$$\psi_{n_x n_y n_z} = \left(\frac{8}{abc}\right)^{1/2} \sin \frac{n_x \pi x}{a} \sin \frac{n_y \pi y}{b} \sin \frac{n_z \pi z}{c} \tag{5-11}$$

and the energy eigenvalues become

$$\varepsilon_{n_x n_y n_z} = \frac{\hbar^2 k^2}{2m} = \frac{\hbar^2}{2m} (k_x^2 + k_y^2 + k_z^2)$$

$$= \frac{\hbar^2 \pi^2}{2m} \left(\frac{n_x^2}{a^2} + \frac{n_y^2}{b^2} + \frac{n_z^2}{c^2}\right). \tag{5-12}$$

Note that the normalization in (5-11) is proportional to (volume of box)$^{-1/2}$.

In general, this problem does not contain a degeneracy. However, we can produce a degeneracy by increasing the spatial symmetry. For example, if we set $a = b = c$, then the eigenvalues take the form

$$\varepsilon_{n_x n_y n_z} = \frac{\hbar^2 \pi^2}{2ma^2} (n_x^2 + n_y^2 + n_z^2). \tag{5-13}$$

Now we observe that it is not the values of n_x, n_y, and n_z individually that determine the energy but rather the sum of their squares. Thus while the ground state ε_{111} is still nondegenerate with an energy $\varepsilon_{111} = [\hbar^2\pi^2/2ma^2]$ (3), the three first excited states ψ_{112}, ψ_{121}, and ψ_{211} are now triply degenerate, each with the same energy $\varepsilon_{112} = \varepsilon_{121} = \varepsilon_{211} = [\hbar^2\pi^2/2ma^2]$ (6). The degeneracy may be greater for states of higher energy. Because of the degeneracy, the linear combination $\psi' = \alpha\psi_{112} + \beta\psi_{121} + \gamma\psi_{211}$ is still another energy eigenfunction corresponding to the first excited state. Also, since the $\psi_{n_x n_y n_z}(\mathbf{r})$ are not unique, their mutual orthogonality must be verified and cannot be taken for granted. These box eigenfunctions are not momentum eigenfunctions since $[\hat{\mathbf{p}}, \mathscr{H}] \neq 0$ for the box Hamiltonian. Whenever the Hamiltonian contains $\hat{\mathbf{r}}$ [via the potential $\hat{V}(\hat{\mathbf{r}})$], it is incompatible with linear momentum.

The eigenfunctions $\psi_{n_x n_y n_z}(\mathbf{r})$ in the limit a, b, $c \to \infty$ should correspond to those for free particles and we expect the sine functions in (5-11) to behave as *box-normalized* plane waves,‡ that is,

$$\psi_{n_x n_y n_z}(\mathbf{r}) \to \frac{1}{(abc)^{1/2}} e^{ik_x x} e^{ik_y y} e^{ik_z z} = \frac{1}{(abc)^{1/2}} e^{i\mathbf{k}\cdot\mathbf{r}}$$

$$= \frac{1}{\sqrt{\mathscr{V}}} e^{i\mathbf{k}\cdot\mathbf{r}}. \tag{5-14}$$

Because free particle eigenfunctions are part of the *continuous* spectrum the sum over states becomes an integral, that is,

$$\sum_{n_x n_y n_z > 0} \to \int_{\text{positive } n_i} dn_x \, dn_y \, dn_z. \tag{5-15}$$

Using (5-10), the integral may be written

$$\int_{\substack{n_x n_y n_z \\ \text{positive}}} dn_x \, dn_y \, dn_z = \frac{abc}{\pi^3} \int_{\substack{k_x k_y k_z \\ \text{positive}}} dk_x \, dk_y \, dk_z = \frac{\mathscr{V}}{\pi^3} \int_{\substack{\mathbf{k} \text{ in} \\ \text{first octant}}} d\mathbf{k}$$

$$= \frac{\mathscr{V}}{8\pi^3} \int_{\substack{\mathbf{k} \text{ over} \\ \text{all space}}} d\mathbf{k}. \tag{5-16}$$

We can convert the integral over states to one over energy by transforming to polar coordinates in $\mathbf{k}$ space, obtaining

$$\frac{\mathscr{V}}{8\pi^3} 4\pi \int_0^\infty k^2 \, dk = \int_0^\infty \rho(\varepsilon) \, d\varepsilon.$$

Using the relation $\hbar^2 k^2/2m = \varepsilon$, $\rho(\varepsilon)$ can be calculated as

$$\rho(\varepsilon) = \frac{(2m)^{3/2}\mathscr{V}}{4\pi^2\hbar^3} \sqrt{\varepsilon}. \tag{5-17}$$

---

‡ In this representation of free particles, we are replacing the boundary conditions which require that the eigenfunctions vanish at infinity with so-called "periodic" boundary conditions.

The function $\rho(\varepsilon)$ is a measure of the dependence of the degeneracy density (number of degenerate states per energy interval) on ε for a free particle and is called the *density of states*. We shall make extensive use of (5-17) in our discussion of the theory of scattering.

IV   The Anisotropic Oscillator

The three-dimensional anisotropic oscillator consists of a particle bound to the origin by an elastic force where the elastic constants depend on the direction in which the particle is displaced from the origin. The Schroedinger equation for this potential is

$$\left\{-\frac{\hbar^2}{2m}\nabla^2 + \frac{1}{2}m\omega_x^2 x^2 + \frac{1}{2}m\omega_y^2 y^2 + \frac{1}{2}m\omega_z^2 z^2\right\}\psi_i = \varepsilon_i\psi_i \qquad (5\text{-}18)$$

where $\omega_x^2 = K_x/m$, $\omega_y^2 = K_y/m$, and $\omega_z^2 = K_z/m$. Again, a separation will result if we use a solution of the form $\psi_i = X(x)Y(y)Z(z)$. In this case we obtain the three equations

$$-\frac{\hbar^2}{2m}X'' + \frac{1}{2}m\omega_x^2 x^2 X = \varepsilon_x X$$

$$-\frac{\hbar^2}{2m}Y'' + \frac{1}{2}m\omega_y^2 y^2 Y = \varepsilon_y Y$$

and

$$-\frac{\hbar^2}{2m}Z'' + \frac{1}{2}m\omega_z^2 z^2 Z = \varepsilon_z Z$$

with $\varepsilon_i = \varepsilon_x + \varepsilon_y + \varepsilon_z$. The solutions have already been obtained in the one-dimensional case [see (4-47)]. The total eigenfunctions and eigenvalues of (5-18) may be written as

$$\psi_{n_x n_y n_z}(x,\,y,\,z) = X_{n_x} Y_{n_y} Z_{n_z} = \left[\frac{\alpha_x \alpha_y \alpha_z}{\pi^3}\right]^{1/4}\left[\frac{1}{2^{n_x+n_y+n_z}n_x!\,n_y!\,n_z!}\right]^{1/2}$$

$$\times \exp[-\tfrac{1}{2}(\alpha_x x^2 + \alpha_y y^2 + \alpha_z z^2)]\,H_{n_x}\big((\alpha_x)^{1/2}x\big)$$
$$\times H_{n_y}\big((\alpha_y)^{1/2}y\big)\,H_{n_z}\big((\alpha_z)^{1/2}z\big) \qquad (5\text{-}19)$$

and

$$\varepsilon_{n_x n_y n_z} = (n_x + \tfrac{1}{2})\hbar\omega_x + (n_y + \tfrac{1}{2})\hbar\omega_y + (n_z + \tfrac{1}{2})\hbar\omega_z$$
$$(n_x,\,n_y,\,n_z = 0,\,1,\,2,\,3,\,\ldots). \qquad (5\text{-}20)$$

The absence of a degeneracy is to be attributed to a lack of symmetry in the Hamiltonian. To verify this, we consider the special case of an isotropic oscillator $(K_x = K_y = K_z)$ for which the energy eigenvalues become

$$\varepsilon_{n_x n_y n_z} = (n_x + n_y + n_z)\hbar\omega + \tfrac{3}{2}\hbar\omega. \qquad (5\text{-}21)$$

We now observe that the energy depends on the sum of the integers and that any set of quantum numbers having the same sum will represent states of the same energy. Thus, for example, while ψ_{000} is nondegenerate, the states ψ_{100}, ψ_{010}, and ψ_{001} are triply degenerate. The degeneracy of the level $\varepsilon_{\bar{n}}$ where $\bar{n} = n_x + n_y + n_z$ equals the number of ways three nonnegative integers may be chosen to total $\bar{n}$, that is,

$$\text{degeneracy of the } \bar{n}\text{th level} = \tfrac{1}{2}(\bar{n} + 1)(\bar{n} + 2). \qquad (5\text{-}22)$$

The isotropic oscillator $\left(V = \tfrac{1}{2}K(x^2 + y^2 + z^2) = \tfrac{1}{2}Kr^2\right)$ is an example of a central force problem (force directed along the radius vector) and hence it is also separable in spherical polar coordinates. We shall see shortly that solution in spherical coordinates leads to a *different set of degenerate eigenfunctions from that in* (5-19), although the energy spectrum is the same. We shall demonstrate below that the degenerate spherical eigenfunctions are at most linear combinations of degenerate Cartesian functions. Therefore, when solving for the eigenstates of the isotropic oscillator, we must be more specific with regard to which set of states is being sought.

V Curvilinear Coordinates

There are many physically important problems where the Schroedinger equation does not separate in Cartesian coordinates. Among the most important is that of the Coulomb potential associated with the hydrogen atom which is separable in spherical and parabolic coordinates. We briefly consider some aspects of generalized curvilinear coordinates.

A set of coordinates is defined by the transformation

$$q_i = q_i(x, y, z) \qquad (i = 1, 2, 3, \ldots).$$

If the square of a differential element of length involves only terms dq_i^2 and not cross terms $dq_i\, dq_j$, then the coordinate system is said to be orthogonal and we may set

$$dl^2 = dx^2 + dy^2 + dz^2 = h_1^2\, dq_1^2 + h_2^2\, dq_2^2 + h_3^2\, dq_3^2. \qquad (5\text{-}23)$$

The parameters h_1, h_2, and h_3 are called *metric coefficients* and each is a function of the q_i.

Once the h's have been determined, the Laplacian may be expressed as

$$\nabla^2 = \frac{1}{h_1 h_2 h_3} \left\{ \frac{\partial}{\partial q_1} \frac{h_2 h_3}{h_1} \frac{\partial}{\partial q_1} + \frac{\partial}{\partial q_2} \frac{h_1 h_3}{h_2} \frac{\partial}{\partial q_2} + \frac{\partial}{\partial q_3} \frac{h_1 h_2}{h_3} \frac{\partial}{\partial q_3} \right\} \quad (5\text{-}24)$$

and a differential volume element becomes

$$d\mathbf{r} = dx\, dy\, dz = h_1 h_2 h_3\, dq_1\, dq_2\, dq_3 . \quad (5\text{-}25)$$

The transformations to spherical polar and cylindrical polar coordinates are, for example (Figure 5-1),

$$
\begin{array}{ccc}
x = r \sin\theta \cos\phi & & x = \rho \cos\phi \\
y = r \sin\theta \sin\phi & \text{and} & y = \rho \sin\phi \\
z = r \cos\theta & & z = z.
\end{array}
\quad (5\text{-}26)
$$

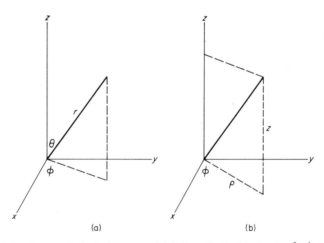

(a)   (b)

Figure 5-1   Geometrical significance of (a) the spherical polar (r, θ, ϕ) and (b) the cylindrical polar (ρ, ϕ, z) variables.

A simple calculation using (5-23) and (5-26) gives

$$
\begin{array}{ccc}
h_r = 1 & & h_\rho = 1 \\
h_\theta = r & \text{and} & h_\phi = \rho \\
h_\phi = r \sin\theta & & h_z = 1
\end{array}
$$

and the respective Laplacians and volume elements become

$$\nabla^2 = \frac{1}{r^2 \sin\theta} \left\{ \frac{\partial}{\partial r} \left(r^2 \sin\theta \frac{\partial}{\partial r} \right) + \frac{\partial}{\partial \theta} \left(\sin\theta \frac{\partial}{\partial \theta} \right) + \frac{\partial}{\partial \phi} \left(\frac{1}{\sin\theta} \frac{\partial}{\partial \phi} \right) \right\}$$

$$d\mathbf{r} = r^2 \sin\theta\, dr\, d\theta\, d\phi$$

and

$$\nabla^2 = \frac{1}{\rho}\left\{ \frac{\partial}{\partial\rho}\left(\rho\,\frac{\partial}{\partial\rho} \right) + \frac{\partial}{\partial\phi}\left(\frac{1}{\rho}\frac{\partial}{\partial\phi} \right) + \frac{\partial}{\partial z}\left(\rho\,\frac{\partial}{\partial z} \right) \right\}$$

$$d\mathbf{r} = \rho\,d\rho\,d\phi\,dz.$$

There are many orthogonal curvilinear coordinate systems and a given problem may separate in more than one system.‡ The central force problem is always separable in spherical polar coordinates as we next shall show.

VI   The Central Force Problem $V = V(r)$

The Schroedinger equation for a particle bound by a central force potential is, in spherical coordinates,

$$\left[-\frac{\hbar^2}{2m}\left\{ \frac{1}{r^2}\frac{\partial}{\partial r}\left(r^2\frac{\partial}{\partial r} \right) + \frac{1}{r^2\sin\theta}\frac{\partial}{\partial\theta}\left(\sin\theta\,\frac{\partial}{\partial\theta} \right) + \frac{1}{r^2\sin^2\theta}\left(\frac{\partial^2}{\partial\phi^2} \right) \right\} + V(r) \right]$$

$$\times\,\psi_i(r,\theta,\phi) = \varepsilon_i\psi_i(r,\theta,\phi). \tag{5-27}$$

In addition to the usual bound state requirement, $\psi \xrightarrow[r\to\infty]{} 0$, we must impose *single-valuedness*, that is,

$$\psi_i(r,\theta,\phi) = \psi_i(r,\theta,\phi+2\pi)$$

since the points ϕ and $\phi + 2\pi$ are in fact the *same* physical points.

We substitute the trial solution $\psi_i = R(r)Y(\theta,\phi)$ into (5-27), and simplyfying, we obtain

$$\frac{\hbar^2}{2m}\left\{ \frac{1}{R}\frac{d}{dr}\left(r^2\frac{dR}{dr} \right) \right\} + (\varepsilon_i - V(r))r^2$$

$$= -\frac{\hbar^2}{2m}\frac{1}{Y}\left\{ \frac{1}{\sin\theta}\frac{\partial}{\partial\theta}\left(\sin\theta\,\frac{\partial Y}{\partial\theta} \right) + \frac{1}{\sin^2\theta}\frac{\partial^2 Y}{\partial\phi^2} \right\}. \tag{5-28}$$

Since the variable r is on one side and the variables θ and ϕ on the other, (5-28) can be satisfied for all r, θ, and ϕ if and only if each side is equal to some constant. Choosing that constant§ to be of the form $(\hbar^2/2m)l(l+1)$, we

---

‡ See, for example, P. M. Morse and H. Feshbach, " Methods of Theoretical Physics," Volume I, p. 655. McGraw-Hill, New York, 1953.

§ This choice of the separation constant leads to equations that are familiar to mathematical physicists and are therefore easier to treat.

obtain the two equations

$$\left\{-\frac{\hbar^2}{2m}\frac{1}{r^2}\frac{d}{dr}\left(r^2\frac{d}{dr}\right)+V(r)+\frac{\hbar^2}{2m}\frac{l(l+1)}{r^2}\right\}R(r)=\varepsilon_i\,R(r) \qquad (5\text{-}29)$$

and

$$-\hbar^2\left\{\frac{1}{\sin\theta}\frac{\partial}{\partial\theta}\left(\sin\theta\frac{\partial}{\partial\theta}\right)+\frac{1}{\sin^2\theta}\frac{\partial^2}{\partial\phi^2}\right\}Y=\hbar^2 l(l+1)Y. \qquad (5\text{-}30)$$

We shall return to the physical significance of these equations shortly. First let us continue the separation process by assuming Y to be of the form

$$Y=\Theta(\theta)\Phi(\phi).$$

Equation (5-30) now becomes

$$\hbar^2\left\{\left[\frac{\sin\theta}{\Theta}\frac{d}{d\theta}\left(\sin\theta\frac{d\Theta}{d\theta}\right)\right]+l(l+1)\sin^2\theta\right\}=-\hbar^2\frac{1}{\Phi}\frac{d^2\Phi}{d\phi^2}. \qquad (5\text{-}31)$$

As in the previous case, this equation can only be satisfied if both sides are equal to the same constant. Taking the separation constant‡ to be of the form $\hbar^2 m_l^2$, we obtain the ordinary differential equations

$$-\hbar^2\frac{d^2}{d\phi^2}\Phi_{m_l}=m_l^2\hbar^2\,\Phi_{m_l} \qquad (5\text{-}32)$$

and

$$\hbar^2\left\{\left[\frac{1}{\sin\theta}\frac{d}{d\theta}\left(\sin\theta\frac{d\Theta_{lm_l}}{d\theta}\right)\right]+\left[l(l+1)-\frac{m_l^2}{\sin^2\theta}\right]\Theta_{lm_l}\right\}=0. \qquad (5\text{-}33)$$

Therefore the Schroedinger equation completely separates into (5-29), (5-32), and (5-33), which are ordinary differential equations in the variables r, ϕ, and θ, respectively. The total solution now takes the form

$$\psi_{lm_l}=R(r)Y_{lm_l}(\theta,\,\phi)=R(r)\Theta_{lm_l}(\theta)\Phi_{m_l}(\phi).$$

We now return to the problem of assigning physical significance to the angular equations and the separation constants.

Note that (5-30), (5-32), and (5-33) involve only angular variables. In particular since $V(r)$ does not appear in them, these equations are common to *all* central force problems. It may be verified that the left-hand side of (5-30) is in fact the operator $\hat{L}^2$. Using the transformation from Cartesian to

---

‡ See second footnote at the bottom of page 135.

spherical polar coordinates and the chain rule for differentiation, we find

$$\hat{L}_x = \hat{y}\hat{p}_z - \hat{z}\hat{p}_y = \frac{\hbar}{i}\left(y\frac{\partial}{\partial z} - z\frac{\partial}{\partial y}\right) = i\hbar\left(\sin\phi\frac{\partial}{\partial \theta} + \cot\theta\cos\phi\frac{\partial}{\partial \phi}\right)$$

$$\hat{L}_y = \hat{z}\hat{p}_x - \hat{x}\hat{p}_z = \frac{\hbar}{i}\left(z\frac{\partial}{\partial x} - x\frac{\partial}{\partial z}\right) = i\hbar\left(-\cos\phi\frac{\partial}{\partial \theta} + \cot\theta\sin\phi\frac{\partial}{\partial \phi}\right) \quad (5\text{-}34)$$

$$\hat{L}_z = \hat{x}\hat{p}_y - \hat{y}\hat{p}_x = \frac{\hbar}{i}\left(x\frac{\partial}{\partial y} - y\frac{\partial}{\partial x}\right) = \frac{\hbar}{i}\frac{\partial}{\partial \phi}.$$

Performing the necessary operations, it follows that

$$\hat{L}^2 = \hat{L}_x{}^2 + \hat{L}_y{}^2 + \hat{L}_z{}^2 = -\hbar^2\left[\frac{1}{\sin\theta}\frac{\partial}{\partial \theta}\left(\sin\theta\frac{\partial}{\partial \theta}\right) + \frac{1}{\sin^2\theta}\frac{\partial^2}{\partial \phi^2}\right].$$

Since (5-30) involves only angular differentials, inserting $R(r)$ on both sides leaves this equation unaffected and it becomes

$$\hat{L}^2 R Y_{lm_l} = \hbar^2 l(l+1) R Y_{lm_l}$$

or

$$\hat{L}^2 \psi_{lm_l} = \hbar^2 l(l+1)\psi_{lm_l}. \quad (5\text{-}35)$$

Furthermore, since

$$\hat{L}_z = \frac{\hbar}{i}\frac{\partial}{\partial \phi}$$

it follows that (5-32) is

$$\hat{L}_z{}^2 \Phi_{m_l} = m_l{}^2 \hbar^2 \Phi_{m_l}$$

or, multiplying both sides by $R\Theta_{lm_l}$,

$$\hat{L}_z{}^2 \psi_{lm_l} = m_l{}^2 \hbar^2 \psi_{lm_l}. \quad (5\text{-}36)$$

We are thus led to the following conclusion: *The process of separation of variables of any central force problem in spherical coordinates always leads to energy eigenfunctions which are also simultaneous eigenfunctions of $\hat{L}^2$ and $\hat{L}_z{}^2$. The respective eigenvalues are $l(l+1)\hbar^2$ and $m_l{}^2\hbar^2$, where l and m_l remain to be determined.*

Turning to the radial equation, we observe that it is equivalent to a one-dimensional (radial direction) Schroedinger equation where the effective potential is due to the central force plus a (repulsive) centrifugal potential (Figure 5-2) associated with angular momentum, that is,

$$V_{\text{eff}_l}(r) = V(r) + \frac{\hbar^2 l(l+1)}{2mr^2}. \quad (5\text{-}37)$$

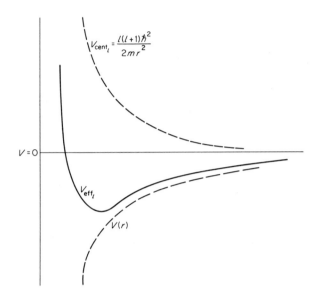

Figure 5-2   The total effective radial potential (V_{eff_l}) and the centrifugal (V_{cent_l}) and central force ($V(r)$) potentials.

The first term in (5-29) can be identified with the radial kinetic energy,‡

$$\frac{\hat{p}_r^{\,2}}{2m} = -\frac{\hbar^2}{2m}\frac{1}{r^2}\frac{d}{dr}\left(r^2\frac{d}{dr}\right). \qquad (5\text{-}38)$$

Since all central force problems have (5-32) and (5-33) in common, it is worthwhile to study their solutions first. In doing so we are actually quantizing angular momentum. The possibility of a central force Hamiltonian having

---

‡ Note that

$$\hat{p}_r = \left(\frac{\hat{\mathbf{r}}}{\hat{r}}\right)\cdot\hat{\mathbf{p}} \to \frac{\hbar}{i}\frac{\partial}{\partial r}$$

is not Hermitian because it is generated by a product of noncommuting Hermitian operators. An acceptable representation for $\hat{p}_r$ is the Hermitian form (see Problem 5-13)

$$\hat{p}_r = \frac{1}{2}\left[\left(\frac{\hat{\mathbf{r}}}{\hat{r}}\right)\cdot\mathbf{p} + \mathbf{p}\cdot\left(\frac{\hat{\mathbf{r}}}{\hat{r}}\right)\right] \to \frac{\hbar}{i}\left(\frac{\partial}{\partial r} - \frac{1}{r}\right)$$

from which it follows that

$$\frac{\hat{p}_r^{\,2}}{2m} = -\frac{\hbar^2}{2m}\frac{1}{r^2}\frac{\partial}{\partial r}\left(r^2\frac{\partial}{\partial r}\right).$$

common eigenstates with $\hat{L}^2$ and $\hat{L}_z$ could have been foreseen from the commutation relations

$$[\mathscr{H}_{cf}, \hat{L}^2] = [\mathscr{H}_{cf}, \hat{L}_z] = 0. \tag{5-39}$$

The verification of these relations is left as an exercise.

VII   Quantization of Angular Momentum

We have just seen that the separation of variables of a central force problem in spherical coordinates always leads to the angular equation

$$-\hbar^2\left\{\frac{1}{\sin\theta}\frac{\partial}{\partial\theta}\left(\sin\theta\frac{\partial}{\partial\theta}\right) + \frac{1}{\sin^2\theta}\frac{\partial^2}{\partial\phi^2}\right\}Y_{lm_l}(\theta,\phi) = \hbar^2 l(l+1)Y_{lm_l}(\theta,\phi) \tag{5-40}$$

which upon further separation, $Y_{lm} = \Theta_{lm_l}(\theta)\Phi_{m_l}(\phi)$, gives

$$\sin\theta\frac{d}{d\theta}\left(\sin\theta\frac{d\Theta_{lm_l}}{d\theta}\right) + (-m_l^2 + l(l+1)\sin^2\theta)\Theta_{lm_l} = 0 \tag{5-41}$$

and

$$-\hbar^2\frac{d^2}{d\phi^2}\Phi_{m_l} = m_l^2\hbar^2\Phi_{m_l}. \tag{5-42}$$

Furthermore (5-40) and (5-42) represent eigenvalue equations for $\hat{L}^2$ and $\hat{L}_z$ (or more precisely $\hat{L}_z^2$) with respective quantum numbers l (called orbital) and m_l (called azimuthal or "magnetic"). Once the permissible values of l and m_l have been determined, the eigenvalues of $\hat{L}^2$ and $\hat{L}_z$ may be written as $l(l+1)\hbar^2$ and $m_l\hbar$ respectively.

We observe that any solution to

$$\hat{L}_z\Phi_{m_l} = \frac{\hbar}{i}\frac{d}{d\phi}\Phi_{m_l} = m_l\hbar\Phi_{m_l} \tag{5-43}$$

is automatically a solution to (5-42). The solution to (5-43) gives as eigenfunctions of $\hat{L}_z$, $\Phi_{m_l} \propto e^{im_l\phi}$. However the *single-valuedness* of ψ requires that

$$\Phi_{m_l}(\phi + 2\pi) = \Phi_{m_l}(\phi)$$
$$e^{im_l(\phi + 2\pi)} = e^{im_l\phi}$$

or

$$e^{im_l 2\pi} = 1$$

from which it follows that $m_l = 0, \pm1, \pm2, \ldots$. The eigenvalues of $\hat{L}_z$ are therefore $0, \pm\hbar, \pm2\hbar, \ldots$.

Next we turn our attention to (5-41) which may be transformed, using $u = \cos\theta$, to

$$\frac{d}{du}\left[(1-u^2)\frac{d}{du}\,\Theta_{lm_l}(u)\right] + \left(l(l+1) - \frac{m_l^2}{1-u^2}\right)\Theta_{lm_l}(u) = 0. \qquad (5\text{-}44)$$

Equation (5-44) is the associated Legendre equation and its solutions are associated Legendre functions. The substitution

$$\Theta_{lm_l} = (1-u^2)^{|m_l|/2}\,\frac{d^{|m_l|}}{du^{|m_l|}}\,P_l(u) \qquad (5\text{-}45)$$

simplifies (5-44) to the ordinary (m_l-independent) Legendre equation

$$\frac{d}{du}\left[(1-u^2)\frac{dP_l(u)}{du}\right] + l(l+1)P_l(u) = 0 \qquad (5\text{-}46)$$

where the P_l are ordinary Legendre functions. *Equation (5-46) must be solved subject to the finiteness requirement of the eigenfunctions.* Therefore, we expect

$$P_l(u) = \text{finite} \qquad \text{for} \quad -1 \leqslant u \leqslant 1 \quad (0 \leqslant \theta \leqslant \pi).$$

The approach used here is quite similar to the solution of the Hermite equation associated with the oscillator eigenfunctions. We shall attempt a series solution of the form

$$P_l(u) = \sum_{j=0}^{\infty} a_j u^j$$

which upon substitution into (5-46) gives the two-term recursion formula

$$a_{j+2} = \frac{j(j+1) - l(l+1)}{(j+1)(j+2)}\,a_j. \qquad (5\text{-}47)$$

Again we generate either an even or an odd series by choosing a_0 and a_1 arbitrarily. Unfortunately, as may be verified, both series solutions diverge at $u = \pm 1$, making them unacceptable as eigenfunctions. If however l is a positive integer or zero, then either the even or odd series (according to the value of l) truncates. For even l, we obtain an even polynomial of degree l. Similarly, an odd polynomial of degree l results when l is odd. *Since the polynomials remain finite at $u = \pm 1$ they are acceptable eigenfunctions.*

It is conventional to choose $a_0^{(l)}$ and $a_1^{(l)}$ so that $P_l(1) \equiv 1$, in which case the polynomials are Legendre polynomials. The first few are

$$\begin{array}{ll}
P_0(u) = 1 & P_3(u) = \tfrac{1}{2}(5u^3 - 3u) \\
P_1(u) = u & P_4(u) = \tfrac{1}{8}(35u^4 - 30u^2 + 3) \\
P_2(u) = \tfrac{1}{2}(3u^2 - 1) & P_5(u) = \tfrac{1}{8}(63u^5 - 70u^3 + 15u).
\end{array} \qquad (5\text{-}48)$$

The (ordinary) Legendre polynomials have the following useful properties:

(1) $(1 - 2uz + z^2)^{-1/2} = \sum_{l=0}^{\infty} P_l(u)z^l$ (generating function)

(2) $\qquad P_l(u) = \dfrac{1}{2^l l!} \dfrac{d^l}{du^l}(u^2 - 1)^l \qquad$ (Rodrigues' formula)

(3) $\qquad (l + 1)P_{l+1}(u) - (2l + 1)uP_l(u) + lP_{l-1}(u) = 0$

(4) $\qquad (u^2 - 1)P_l'(u) = luP_l(u) - lP_{l-1}(u)$

(5) $\qquad \displaystyle\int_{-1}^{1} P_l(u)P_{l'}(u)\,du = \dfrac{2}{2l + 1}\delta_{ll'} \qquad$ (orthogonality). $\qquad$ (5-49)

Using (5-45), the *associated* Legendre functions‡ become

$$\Theta_{lm_l} = P_l^{m_l} = (1 - u^2)^{|m_l|/2}\,\frac{d^{\,|m_l|}}{du^{|m_l|}}\,P_l(u). \qquad (5\text{-}50)$$

The associated functions have the orthonormalization property

$$\int_{-1}^{1} P_l^{m_l}(u)P_l^{m_l}(u)\,du = \frac{2}{2l + 1}\frac{(l + |m_l|)!}{(l - |m_l|)!}\delta_{ll'}.$$

Since the $|m_l|$th derivative of an lth-degree polynomial vanishes unless $|m_l| \leqslant l$ it follows from (5-50) that *an eigenfunction will exist only if the magnetic number does not exceed the orbital index.*§

The (normalized) spherical harmonics are defined as

$$Y_{lm_l}(\theta,\,\phi) = \left[\frac{2l + 1}{4\pi}\frac{(l - |m_l|)!}{(l + |m_l|)!}\right]^{1/2}P_l^{m_l}(u = \cos\theta)e^{im_l\phi} \qquad (5\text{-}51)$$

where $l = 0, 1, 2, \ldots$ and $m_l = 0, \pm 1, \pm 2, \ldots, \pm l$. They represent eigenfunctions of $\hat{L}^2$ and $\hat{L}_z$ with the respective eigenvalues $l(l + 1)\hbar^2$ and $m_l\hbar$ and have the normalization property

$$\int_0^{2\pi}\int_0^{\pi} Y_{lm_l}^*(\theta,\,\phi)Y_{l'm_{l'}}(\theta,\,\phi)\sin\theta\,d\theta\,d\phi = \delta_{ll'}\,\delta_{m_l m_{l'}}. \qquad (5\text{-}52)$$

The total eigenfunction for any central force problem takes the form

$$\psi_{nlm_l} = R_{nl}(r)Y_{lm_l}(\theta,\,\phi). \qquad (5\text{-}53)$$

We are anticipating a third quantum number n whose significance has not yet been established. Thus the eigenfunctions generated in this manner are simultaneous eigenfunctions of our central force Hamiltonian $\mathcal{H}_{cf}$ and the operators $\hat{L}^2$ and $\hat{L}_z$. At this point the student may ask, since $\mathcal{H}_{cf}$ also commutes with $\hat{L}_x$ and $\hat{L}_y$, why does an eigenfunction of $\hat{L}_z$ evolve? Equivalently, what is so special about the z axis in a physical problem that is spherically symmetric? The answer is that it is the mathematics rather than

---

‡ Due to the factor $(1 - u^2)^{|m_l|/2}$, the Θ_{lm_l} are not polynomials.

§ This restriction corresponds to the classical condition that any component of a vector (L_z) cannot exceed its magnitude (L).

the physics that singles out the z direction. Looking at the transformation to spherical coordinates (5-26), we observe that the angle θ is defined by the inclination to the z axis. Now strictly speaking, the eigenfunction ψ_{nlm_l} is an eigenfunction of $\hat{L}^2$ and one component of $\hat{L}$ which may be along any direction in space. For mathematical convenience that direction is taken to be the z axis.

From the Poisson bracket relation (2-80), it follows that the commutation relation for the components of angular momentum must be

$$[\hat{L}_i, \hat{L}_j] = i\hbar\hat{L}_k \qquad \text{where} \quad i, j, k = x, y, z \text{ in cyclic order.} \qquad (5\text{-}54)$$

The incompatability of any two components of $\hat{L}$ explains why eigenfunctions of only one component may be simultaneously quantized with $\hat{L}^2$ and $\mathscr{H}_{cf}$. Although $\mathscr{H}_{cf}$ commutes with $\hat{L}_x$, $\hat{L}_y$, and $\hat{L}_z$, the components of $\hat{L}$ do not commute among themselves.

The central force eigenfunctions $\psi_{nlm_l}(r, \theta, \phi)$ correspond to eigenkets with the following properties:

$$\mathscr{H}|n, l, m_l\rangle = \varepsilon_{nl}|n, l, m_l\rangle$$
$$\hat{L}^2|n, l, m_l\rangle = l(l+1)\hbar^2|n, l, m_l\rangle \qquad (l = 0, 1, 2, \ldots)$$

and $\qquad\qquad\qquad\qquad\qquad\qquad\qquad\qquad\qquad\qquad\qquad\qquad$ (5-55)

$$\hat{L}_z|n, l, m_l\rangle = m_l\hbar|n, l, m_l\rangle \qquad\qquad (m_l = 0, \pm 1, \pm 2, \ldots, \pm l).$$

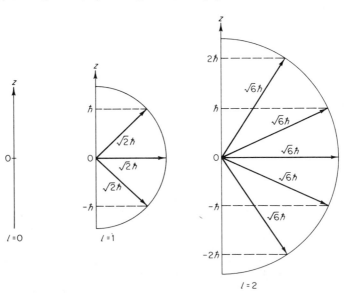

Figure 5-3 Quantization of $\hat{L}$ and $\hat{L}_z$. Note that the vector $\hat{L}$ itself never coincides with the z axis.

The radial equation (5-29) does not contain m_l; hence the energy eigenvalues can depend at most on the quantum numbers n and l. Thus we have an "orientational" or spatial degeneracy with respect to magnetic quantum states. This orientational degeneracy of the level ε_{nl} is at least $2l + 1$ (since there are this many values of m_l for a fixed l) and may be attributed to the spherical symmetry of the potential (Figure 5-3). It follows that any linear combination of magnetic substates is also an eigenstate of $\mathcal{H}$ and $\hat{L}^2$, that is,

$$|n, l\rangle' = \sum_{m_l = -l}^{l} a_{m_l} |n, l, m_l\rangle$$

or

$$\psi'_{nl} = \sum_{m_l = -l}^{l} a_{m_l} \psi_{nlm_l}.$$

However, since we have "mixed" m_l values, the new eigenstate is no longer an eigenstate of $\hat{L}_z$.

The third or principal quantum number n (and consequently ε_{nl}) is to be determined by the radial equation for the particular central force problem. We consider the free particle, isotropic oscillator, and Coulomb problem as examples.

VIII   The Free Particle (Spherical Coordinates)

The free particle eigenfunction was obtained in cartesian coordinates and shown to be of the form $e^{i\mathbf{k} \cdot \mathbf{r}}$. Now we shall consider the free particle eigenfunctions that evolve using spherical coordinates. The wave equation (5-27) is to be solved with $V(r) = 0$. There are three quantum numbers for each energy eigenstate, k, l, and m_l. The solution must be of the form $\psi_{k l m_l} = R_{kl}(r) Y_{lm_l}(\theta, \phi)$ where the continuous quantum parameter, k, is used in place of n. The radial equation (5-29) for the free particle ($V = 0$) becomes

$$\frac{1}{r^2} \frac{d}{dr}\left(r^2 \frac{dR_{kl}}{dr}\right) + \left[k^2 - \frac{l(l + 1)}{r^2}\right] R_{kl} = 0 \qquad (5\text{-}56)$$

where $k^2 = 2m\varepsilon_k/\hbar^2$. Using the change of variable $\rho = kr$ the above equation becomes the *spherical Bessel equation*

$$\frac{1}{\rho^2} \frac{d}{d\rho}\left(\rho^2 \frac{dR_l(\rho)}{d\rho}\right) + \left(1 - \frac{l(l + 1)}{\rho^2}\right) R_l(\rho) = 0 \qquad (5\text{-}57)$$

with the general solution

$$R_l(kr) = A_l j_l(kr) + B_l n_l(kr)$$

where j_l and n_l are respectively termed lth-order spherical Bessel‡ and spherical Neumann functions. We again require that ψ be finite everywhere. The Neumann functions have the property $n_l(kr) \xrightarrow[r \to 0]{} \infty$ and are therefore unacceptable solutions to our problem. Therefore, the free particle eigenfunctions and eigenvalues take the form

$$\psi_{klm_l} = j_l(kr)Y_{lm_l}(\theta, \phi) \qquad (5\text{-}58)$$

with $\varepsilon_k = \hbar^2 k^2/2m$. Note that the energy depends only on the principal parameter k; hence the degeneracy involves both l and m_l. It should be possible to take linear combinations of states of the same k but different l and m_l and generate the Cartesian solution found earlier, that is,

$$e^{i\mathbf{k} \cdot \mathbf{r}} = \sum_{l=0}^{\infty} \sum_{m_l=-l}^{l} a_{lm_l}(k)j_l(kr)Y_{lm_l}(\theta, \phi) \qquad \text{where} \quad |\mathbf{k}| = k.$$

For the special case where $\mathbf{k}$ is along the z axis, the coefficients can be evaluated so as to give the representation

$$e^{ikz} = \sum_{l=0}^{\infty} (2l + 1)i^l j_l(kr)P_l(\cos \theta). \qquad (5\text{-}59)$$

The spherical free particle eigenfunctions are simultaneous functions of $\hat{L}^2$ and $\hat{L}_z$ (rather than of $\hat{\mathbf{p}}$); the importance of the coordinate system used in obtaining energy eigenfunctions should by now be clear.

IX   The Isotropic Oscillator

The radial equation (5-29) for this problem ($V = \frac{1}{2}Kr^2 = \frac{1}{2}m\omega^2 r^2$) takes the form

$$\frac{1}{r^2}\frac{d}{dr}\left(r^2 \frac{dR_{nl}}{dr}\right) + \frac{2m}{\hbar^2}\left(\varepsilon_{nl} - \frac{1}{2}m\omega^2 r^2 - \frac{l(l+1)}{r^2}\right)R_{nl} = 0. \qquad (5\text{-}60)$$

The solution is tedious and a detailed description is not particularly illuminating. We shall merely outline the procedures involved. First we set $\rho = \alpha r^2$ where $\alpha = m\omega/\hbar$ and we write

$$R_{nl} = \rho^{(l+1)/2}e^{-\rho/2}\frac{L_n^{l+1/2}}{\rho^{1/2}}.$$

---

‡ These Bessel functions are related to the cylindrical Bessel functions by

$$j_l(kr) = \left(\frac{\pi}{2kr}\right)^{1/2} J_{l+1/2}(kr).$$

Next we set the energy equal to $\varepsilon_{nl} = \hbar\omega(2n + l + \frac{3}{2})$. Now (5-60) takes the form of the *confluent hypergeometric equation* [see (5-68)],

$$\rho \frac{d^2}{d\rho^2} L_n^{l+1/2}(\rho) + \left[\left(l + \frac{1}{2}\right) + 1 - \rho\right] \frac{d}{d\rho} L_n^{l+1/2}(\rho) + n L_n^{l+1/2}(\rho) = 0. \quad (5\text{-}61)$$

The solutions diverge as $r \to \infty$ unless n is an integer ($n = 0, 1, 2, \ldots$) in which case the solutions $L_n^{l+1/2}$ become Laguerre polynomials (of half-integral order). The first few are

$$L_0^{1/2} = 1 \qquad L_1^{1/2} = \tfrac{3}{2} - \rho$$

$$L_0^{3/2} = 1 \qquad L_1^{3/2} = \tfrac{5}{2} - \rho.$$

The oscillator eigenfunctions and eigenvalues‡ become

$$\psi_{nlm_l}(r, \theta, \phi) \propto \frac{r^{l+1} e^{-\alpha r^2/2} L_n^{l+1/2}(\alpha r^2) Y_{lm_l}(\theta, \phi)}{r} \quad (5\text{-}62)$$

and

$$\varepsilon_{nl} = \varepsilon_{\bar{n}} = (\bar{n} + \tfrac{3}{2})\hbar\omega \quad (5\text{-}63)$$

where $\bar{n} = 2n + l = 0, 1, 2, \ldots$. The eigenvalues are in complete agreement with those obtained using Cartesian coordinates. The two sets of quantum numbers are related by $n_x + n_y + n_z = \bar{n} = 2n + l$. From (5-62) the ground state eigenfunction ($n = l = m_l = 0$) is

$$\psi_{000} \propto e^{-\alpha r^2/2} L_0^{1/2} Y_{00} \propto e^{-\alpha r^2/2} = e^{-\alpha(x^2+y^2+z^2)/2} \quad (5\text{-}64)$$

and has an energy $\varepsilon_{000} = \tfrac{3}{2}\hbar\omega$. It is nondegenerate and unique and is therefore identical (except for normalization) to the Cartesian result in (5-19).

The first excited state ($n = 0$, $l = 1$) is triply degenerate with ψ_{010}, ψ_{011}, and ψ_{01-1} all having the same energy $\varepsilon_{\bar{n}=1} = \tfrac{5}{2}\hbar\omega$. Again it should be possible to construct the degenerate Cartesian states from spherical eigenfunctions. It happens that $\psi_{010}(r, \theta, \phi)$ and $\psi_{001}(x, y, z)$ are identical, that is,

$$\psi_{010}(r, \theta, \phi) = e^{-\alpha r^2/2} r \cos\theta = e^{-\alpha(x^2+y^2+z^2)/2} z = \psi_{001}(x, y, z).$$

On the other hand, we find that $\psi_{100}(x, y, z)$ is a linear combination of

---

‡ We may express (5-62) in terms of $\bar{n}$ as

$$\psi_{\bar{n}lm} = r^l e^{-\alpha r^2/2} L_{(\bar{n}-l)/2}^{l+1/2}(\alpha r^2) Y_{lm_l}(\theta, \phi) \qquad \text{with} \quad \varepsilon_{\bar{n}} = (\bar{n} + \tfrac{3}{2})\hbar\omega.$$

In this form it becomes clear that the energy depends on the single quantum number $\bar{n}$ and *not* on l. Such an occurrence, for a central force problem, is known as an "accidental" degeneracy and is due to some special symmetry other than the spherical symmetry of the potential. This symmetry usually results in the fact that the Schroedinger equation for the potential separates in some coordinate system other than spherical coordinates.

Table 5-1

Classification of the First Few Oscillator States

State	Energy	Cartesian eigenfunction	Spherical eigenfunction	Degeneracy
$\bar{n} = 2n + l$ $\bar{n} = n_x + n_y + n_z$	$\varepsilon_{\bar{n}} = (\bar{n} + \frac{1}{2})\hbar\omega$	$\psi_{n_x n_y n_z}$	ψ_{nlm_l}	$\frac{1}{2}(\bar{n} + 1)(\bar{n} + 2)$
ground state $\bar{n} = 0$	$\frac{1}{2}\hbar\omega$	ψ_{000}	ψ_{000}	1
first excited state $\bar{n} = 1$	$\frac{3}{2}\hbar\omega$	$\psi_{100}, \psi_{010}, \psi_{001}$	$\psi_{011}, \psi_{01-1}, \psi_{010}$	3
second excited state $\bar{n} = 2$	$\frac{5}{2}\hbar\omega$	$\psi_{110}, \psi_{101}, \psi_{011}$ $\psi_{200}, \psi_{020}, \psi_{002}$	$\psi_{022}, \psi_{02-2}, \psi_{021}$ $\psi_{02-1}, \psi_{020}, \psi_{100}$	6

$\psi_{011}(r, \theta, \phi)$ and $\psi_{01-1}(r, \theta, \phi)$ as can be seen from

$$\psi_{100}(x, y, z) = e^{-\alpha(x^2+y^2+z^2)/2}x = e^{-\alpha r^2/2}r \sin \theta \cos \phi$$
$$= e^{-\alpha r^2/2} \sin \theta \, \tfrac{1}{2}(e^{i\phi} + e^{-i\phi})$$
$$= \tfrac{1}{2}[\psi_{011}(r, \theta, \phi) + \psi_{01-1}(r, \theta, \phi)].$$

A similar calculation shows that

$$\psi_{010}(x, y, z) = \frac{1}{2i}[\psi_{011}(r, \theta, \phi) - \psi_{01-1}(r, \theta, \phi)].$$

While the eigenfunctions and the quantum numbers vary with the coordinate system, the energy eigenvalues and the degeneracy of a level are always the same. The relationship between the spherical and Cartesian eigenfunctions for an oscillator have been summarized in Table 5-1.

X   Bound States of an Attractive Coulomb Potential $(V = -K/r)$

The Coulomb problem is perhaps the most important example of a central force problem that we shall consider. The hydrogenic electron experiences a potential of this form due to the presence of the proton in the nucleus. The solution to this problem therefore gives a rather accurate description of the behavior of the simplest of atoms and with some generalization provides the framework for the entire theory of atomic physics. Yet, in a certain sense, this Coulomb potential is a most peculiar one. We shall see that its bound states $(\varepsilon < 0)$ have eigenvalues which are independent of l. This interesting "accidental" degeneracy, as in the case of the isotropic oscillator, is attributable to an intrinsic symmetry (in addition to the usual spatial symmetry) associated with a $1/r$ potential. Furthermore, its unbound states lead to a quantum scattering cross section which is identical to the classical Rutherford result‡ and which leads to a divergent total cross section. Here the long range of the Coulomb potential is directly responsible for this divergence.

We restrict our present discussion to the bound states for which the discrete energy eigenvalues are negative $(\varepsilon_n < 0)$. The radial equation for the Coulomb potential (5-29) becomes§

$$\frac{1}{r^2}\frac{d}{dr}\left(r^2\frac{dR_{nl}}{dr}\right) + \frac{2m}{\hbar^2}\left(\varepsilon_n + \frac{K}{r} - \frac{l(l+1)}{r^2}\right)R_{nl} = 0. \qquad (5\text{-}65)$$

---

‡ See H. Goldstein, "Classical Mechanics," p. 84. Addison-Wesley, Reading, Massachusetts, 1950.

§ We shall assume an "accidental" degeneracy, $\varepsilon_{nl} = \varepsilon_n$. This assumption will be borne out by the results.

We seek solutions which satisfy the boundary condition $R_{nl}(r) \xrightarrow[r=\infty]{} 0$. Making the substitution $\rho = 2\zeta_n r$, where $\zeta_n = (-2m\varepsilon_n/\hbar^2)^{1/2}$ and $R_{nl} = y_{nl}(\rho)/\rho$, (5-65) takes the simpler form

$$\left\{ \frac{d^2}{d\rho^2} - \frac{l(l+1)}{\rho^2} + \frac{\gamma_n}{\rho} - \frac{1}{4} \right\} y_{nl}(\rho) = 0 \qquad (5\text{-}66)$$

where $\gamma_n = mK/\zeta_n\hbar^2$.

We assume that a solution exists of the form

$$y_{nl}(\rho) = \rho^{l+1} e^{-\rho/2} F_{nl} \qquad (5\text{-}67)$$

where F is a power series in the variable ρ. Substituting (5-67) into (5-66) and simplifying, we obtain the *confluent hypergeometric equation*

$$\left\{ \rho \frac{d^2}{d\rho^2} + (\beta - \rho) \frac{d}{d\rho} - \alpha \right\} F(\alpha|\beta|\rho) = 0 \qquad (5\text{-}68)$$

with $\alpha = l + 1 - \gamma_n$ and $\beta = 2(l+1) = 2, 4, 6 \ldots$. The series solution $F(\alpha|\beta|\rho)$ which is well-behaved at the origin is called the *confluent hypergeometric series*. Substituting $F = \sum_{j=0}^{\infty} a_j(\alpha, \beta)\rho^j$ into (5-68), we obtain a two-term recursion formula relating a_{j+1} to a_j. Choosing $a_0 = 1$, the remaining a_j may be calculated giving

$$F(\alpha|\beta|\rho) = 1 + \frac{\alpha}{\beta}\rho + \frac{\alpha(\alpha+1)}{2!\beta(\beta+1)}\rho^2 + \frac{\alpha(\alpha+1)(\alpha+2)}{3!\beta(\beta+1)(\beta+2)}\rho^3 + \cdots$$

as a particular solution.

Now it may be verified that this infinite series diverges at infinity and leads to unacceptable eigenfunctions. However for the special case where α is a negative integer or zero, the series truncates to a polynomial of degree $q = -\alpha$ and represents an acceptable eigenfunction. Therefore acceptable solutions exist when $q = -\alpha = \gamma_n - (l+1) = 0, 1, 2, \ldots$. *Since l is a nonnegative integer it follows that* $\gamma_n = n = 1, 2, 3, \ldots$ *with the auxiliary condition that* $l + 1 \leqslant n$. The (negative) energy eigenvalues immediately follow as

$$\gamma_n^2 = n^2 = \frac{m^2 K^2}{\zeta_n^2 \hbar^4} = -\frac{mK^2}{2\hbar^2 \varepsilon_n} \qquad (5\text{-}69)$$

or

$$\varepsilon_n = -\frac{mK^2}{2\hbar^2 n^2} \qquad (n = 1, 2, 3, \ldots).$$

Equation (5-69) reflects the accidental degeneracy. There was no *a priori* reason for this degeneracy to arise; it comes from the special mathematical nature of the acceptable solutions. Apparently there is a higher symmetry (in addition to the spherical symmetry) associated with a $1/r$ potential.‡

Setting the integer $p = \beta - 1 = 2l + 1$, the confluent hypergeometric equation (5-68) becomes the *associated Laguerre equation*

$$\left\{ \rho \frac{d^2}{d\rho^2} + \left[(p+1) - \rho \right] \frac{d}{d\rho} + q \right\} L_q^p(\rho) = 0 \qquad (5\text{-}70)$$

where the polynomial solutions are the associated Laguerre polynomials (integral order)

$$L_q^p(\rho) = L_{n-(l+1)}^{2l+1}(\rho).$$

The first few are:

$$
\begin{array}{lll}
L_0^0 = 1 & L_1^0 = 1 - \rho & L_2^0 = 1 - 2\rho + \tfrac{1}{2}\rho^2 \\
& L_1^1 = 2 - \rho & L_2^1 = 3 - 3\rho + \tfrac{1}{2}\rho^2 \\
& L_1^2 = 3 - \rho & L_2^2 = 6 - 4\rho + \tfrac{1}{2}\rho^2.
\end{array}
\qquad (5\text{-}71)
$$

They have the following properties:

(1) $L_p^q(z) = (-1)^q \dfrac{d^q}{dz^q} L_{p+q}^0(z)$

$$= \frac{e^z z^{-q}}{p!} \frac{d^p}{dz^p} (z^{p+q} e^{-z}) \qquad (p, q = 0, 1, 2, 3, \ldots)$$

(Rodrigues' formula)

(2) $\dfrac{e^{-zt/(1-t)}}{(1-t)^{q+1}} = \displaystyle\sum_{p=0}^{\infty} t^p L_p^q(z)$ (generating function)

(3) $zL_p^{q+1}(z) = (p + q + 1)L_p^q(z) - (p + 1)L_{p+1}^q(z)$

(4) $zL_p^{q+1}(z) = -(p - z)L_p^q(z) + (p + q)L_{p-1}^q(z)$

(5) $\displaystyle\int_0^{\infty} dz\, z^q e^{-z} L_p^q(z) L_r^q(z) = \frac{[(p+q)!]^3}{p!} \delta_{pr}$

$$(5\text{-}72)$$

(orthogonality).

---

‡ This symmetry is related to the fact that the Schroedinger equation for a $1/r$ potential is separable in both spherical polar *and* parabolic coordinates.

Gathering our results and normalizing, we find the bound energy eigenfunctions for a Coulomb potential to be

$$\psi_{nlm_l}(r, \theta, \phi) = \left[\frac{2l+1}{4\pi} \frac{(l-|m_l|)!}{(l+|m_l|)!} \right]^{1/2} P_l^{m_l}(\cos \theta) e^{im_l \phi}$$

$$\cdot \left[\frac{(n-l-1)!}{2n[(n+l)!]^3} \left(\frac{2}{na} \right)^3 \right]^{1/2} \left(\frac{2r}{na} \right)^l e^{-r/na} L_{n-(l+1)}^{2l+1} \left(\frac{2r}{na} \right)$$

$$(n = 1, 2, 3, \ldots; \quad l = 0, 1, 2, \ldots, n-1;$$

$$m_l = 0, \pm 1, \ldots, \pm l) \quad (5\text{-}73)$$

where we have used the abbreviation $a = \hbar^2/mK$. The first few bound Coulomb eigenfunctions are:

$$\psi_{100} = \left(\frac{1}{\pi a^3} \right)^{1/2} e^{-r/a}$$

$$\psi_{200} = \frac{1}{4} \left(\frac{1}{2\pi a^3} \right)^{1/2} \left(2 - \frac{r}{a} \right) e^{-r/2a}$$

$$\psi_{210} = \frac{1}{4} \left(\frac{1}{2\pi a^3} \right)^{1/2} \left(\frac{r}{a} \right) e^{-r/2a} \cos \theta$$

$$\psi_{21\pm1} = \frac{1}{8} \left(\frac{1}{\pi a^3} \right)^{1/2} \left(\frac{r}{a} \right) e^{-r/2a} \sin \theta \, e^{\pm i\phi}$$

$$\psi_{300} = \frac{1}{81} \left(\frac{1}{3\pi a^3} \right)^{1/2} \left(27 - \frac{18r}{a} + \frac{2r^2}{a^2} \right) e^{-r/3a}$$

$$\psi_{310} = \frac{1}{81} \left(\frac{2}{\pi a^3} \right)^{1/2} \left(6 - \frac{r}{a} \right) \left(\frac{r}{a} \right) e^{-r/3a} \cos \theta \qquad (5\text{-}74)$$

$$\psi_{31\pm1} = \frac{1}{81} \left(\frac{1}{\pi a^3} \right)^{1/2} \left(6 - \frac{r}{a} \right) \left(\frac{r}{a} \right) e^{-r/3a} \sin \theta \, e^{\pm i\phi}$$

$$\psi_{320} = \frac{1}{81} \left(\frac{1}{6\pi a^3} \right)^{1/2} \left(\frac{r}{a} \right)^2 e^{-r/3a} (3 \cos^2 \theta - 1)$$

$$\psi_{32\pm1} = \frac{1}{81} \left(\frac{1}{\pi a^3} \right)^{1/2} \left(\frac{r}{a} \right)^2 e^{-r/3a} \sin \theta \cos \theta \, e^{\pm i\varphi}$$

$$\psi_{32\pm2} = \frac{1}{162} \left(\frac{1}{\pi a^3} \right)^{1/2} \left(\frac{r}{a} \right)^2 e^{-r/3a} \sin^2 \theta \, e^{\pm 2i\phi}.$$

XI   The Hydrogen Atom‡

The above results can be applied to the hydrogen atom in which the hydrogenic electron experiences the Coulomb potential of a stationary proton in the nucleus. Setting $K = e^2$, we find $a = \hbar^2/mK = \hbar^2/me^2$ to be the first Bohr radius. The energy eigenvalues take the form

$$\varepsilon_n = -\frac{me^4}{2\hbar^2 n^2} = \frac{\varepsilon_1}{n^2} \tag{5-75}$$

where $\varepsilon_1 = -me^4/2\hbar^2 \simeq -13.6$ eV in complete agreement with Bohr's results. It is remarkable that Bohr obtained (5-75) with so elementary a theory.

As a result of the accidental degeneracy, the nth level is now $\sum_{l=0}^{n-1} (2l + 1) = n^2$-fold degenerate. We shall on occasion use the spectroscopic notation given in Table 5-2, labeling a hydrogenic state according to

Table 5-2

Spectroscopic Notation

l value	letter (lower case)
0	s (sharp)
1	p (principal)
2	d (diffuse)
3	f (fundamental)
4	g

its l value. The accidental degeneracy may be compared to the classical case where different ellipses (different L values) have the same energies provided they have the same major axes. The spatial degeneracy corresponds to the fact that the plane of the classical orbit may be rotated without affecting its energy. While the concept of a well-defined orbit is inappropriate to quantum theory, it is nevertheless useful to picture each state (n, l) by an orbit. Since s states $(l = 0)$ have zero angular momentum, they should be pictured as degenerate ellipses (straight lines). As l increases for fixed n, we expect the ellipse to become quasi-circular; we therefore picture the states $n, l = n - 1$ as circles. All other states are ellipses of varying eccentricities with those states of the same n having equal major axes (Figure 5-4).

---

‡ The results presented here contain only gross structure. The smaller corrections due to the spin of the electron and the motion of the proton will be considered in Chapters 6 and 10.

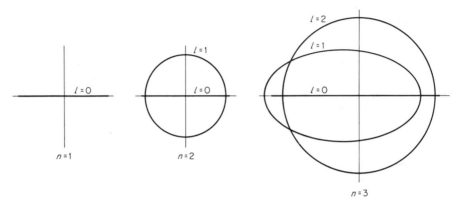

Figure 5-4   Some semiclassical orbits for the states *n, l*.

The spatial quantization (m_l) restricts the inclination of the plane of the classical orbit to the *z* axis. The state ψ_{nlm_l} contains no orientational quantization with respect to the *x* and *y* axes. Thus the normal to the plane may lie anywhere on a cone about the *z* axis (Figure 5-5).

The probability of finding the hydrogenic electron in a given volume

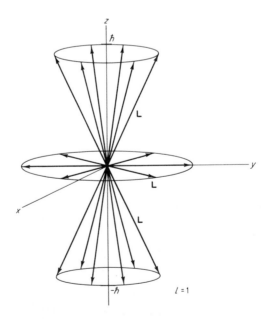

Figure 5-5   The "statistical" cones associated with spatial quantization of angular momentum.

element of space is

$$\mathscr{P}_{nlm_l}\, d\mathbf{r} = R_{nl}^2 |\,Y_{lm_l}|^2 r^2\, dr\, d\Omega$$

where $d\Omega = \sin\theta\, d\theta\, d\phi$. This may be decomposed into an angular density $|\,Y_{lm_l}|^2$ and a radial density $R_{nl}^2 r^2$. The former is associated with the probability of finding the electron within a solid angle $d\Omega$ about the origin (proton). Note that the angular density is

$$|\,Y_{lm_l}|^2 = \Theta_{lm_l}^2$$

and depends only on the polar angle θ and not on the azimuthal angle ϕ. The angular density is conveniently plotted on a polar diagram (Figure 5-6).

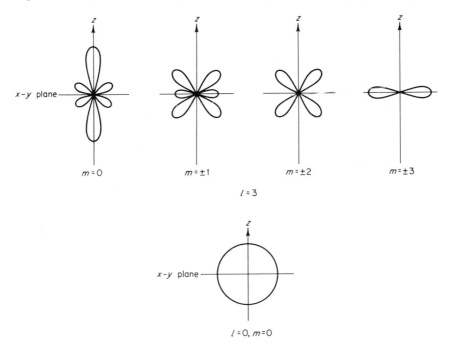

Figure 5-6 The polar plots for the angular densities $|\,Y_{lm_l}|^2$ associated with the states $l = 0$ and $l = 3$.

A particle in the state $l = 3$, $m_l = 0$ is most likely to be found at $\theta = 0$ or $\theta = \pi$. In the state $l = 3$, $m_l = 3$ the maximum probability occurs at $\theta = \pi/2$ (xy plane).

The radial probability density $R_{nl}^2 r^2$ is related to the probability of finding the electron in a spherical shell of thickness dr at a distance r from the origin (Figure 5-7). Note that the radial position of maximum probability for the ground state ψ_{100} occurs at $r_{max} = a$ which corresponds to Bohr's result.

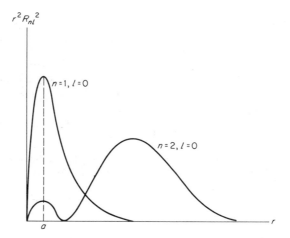

Figure 5-7   Radial probability densities for the states ψ_{100} and ψ_{200} in hydrogen.

However, of greater importance is the *average* position with respect to the origin

$$\langle r \rangle_{nl} = \int_0^\infty R_{nl} r R_{nl} r^2 \, dr.$$

Integrals of the type

$$\langle r^\nu \rangle_{nl} = \int_0^\infty R_{nl} r^\nu R_{nl} r^2 \, dr$$

are rather difficult to evaluate. We list a few results‡:

$$\langle r \rangle_{nl} = n^2 a \left[1 + \frac{1}{2} \left\{ 1 - \frac{l(l+1)}{n^2} \right\} \right]$$

$$\langle r^2 \rangle_{nl} = \frac{1}{2} n^2 [5n^2 + 1 - 3l(l+1)] a^2$$

$$\langle r^{-1} \rangle_{nl} = \frac{1}{n^2 a} \tag{5-76}$$

$$\langle r^{-2} \rangle_{nl} = \frac{1}{n^3 (l + \frac{1}{2}) a^2}.$$

We can ask the following question: How well is the position of the electron in an orbit localized radially? For this we must calculate the uncertainty Δr_n

‡ See H. A. Bethe and E. E. Salpeter, "Quantum Mechanics of One and Two Electron Atoms," p. 17. Springer-Verlag, Berlin, 1957.

about the expectation value $\langle r \rangle_n$. Using the definitions introduced in Chapter 3, and the results in (5-76), Δr_n and $\langle r \rangle_n$ may be readily calculated.

It is interesting to estimate the limit of the ratio

$$\lim_{n,l \to \infty} \frac{\Delta r_{nl}}{\langle r \rangle_{nl}} = \lim_{n,l \to \infty} \frac{(\langle r^2 \rangle_{nl} - \langle r \rangle_{nl}^2)^{1/2}}{\langle r \rangle_{nl}}.$$

Setting l equal to its maximum value, $l_{max} = n - 1$, and using (5-76) we obtain, in the limit $n \to \infty$,

$$\frac{\Delta r_{n, n-1}}{\langle r \rangle_{n, n-1}} = \frac{1}{(2n+1)^{1/2}} \xrightarrow[n \to \infty]{} 0.$$

Thus the probability densities for large n and l become well-defined about the Bohr radii and correspond to the quasi-circular orbits of Bohr's semiclassical theory.

The expectation value of $V = -e^2/r$ may be obtained using (5-76), which gives

$$\langle V \rangle_{nl} = \left\langle \frac{-e^2}{r} \right\rangle_{nl} = -e^2 \langle r^{-1} \rangle_{nl} = -\frac{e^2}{n^2 a} = 2\varepsilon_n.$$

Since $\langle T + V \rangle_{nl} = \langle \mathcal{H} \rangle_{nl} = \varepsilon_n$, it follows that

$$\langle T \rangle_{nl} = \varepsilon_n - \langle V \rangle_{nl} = -\varepsilon_n = -\tfrac{1}{2} \langle V \rangle_{nl}. \tag{5-77}$$

This is also the case in classical mechanics (from the virial theorem) where the expectation values are replaced by time averages, that is, $\bar{T} = -\tfrac{1}{2}\bar{V}$ (see Problems 2-3 and 4-10).

XII Parity and the Central Force Problem

Since any central force Hamiltonian is invariant upon reflection through the origin $[\mathcal{H}_{cf}(-\hat{r}) = \mathcal{H}_{cf}(\hat{r})]$, it commutes with the parity operator. Consequently there must exist at least one set of central force energy eigenfunctions of definite parity. Whether the states $\psi_{nlm_l}(r, \theta, \phi)$ are such a set will be investigated next.

A reflection through the origin $\mathbf{r} \to -\mathbf{r}$ is represented by $x \to -x$, $y \to -y$, $z \to -z$, or in spherical coordinates by $r \to r$, $\theta \to \pi - \theta$, and $\phi \to \phi + \pi$ (Figure 5-8). The replacement $\theta \to \pi - \theta$ is equivalent to $u \to -u$, where $u = \cos \theta$. Since the radial part of R_{nl} remains unaffected by a reflection, we consider only the behavior of the spherical harmonics

$$Y_{lm_l}(\theta, \phi) \propto (1 - u^2)^{|m_l|/2} \frac{d^{|m_l|}}{du^{|m_l|}} P_l(u) e^{im_l\phi}.$$

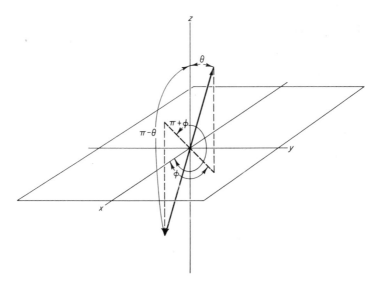

Figure 5-8 The parity (reflection) operation in spherical coordinates.

Under a reflection the ϕ part becomes

$$e^{im_l\phi} \rightarrow e^{im_l(\phi + \pi)} = e^{im_l\pi}e^{im_l\phi} = (-1)^{m_l}e^{im_l\phi}$$

that is, it changes sign only if m_l is odd. The only term in Y_{lm_l} affected by the reflection $u \rightarrow -u$ is

$$\frac{d^{|m_l|}}{du^{|m_l|}} P_l(u).$$

Since this is an even or odd polynomial according to the value $l - m_l$, it will change sign only if $l - m_l$ is odd. We consider two cases:

Case (a)   l is even If m_l is odd, then the ϕ part changes sign. But since $l - m_l$ is also odd, the θ part also changes sign, leaving the eigenfunction unchanged. Similarly, if m_l is even, no change occurs in ψ_{nlm_l}.

Case (b)   l is odd If m_l is odd, again the ϕ part changes sign. However since $l - m_l$ is now even the θ part remains unchanged and Y_{lm_l} therefore has a net change in sign. A similar result is obtained if m_l is even.

We conclude that the central force eigenstates obtained in spherical coordinates are of definite parity, being odd or even according to odd or even values of l respectively. Thus in addition to the usual eigenvalue properties these states have the property

$$\hat{P}|n, l, m_l\rangle = (-1)^l|n, l, m_l\rangle$$

or   (5-78)

$$\psi_{nlm_l}(-\mathbf{r}) = (-1)^l\psi_{nlm_l}(\mathbf{r}).$$

XIII   The Effect of a Uniform Magnetic Field on the Central Force Problem

We now wish to show that the reduction in symmetry produced by a **B** field results in a removal of the orientational degeneracy (with respect to m_l). The magnetic field is derivable from a vector potential **A** using the relation

$$\mathbf{B} = \nabla \times \mathbf{A}. \tag{5-79}$$

It is **A** rather than **B** that enters directly into the classical Hamiltonian, which according to (2-62) is

$$\mathcal{H} = \frac{(\mathbf{p} - (q/c)\mathbf{A})^2}{2m} + V(r) \tag{5-80}$$

where q is the charge of our particle and c is the speed of light. If the magnetic field is uniform, that is, its spatial derivatives vanish, then (5-79) may be inverted to give $\mathbf{A} = \frac{1}{2}\mathbf{B} \times \mathbf{r}$. This may be verified by taking the curl of both sides obtaining (for constant **B**)

$$\nabla \times \mathbf{A} = \nabla \times (\tfrac{1}{2}\mathbf{B} \times \mathbf{r}) = \tfrac{1}{2}[(\nabla \cdot \mathbf{r})\mathbf{B} - (\nabla\mathbf{r}) \cdot \mathbf{B}] = \tfrac{1}{2}[3\mathbf{B} - \mathbf{B}] = \mathbf{B}$$

which is the desired result.

Now the Hamiltonian (5-80) takes the form

$$\mathcal{H} = \frac{p^2}{2m} + V(r) - \frac{q}{mc}\mathbf{p} \cdot \mathbf{A} + \frac{q^2}{2mc^2}A^2$$
$$= \frac{p^2}{2m} + V(r) - \frac{q}{2mc}\mathbf{p} \cdot \mathbf{B} \times \mathbf{r} + \frac{q^2}{2mc^2}\left|\frac{1}{2}\mathbf{B} \times \mathbf{r}\right|^2. \tag{5-81}$$

In practice the magnetic field is usually sufficiently weak‡ to drop the quadratic term in **B**. Rearranging the triple product, (5-81) becomes

$$\mathcal{H} = \frac{p^2}{2m} + V(r) - \frac{q}{2mc}(\mathbf{r} \times \mathbf{p}) \cdot \mathbf{B} = \frac{p^2}{2m} + V(r) - \boldsymbol{\mu}_l \cdot \mathbf{B} \tag{5-82}$$

---

‡ This approximation is valid if the orbital energy is large compared with the magnetic energy associated with the external **B** field. For a discussion of the quadratic Zeeman effect see H. A. Bethe and R. W. Jackiw, "Intermediate Quantum Mechanics," Benjamin, New York, 1968, p. 122.

where

$$\boldsymbol{\mu}_l = \frac{q}{2mc} \, \mathbf{r} \times \mathbf{p} = \frac{q}{2mc} \, \mathbf{L}$$

is the orbital magnetic dipole moment of the charge. Thus for weak fields, the dynamical system carries a magnetic dipole associated with the orbital angular momentum. It is convenient to make a trivial generalization of (5-82) by writing

$$\boldsymbol{\mu}_l = g_l \frac{q}{2mc} \, \mathbf{L} \qquad (5\text{-}83)$$

where $g_l = 1$ is called the orbital gyromagnetic or "g" factor.‡

For convenience, we shall orient our coordinate system so that the **B** field is along the z axis. The z direction is now a well-defined physical direction and is no longer purely mathematical.

The quantum-mechanical Hamiltonian becomes

$$\mathscr{H} = \mathscr{H}_{\mathrm{cf}} - \frac{g_l q}{2mc} \, B\hat{L}_z \qquad (5\text{-}84)$$

where

$$\mathscr{H}_{\mathrm{cf}} = \frac{\hat{p}^2}{2m} + V(r).$$

To find the eigenfunctions and eigenvalues of this new Hamiltonian we observe that $[\mathscr{H}, \mathscr{H}_{\mathrm{cf}}] = 0$. This means that the two Hamiltonians have a common eigenfunction set. Perhaps the eigenfunctions of $\mathscr{H}_{\mathrm{cf}}$ already obtained are also eigenfunctions of $\mathscr{H}$. We confirm this by writing

$$\mathscr{H}\psi_{nlm_l} = \mathscr{H}_{\mathrm{cf}}\psi_{nlm_l} - \frac{g_l q B}{2mc} \, \hat{L}_z \psi_{nlm_l}$$

$$= \left(\varepsilon_{nl} - \frac{g_l q B m_l \hbar}{2mc} \right) \psi_{nlm_l}. \qquad (5\text{-}85)$$

Thus the functions ψ_{nlm_l} are also eigenfunctions§ of $\mathscr{H}$ but with modified eigenvalues which are given by

$$\varepsilon_{nlm_l} = \varepsilon_{nl} - m_l \mu_{\mathrm{B}} B \qquad (g_l = 1).$$

---

‡ The quotient $\mu_l/L = g_l q/2mc$ is called the gyromagnetic ratio.

§ Note that the eigenfunctions remain unaffected by the magnetic field. In particular, the states "conserve" their parity characteristics. This implies that the magnetic term in the Hamiltonian, that is, $qL_z B/2mc$, must be reflection invariant (that is, commute with the parity operator). Since **B** is a constant, L must be reflection invariant. This is indeed the case since under reflections we have $\mathbf{r} \to -\mathbf{r}$ and $\mathbf{p} \to -\mathbf{p}$ so that $\mathbf{L} \to -\mathbf{r} \times -\mathbf{p} = \mathbf{L}$.

The parameter $\mu_B = q\hbar/2mc$ is called *a Bohr magneton* and is a character-istic constant of a particle. The value of the Bohr magneton for the electron is

$$\mu_B = \frac{e\hbar}{2mc} \simeq 0.927 \times 10^{-20} \quad \text{erg/gauss.} \tag{5-86}$$

The energy now depends on m_l as well as on n and l; the magnetic field thus removes the orientational degeneracy. The origin of the term "magnetic quantum number" for m_l is now obvious. For the hydrogenic electron, the accidental degeneracy (with respect to l) remains. By setting $q = -e$, the Bohr levels are shifted by

$$\varepsilon_{nm_l} = \frac{\varepsilon_1}{n^2} + \frac{m_l e\hbar B}{2mc}. \tag{5-87}$$

This shift is known as the normal Zeeman effect (Figure 5-9). It is noteworthy that the energy shifts of the magnetic sublevels occur in multiples of

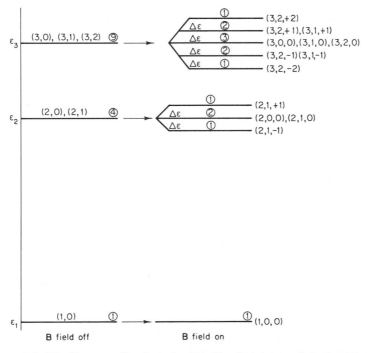

Figure 5-9 The Zeeman effect in hydrogen. The degeneracy of the level is circled. Note that the **B** field does not remove the degeneracy completely; the accidental (*l*) degener-acy remains.

$$\Delta\varepsilon(\text{mag}) = \frac{e\hbar B}{2mc}.$$

The classical ($\hbar$-independent) quantity

$$\omega_{\text{L}} = \frac{\Delta\varepsilon(\text{mag})}{\hbar} = \frac{eB}{2mc}$$

is the Larmor frequency of orbital precession obtained in the classical analysis of (5-80).‡

The eigenstates of $\mathscr{H}$ are now simultaneous eigenstates of $\hat{L}^2$ and the component of **L** along **B** (in this case L_z). The other components L_x and L_y are not precise; it may be verified that in these states

$$\langle L_x \rangle_{nlm_l} = \langle L_y \rangle_{nlm_l} = 0 \qquad (5\text{-}88)$$

so that on the average there is zero angular momentum transverse to **B**. Equivalently **L** is statistically random on a cone about **B**§ (Figure 5-10).

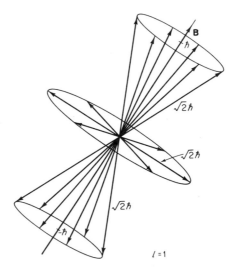

Figure 5-10   The distribution of **L** on cones about **B**. Classically the vector **L** precesses about **B** with a frequency $\omega_{\text{L}} = eB/2mc$ (Larmor precession).

---

‡ For a discussion of Larmor precession, see K. R. Symon, "Mechanics," 3rd ed., p. 284. Addison-Wesley, Reading, Massachusetts, 1971.

§ From a classical standpoint we could say that **L** precesses about **B**; however, quantum-mechanically the state is stationary. Thus the cone is statistical rather than dynamical in nature.

When emission lines in hydrogen are observed with high-resolution spectroscopes, the line shifts are not in quantitative agreement with the normal Zeeman effect. This discrepancy will be resolved in Chapter 6 by postulating that in addition to an orbital magnetic moment, the electron carries a magnetic dipole generated by an intrinsic electron spin. Before discussing spin, we shall attempt to obtain the angular momentum eigenvalues from the ladder method.

This digression has two important objectives. First, it again shows the general nature of the vector-operator formalism. More important for our purposes, it leads to the conclusion that based on commutator algebra alone, the quantum numbers of angular momentum may be *half-integral* as well as *integral*. In the case of orbital angular momentum where a classical analog and consequently an eigenfunction exists, we must rule out half-integers.‡ However, in a purely quantum-mechanical case such as intrinsic electron spin, half-integral momenta are consistent with the fundamental commutation relations for the components of angular momentum, as we now show.

XIV   The Ladder Method

We wish to obtain the eigenvalues and eigenvectors of angular momentum using only the commutation relations

$$[\hat{L}_i, \hat{L}_j] = i\hbar\hat{L}_k \qquad (ijk = xyz \text{ in cyclic order}). \qquad (5\text{-}89)$$

These relations may be obtained as a generalization of the Poisson bracket (2-80) or they may be obtained directly using the commutation relations for $\hat{\mathbf{r}}$ and $\hat{\mathbf{p}}$.

We concern ourselves with the eigenvalue equations

$$\hat{L}^2|l, m_l\rangle = \hbar^2\lambda_l|l, m_l\rangle \qquad (5\text{-}90a)$$

$$\hat{L}_z|l, m_l\rangle = \hbar\lambda_{m_l}|l, m_l\rangle. \qquad (5\text{-}90b)$$

where λ_l and λ_{m_l} are to be determined. The expectation values $\langle L_z^2\rangle$, $\langle L_y^2\rangle$, and $\langle L_z^2\rangle$ are all nonnegative, from which it follows that

$$\langle L^2\rangle = \langle L_x^2\rangle + \langle L_y^2\rangle + \langle L_z^2\rangle \geqslant \langle L_z^2\rangle$$

or in particular

$$\langle l, m_l|\hat{L}^2|l, m_l\rangle \geqslant \langle l, m_l|\hat{L}_z^2|l, m_l\rangle.$$

‡ Recall that it was the single-valuedness requirement in ψ_{nlm_l} that led to the integral m_l.

Using (5-90) it follows that the eigenvalue parameters have the restriction

$$\lambda_l \geqslant \lambda_{m_l}^2. \tag{5-91}$$

We introduce the (non-Hermitian) operator $\hat{L}^+ = \hat{L}_x + i\hat{L}_y$ and its adjoint $\hat{L}^- = \hat{L}_x - i\hat{L}_y$. Operating with $\hat{L}^\pm$ on (5-90a) and noting that $[\hat{L}^\pm, \hat{L}^2] = 0$, we find

$$\hat{L}^\pm \hat{L}^2 |l, m_l\rangle = \hbar^2 \lambda_l \hat{L}^\pm |l, m_l\rangle$$

or

$$\hat{L}^2 \{\hat{L}^\pm |l, m_l\rangle\} = \hbar^2 \lambda_l \{\hat{L}^\pm |l, m_l\rangle\}$$

so that the vector in brackets is still an eigenket of $\hat{L}^2$ with the *same* eigenvalue $\hbar^2 \lambda_l$.

Now we operate with $\hat{L}^\pm$ on (5-90b) and obtain

$$(\hat{L}_x \pm i\hat{L}_y)\hat{L}_z |l, m_l\rangle = \lambda_m \hbar(\hat{L}_x \pm i\hat{L}_y)|l, m_l\rangle. \tag{5-92}$$

Using the commutation relations $\hat{L}_x \hat{L}_z = \hat{L}_z \hat{L}_x - i\hbar\hat{L}_y$ and $\hat{L}_y \hat{L}_z = \hat{L}_z \hat{L}_y + i\hbar\hat{L}_x$, (5-92) may be simplified to

$$(\hat{L}_z \hat{L}^\pm \mp \hbar\hat{L}^\pm)|l, m_l\rangle = \lambda_{m_l} \hbar\hat{L}^\pm |l, m_l\rangle$$

or, transposing,

$$\hat{L}_z \{\hat{L}^\pm |l, m_l\rangle\} = (\lambda_{m_l} \pm 1)\hbar\{\hat{L}^\pm |l, m_l\rangle\}. \tag{5-93}$$

Thus the kets within the brackets are eigenvectors of $\hat{L}_z$ with the adjacent eigenvalues $\lambda_{m_l} \pm 1$. It may be verified that the properly normalized ket is

$$|l, m_l \pm 1\rangle = \frac{1}{[\{\lambda_l - \lambda_{m_l}(\lambda_{m_l} \pm 1)\}\hbar^2]^{1/2}} \hat{L}^\pm |l, m_l\rangle. \tag{5-94}$$

The operators $\hat{L}^+$ and $\hat{L}^-$ are respectively raising and lowering operators, and we conclude that the spectrum of $\hat{L}_z$ is integrally spaced. However from (5-91) it follows that λ_{m_l} is bounded. From considerations of spatial symmetry we may assume that the integrally spaced values of λ_{m_l} are symmetric with respect to zero and are bounded by λ_{m_0} according to

$$-\lambda_{m_0}, \quad -\lambda_{m_0} + 1, \quad \ldots, \quad \lambda_{m_0} - 1, \quad \lambda_{m_0}. \tag{5-95}$$

However it is impossible to have a set of numbers integrally spaced which will satisfy (5-95) unless λ_{m_0} is either an *integer* or a *half-integer*. For example, if $\lambda_{m_0} = +\frac{2}{3}$ then it is impossible to reach $-\frac{2}{3}$ in integral steps. Thus we set $\lambda_{m_0} = l$ where l is either integral or half-integral; it then follows that $\lambda_{m_l} = m_l$ where m_l is integral or half-integral according to the value of l. Equation (5-95) now reads

$$-l \leqslant m_l \leqslant l \qquad (l, |m_l| = 0, 1, 2, \ldots \quad \text{or} \quad l, |m_l| = \tfrac{1}{2}, \tfrac{3}{2}, \ldots). \tag{5-96}$$

The ladder method leads to the additional result that based on (5-89), l and m_l may be half-integral as well as integral. However in the case where $\hat{L}$ has a classical analog (that is, $\hat{L}$ is a canonical function of $\hat{r}$ and $\hat{p}$) and consequently has an eigenfunction ψ_{lm_l}, *single-valuedness rules out half-integral values. In the case of a purely quantum-mechanical angular momentum (for example, spin), we shall see that half-integral values are not only permissible but actually necessary.*

To find the value of λ_l we observe that when $\lambda_{m_l} = \lambda_{m_0} = l$ then a raising operation does not produce a new state, that is,

$$\hat{L}^+ |l, l\rangle = 0. \tag{5-97}$$

Using the identity $\hat{L}^2 = \hat{L}_x{}^2 + \hat{L}_y{}^2 + \hat{L}_z{}^2 = \hat{L}^-\hat{L}^+ + \hat{L}_z{}^2 + \hbar\hat{L}_z$, we have

$$\hat{L}^2 |l, l\rangle = \hbar^2 \lambda_l |l, l\rangle$$

or

$$(\hat{L}^-\hat{L}^+ + \hat{L}_z{}^2 + \hbar\hat{L}_z)|l, l\rangle = \hbar^2 \lambda_l |l, l\rangle.$$

Using (5-97) and (5-90b) we find

$$(l^2\hbar^2 + l\hbar^2)|l, l\rangle = \hbar^2 \lambda_l |l, l\rangle$$

or

$$\lambda_l = l(l + 1),$$

in agreement with previous results. Once any one eigenvector $|l, m_l\rangle$ is known, the adjacent vectors may be obtained using (5-94), which gives

$$|l, m_l \pm 1\rangle = \frac{1}{\{[l(l + 1) - m_l(m_l \pm 1)]\hbar^2\}^{1/2}} \hat{L}^\pm |l, m_l\rangle.$$

The eigenfunctions may be obtained by expressing (5-97) in wave mechanics and generating a differential equation for $Y_{l,l}$. The adjacent eigenfunctions $Y_{l, l-1}$ are then obtained by successive lowering operations in a manner similar to that used in Chapter 4 for the oscillator. The results here are in agreement with those obtained from wave mechanics.

We have pointed out that the normal Zeeman pattern presents an incomplete picture of the energy levels of the hydrogenic electron in a uniform magnetic field. We have also verified that a nonclassical angular momentum (if it exists) can have either half-integral or integral quantum numbers. These facts along with some experiment data to be discussed lead quite naturally to the existence of *intrinsic spin*—the subject of the next chapter.

Suggested Reading

Arfken, G., "Mathematical Methods for Physicists," 2nd ed. Academic Press, New York, 1970.

Bethe, H. A., and Salpeter, E. E., "Quantum Mechanics of One and Two Electron Atoms," Chapter 1a. Springer-Verlag, Berlin, 1957.
(This reference work represents one of the most comprehensive texts on hydrogen and helium ever written!)

Bohm, D., "Quantum Theory," Chapter 14. Prentice-Hall, Englewood Cliffs, New Jersey, 1951.

Borowitz, S., "Fundamentals of Quantum Mechanics," Chapters 12 and 13. Benjamin, New York, 1967.

Merzbacher, E., "Quantum Mechanics," 2nd ed. Chapter 9. Wiley, New York, 1970.

Morse, P. M., and Feshbach, H., "Methods of Theoretical Physics," Volume I, Chapter 5. McGraw-Hill, New York, 1953.

Pauling, L., and Wilson, E. B., "Introduction to Quantum Mechanics," Chapter 4. McGraw-Hill, New York, 1935.

Saxon, D. S., "Elementary Quantum Mechanics," Chapter 9. Holden-Day, San Francisco, 1964.

Problems

5-1. The state function associated with the state ket $|\beta\rangle$ in the (continuous) momentum representation is defined by

$$\Psi_\beta(\mathbf{p}) = \langle \mathbf{p}|\beta\rangle.$$

Show that $\Psi_\beta(\mathbf{p})$ is the *Fourier transform* of

$$\Psi_\beta(\mathbf{r}) = \langle \mathbf{r}|\beta\rangle$$

that is,

$$\Psi_\beta(\mathbf{p}) = \frac{1}{(2\pi\hbar)^{3/2}} \int e^{-i\mathbf{p}\cdot\mathbf{r}/\hbar}\Psi_\beta(\mathbf{r})\,d\mathbf{r}.$$

(Hint: Insert the completeness relation $\hat{1} = \int |\mathbf{r}\rangle\langle\mathbf{r}|\,d\mathbf{r}$ into the definition of $\Psi_\beta(\mathbf{p})$.)

5-2. Show how to find the ground-state energy and eigenfunction for a *finite* box potential, that is, one for which

$$V = 0 \quad \text{for} \ \ 0 \leqslant x \leqslant a, \ \ 0 \leqslant y \leqslant b, \ \ 0 \leqslant z \leqslant c$$

$$V = V_0 \quad \text{otherwise.}$$

5-3. Using the transformation given in (5-26), show that the components of $\hat{\mathbf{L}}$ are given by (5-34).

5-4. Using the commutation relations for the components of $\hat{p}$ and $\hat{r}$, verify the commutation relation

$$[\hat{L}_x, \hat{L}_y] = i\hbar\hat{L}_z.$$

5-5. Prove that $\hat{L}^2$ and $\hat{L}_z$ commute with any central force Hamiltonian.

5-6. Verify that the isotropic oscillator problem is separable in cylindrical polar coordinates and obtain the first few eigenvalues and eigenfunctions. Compare these results with the solutions obtained using Cartesian coordinates.

5-7. Show how to calculate the ground-state energy and eigenfunction for a particle of mass M in a spherical square well potential

$$V = -V_0 \qquad (0 \leqslant r \leqslant R)$$

$$V = 0 \qquad \text{(otherwise)}.$$

(Assume this state to have $l = 0$.)

5-8. Verify that the most *probable* position for the hydrogenic electron in the ground state is at $r = a$ where $a = \hbar^2/me^2$ is the first Bohr radius.

5-9. When a system is in an eigenstate of $\hat{L}_z$, which we label $|nlm_l\rangle$, show that $\langle L_x \rangle = \langle L_y \rangle = 0$. (Hint: Use $[\hat{L}_i, \hat{L}_j] = i\hbar\hat{L}_k$.)

5-10. Check the formula for the degeneracy of the isotropic oscillator, that is, degeneracy $= \frac{1}{2}(\bar{n} + 1)(\bar{n} + 2)$, for the level $\bar{n} = 3$ by enumerating the degenerate Cartesian and spherical eigenfunctions for this level.

5-11. Estimate the ground-state energy and the first Bohr radius in hydrogen by minimizing the energy function,

$$\mathscr{E} = \frac{(\Delta p)^2}{2m} - \frac{e^2}{\Delta r},$$

subject to the restriction $\Delta p \, \Delta r \simeq \hbar$. Compare Δr_{min} and $\mathscr{E}_{min}$ with the correct values.

5-12. Since $\hat{L}_z = (\hbar/i) \, \partial/\partial\phi$ is canonically conjugate to the azimuthal angle ϕ we would expect the following uncertainty relation to hold:

$$\Delta L_z \, \Delta\phi \geqslant \tfrac{1}{2}\hbar.$$

However it is always possible, in theory, to measure L_z with unlimited precision (that is, $\Delta L_z \to 0$). Yet $\Delta\phi$ must always be less than or equal to 2π since this covers the entire range of the azimuthal angle. Explain this apparent contradiction to the uncertainty principle.

5-13. (a)   Show that the radial momentum may be represented by

$$\hat{p}_r = \frac{1}{2}\left[\frac{\hat{\mathbf{r}}}{\hat{r}}\cdot\hat{\mathbf{p}} + \hat{\mathbf{p}}\cdot\frac{\hat{\mathbf{r}}}{\hat{r}}\right] \to \frac{1}{2}\frac{\hbar}{i}\left[\frac{\mathbf{r}}{r}\cdot\nabla + \nabla\cdot\frac{\mathbf{r}}{r}\right] = \frac{\hbar}{i}\left(\frac{\partial}{\partial r} + \frac{1}{r}\right)$$

and that

$$\frac{\hat{p}_r{}^2}{2m} = -\frac{\hbar^2}{2m}\frac{1}{r^2}\frac{\partial}{\partial r}\left(r^2\frac{\partial}{\partial r}\right).$$

(b)   Verify the commutation relation $[\hat{r}, \hat{p}_r] = i\hbar\hat{1}$.

6 | Spin Angular Momentum

Experimental developments during the 1920s provided data which led physicists to the idea that elementary particles such as electrons not only possess wave properties but also may have an intrinsic spin angular momentum. As early as 1922 an experiment by Stern and Gerlach revealed this spin most dramatically. The experiment involved passing a beam of silver atoms through a slightly inhomogeneous magnetic field (Figure 6-1). First, the field's nonuniformity produces a net force on the magnetic dipoles associated with the atoms of the beam. Second, the field provides a direction (for example, the z axis) of spatial quantization. It is left as an exercise (Problem 6-1) to show that the *average* net force on a magnetic dipole is $\langle F_z \rangle = \mu_z \, \partial B_z/\partial z$, where μ_z is the component of the atom's magnetic dipole moment along the z axis.

As discussed in Chapter 5, a magnetic dipole originates with the orbital motion of the electrons about the nucleus. Assuming for simplicity that this moment is due to the single outer electron of each silver atom, then according to (5-83) we expect the magnetic moment to be

$$\mu_z = -\frac{1}{2}\frac{e}{mc}L_z = -\frac{1}{2}\frac{e\hbar}{mc}m_l. \tag{6-1}$$

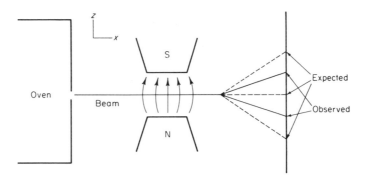

Figure 6-1   The Stern–Gerlach experiment.

The force on each atom therefore depends on its inclination to the z axis, that is, on its magnetic substate. If l is integral, as is always the case with orbital angular momentum, then the number of substates, $2l + 1$, is odd and we would expect the beam to be split in such a way that an odd number of traces appears on a detecting screen. Yet when the experiment was performed, only two traces appeared. The dipole moment of the silver atoms could not have been produced by orbital momentum.

Further evidence suggesting the existence of spin was obtained from fine-structure spectroscopic measurements on the emission spectra of atoms. Goudsmit and Uhlenbeck (1925) resolved the discrepancy between theory and experiment by suggesting that in addition to any orbital angular momentum, the electron possessed an internal spin or intrinsic angular momentum characterized by the quantum numbers s and m_s. These were the respective analogs of l and m_l associated with orbital momentum. Since only two projections were observed in the Stern–Gerlach experiment, the quantum numbers were restricted to $s = \frac{1}{2}$ with $m_s = \pm\frac{1}{2}$. The theory of spin was soon to be formalized by Pauli (1926) and it is his theory that we will now describe.

I   Pauli's Theory of Electron Spin

We emphasize at the outset that the intrinsic spin of a particle has no classical analog. Therefore it is entirely incorrect to visualize the electron as a sphere rotating on its axis. First, the spin operator $\hat{S}$ is not a canonical function of $\hat{r}$ and $\hat{p}$ and the electron has no internal classical structure. Second, if there were a classical limit, according to the correspondence principle, it would be reached when s and m_s are large. But unlike l and m_l in orbital angular momentum, the spin values of elementary particles are always of the order of $\hbar$.

While the theory to be discussed applies to the electron ($s = \frac{1}{2}$), it may be generalized with little difficulty for higher intrinsic spin values.‡

We begin by using the following *postulates*:

(a) Spin is represented by a Hermitian operator $\hat{\mathbf{S}}$ whose components satisfy commutation relations analogous to (5-89),

$$[\hat{S}_i, \hat{S}_j] = i\hbar \hat{S}_k \qquad (i, j, k = x, y, z \text{ in cyclic order}). \tag{6-2}$$

(b) The eigenvalues of $\hat{S}^2$ all have the value $s(s + 1)\hbar^2$ where $s = \frac{1}{2}$. Furthermore the eigenvalues of any one component of $\hat{\mathbf{S}}$ (for example $\hat{S}_z$) are $m_s\hbar$ where $|m_s| \leqslant s$ or $m_s = \pm\frac{1}{2}$.

We may regard postulate (a) as a definition of any angular momentum. Postulate (b) implies that spin is purely quantum-mechanical and that since **r** is not a relevant variable, a wave-mechanical analog does not exist. The two eigenvalues of $\hat{S}_z$ suggest that the spin operators are to be represented in a two-dimensional "spin" space.

We have seen that the eigenvectors of a Hermitian operator may be used as a basis with which to represent other operators and vectors. In the case of spin, it is conventional to use the $\hat{S}_z$ eigenbasis in which $\hat{S}_z$ is displayed as a diagonal matrix,

$$\hat{S}_z \rightarrow \begin{pmatrix} \frac{1}{2}\hbar & 0 \\ 0 & -\frac{1}{2}\hbar \end{pmatrix} \tag{6-3}$$

with orthonormal eigenvectors

$$|m_s\rangle = |\tfrac{1}{2}\rangle = \begin{pmatrix} 1 \\ 0 \end{pmatrix} \qquad \text{and} \qquad |m_s\rangle = |-\tfrac{1}{2}\rangle = \begin{pmatrix} 0 \\ 1 \end{pmatrix} \tag{6-4}$$

as the basis. In this basis the other operators take the form

$$\hat{S}_x \rightarrow \begin{pmatrix} 0 & \frac{1}{2}\hbar \\ \frac{1}{2}\hbar & 0 \end{pmatrix}, \qquad \hat{S}_y \rightarrow \begin{pmatrix} 0 & -\frac{1}{2}i\hbar \\ \frac{1}{2}i\hbar & 0 \end{pmatrix}. \tag{6-5}$$

The commutation relations (6-2) will then be satisfied as can be seen from

$$\begin{pmatrix} 0 & \frac{1}{2}\hbar \\ \frac{1}{2}\hbar & 0 \end{pmatrix}\begin{pmatrix} 0 & -\frac{1}{2}i\hbar \\ \frac{1}{2}i\hbar & 0 \end{pmatrix} - \begin{pmatrix} 0 & -\frac{1}{2}i\hbar \\ \frac{1}{2}i\hbar & 0 \end{pmatrix}\begin{pmatrix} 0 & \frac{1}{2}\hbar \\ \frac{1}{2}\hbar & 0 \end{pmatrix} = i\hbar\begin{pmatrix} \frac{1}{2}\hbar & 0 \\ 0 & -\frac{1}{2}\hbar \end{pmatrix}.$$

---

‡ Particles with half-integral spins are termed fermions, those with integral spins are termed bosons. We shall see (Chapter 9) that bosons and fermions obey entirely different statistics; consequently the thermodynamic behavior of the matter that they constitute is grossly different.

The $\hat{S}^2$ matrix becomes, after some simplication,

$$\hat{S}^2 = \hat{S}_x{}^2 + \hat{S}_y{}^2 + \hat{S}_z{}^2 \rightarrow \tfrac{3}{4}\hbar^2 \begin{pmatrix} 1 & 0 \\ 0 & 1 \end{pmatrix} = \tfrac{3}{4}\hbar^2\hat{1} \tag{6-6}$$

from which the eigenvalues follow as

$$s(s+1)\hbar^2 = \tfrac{1}{2}(\tfrac{1}{2}+1)\hbar^2 = \tfrac{3}{4}\hbar^2$$

as required by the first postulate (Figure 6-2).

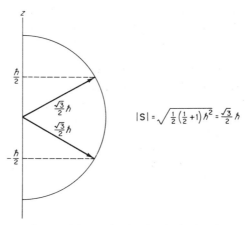

Figure 6-2   Spatial quantization for intrinsic spin, $s = \tfrac{1}{2}$.

It is sometimes convenient to express the spin operators as $\hat{S}_i = \tfrac{1}{2}\hbar\hat{\sigma}_i$ where

$$\hat{\sigma}_x = \begin{pmatrix} 0 & 1 \\ 1 & 0 \end{pmatrix}; \qquad \hat{\sigma}_y = \begin{pmatrix} 0 & -i \\ i & 0 \end{pmatrix}; \qquad \hat{\sigma}_z = \begin{pmatrix} 1 & 0 \\ 0 & -1 \end{pmatrix} \tag{6-7}$$

are the *Pauli* matrices. They have the properties:

(a) $$[\hat{\sigma}_i, \hat{\sigma}_j] = 2i\hat{\sigma}_k$$

(b) $$[\hat{\sigma}_i, \hat{\sigma}_j]_+ = 2\delta_{ij} \tag{6-8}$$

(c) $$\sum_{i=1}^{3} \hat{\sigma}_i{}^2 = 3\hat{1}.$$

From (b) it follows that two different Pauli matrices *anticommute*, that is, $\hat{\sigma}_i\hat{\sigma}_j = -\hat{\sigma}_j\hat{\sigma}_i$.

The eigenvectors and eigenvalues of $\hat{S}_x$ and $\hat{S}_y$ may be obtained (in the $\hat{S}_z$ eigenbasis) using conventional matrix methods. For example, it may be verified directly that

$$\frac{1}{\sqrt{2}}\begin{pmatrix} 1 \\ 1 \end{pmatrix} \qquad \text{and} \qquad \frac{1}{\sqrt{2}}\begin{pmatrix} 1 \\ -1 \end{pmatrix}$$

are eigenvectors of $\hat{S}_x$ with the respective eigenvalues $\frac{1}{2}\hbar$ and $-\frac{1}{2}\hbar$. The eigenvectors of $\hat{S}_z$ which we label $|m_s\rangle$ constitute a complete orthonormal basis and any spin ket may be represented by

$$|\chi\rangle = \sum_{m_s=-\frac{1}{2}}^{\frac{1}{2}} a_{m_s}|m_s\rangle.$$

It is not necessary to label the eigenkets by $|s, m_s\rangle$ since s for a single particle is always fixed (for example, $s = \frac{1}{2}$ for the electron).

Now that a representation for spin has been developed, we wish to generate a total state ket which reflects both spatial and spin properties. To accomplish this we form a *direct*‡ product of the spatial ket and the spin ket, that is,

$$|\text{total}\rangle = |\text{space}\rangle \otimes |\text{spin}\rangle.$$

It is assumed that all spin operators commute with all spatial operators (for example, $\hat{L}$, $\hat{r}$, $\hat{p}$). Furthermore, any operation of a spin operator on $|\text{total}\rangle$ treats the spatial part as a constant and operates only on the spin part. The reverse is true for spatial operators. A particle in a central force field (for example, a hydrogenic electron) may be represented by a total ket

$$|nlm_l m_s\rangle = |nlm_l\rangle \otimes |m_s\rangle \qquad (6\text{-}9)$$

where the new quantum number refers to the spin of the particle. These ket vectors have the properties

$$\mathscr{H}|nlm_l m_s\rangle = \varepsilon_{nl}|nlm_l m_s\rangle$$

$$\hat{L}^2|nlm_l m_s\rangle = l(l+1)\hbar^2|nlm_l m_s\rangle \qquad (l = 0, 1, 2, \ldots)$$

$$\hat{L}_z|nlm_l m_s\rangle = m_l \hbar|nlm_l m_s\rangle \qquad (m_l = 0, \pm 1, \pm 2, \ldots, \pm l)$$

$$\hat{S}^2|nlm_l m_s\rangle = s(s+1)\hbar^2|nlm_l m_s\rangle \qquad (s = \tfrac{1}{2})$$

and

$$\hat{S}_z|nlm_l m_s\rangle = m_s \hbar|nlm_l m_s\rangle \qquad (m_s = \pm\tfrac{1}{2})$$

with

$$\langle n'l'm_l'm_s'|nlm_l m_s\rangle = \delta_{nn'}\delta_{ll'}\delta_{m_l m_l'}\delta_{m_s m_s'}.$$

For brevity, we shall henceforth omit the symbol $\otimes$ and write the direct product as $|\text{space}\rangle|\text{spin}\rangle$. We generate the *eigenspinor* (eigenfunction plus spin) by taking the inner product of $|\text{total}\rangle$ with $\langle \mathbf{r}|$. Since the latter is a

---

‡ This direct product of the spatial and spin kets is not to be confused with the conventional cross product. The direct product is also called the tensor product.

spatial eigenbra, the inner product affects only the spatial part and we obtain

$$\psi_{nlm_lm_s}(r) = \langle \mathbf{r}|nlm_l\rangle|m_s\rangle = \psi_{nlm_l}(r, \theta, \phi)|m_s\rangle. \tag{6-10}$$

Using the $\hat{S}_z$ eigenbasis for $|m_s\rangle$ we find the "spin up" ($m_s = \frac{1}{2}$) and "spin down" ($m_s = -\frac{1}{2}$) eigenspinors to be, respectively,

$$\psi_{nlm_l\frac{1}{2}} = \psi_{nlm_l}\begin{pmatrix}1\\0\end{pmatrix} = \begin{pmatrix}\psi_{nlm_l}\\0\end{pmatrix}$$

and $\tag{6-11}$

$$\psi_{nlm_l-\frac{1}{2}} = \psi_{nlm_l}\begin{pmatrix}0\\1\end{pmatrix} = \begin{pmatrix}0\\\psi_{nlm_l}\end{pmatrix}.$$

The central force states in addition to the $2l + 1$ spatial degeneracy now have a double degeneracy due to spin. In the special case of hydrogen, the accidental degeneracy becomes

$$2\sum_{l=0}^{n-1} 2l + 1 = 2n^2\text{-fold degenerate.}$$

For example, the ground state of hydrogen (Coulomb potential) is doubly degenerate with $\psi_{100\frac{1}{2}}$ and $\psi_{100-\frac{1}{2}}$ having an energy of -13.6 eV.

II   Transformation Properties of Spin Kets— The Total Angular Momentum

In Chapter 3 it was pointed out that the operator $\hat{p}_x$ was the generator of *translations* of a system along the x direction. Similarly, the operator $\hat{L}_z$ is the generator of *rotations* about the z axis. In fact it may be verified that if a particle is in a state $|r, \theta, \phi\rangle$, in which it is precisely at the point r, θ, ϕ, then the operator $\exp(-i\hat{L}_z\bar{\phi}/\hbar)$ produces a state in which the particle has been rotated to $r, \theta, \phi + \bar{\phi}$. Mathematically the operation reads

$$\exp(-i\hat{L}_z\bar{\phi}/\hbar)|r, \theta, \phi\rangle = |r, \theta, \phi + \bar{\phi}\rangle. \tag{6-12}$$

We now consider an arbitrary spatial ket $|\alpha\rangle$. Since the ket basis $|nlm_l\rangle$ is complete as far as spatial kets are concerned, we may make the expansion

$$|\alpha\rangle = \sum_{nlm_l} a_{nlm_l}|nlm_l\rangle.$$

If we rotate the actual system $\bar{\phi}$ degrees about the z axis, the new state becomes

$$|\alpha'\rangle = \exp(-i\hat{L}_z\bar{\phi}/\hbar)|\alpha\rangle$$

$$= \sum_{nlm_l} a_{nlm_l} \exp(-im_l\bar{\phi})|nlm_l\rangle \qquad (m_l = 0, \pm1, \ldots, \pm l). \tag{6-13}$$

Let us suppose that the rotation is $\bar{\phi} = 2\pi$. In that case $e^{-im_l 2\pi} = 1$ and we find $|\alpha'\rangle = |\alpha\rangle$ as expected. Thus single-valuedness is inherent in a spatial ket.

Generalizing to spin, we take the rotation generator (about the z axis) to be $\exp(-i\hat{S}_z\bar{\phi}/\hbar)$. Applying this operator to a general spin ket

$$|\chi\rangle = \sum_{m_s = \frac{1}{2}}^{\frac{1}{2}} a_{m_s}|m_s\rangle$$

we find

$$|\chi'\rangle = \exp(-i\hat{S}_z\bar{\phi}/\hbar)|\chi\rangle = \sum_{m_s = -\frac{1}{2}}^{\frac{1}{2}} a_{m_s}\exp(-im_s\bar{\phi})|m_s\rangle. \qquad (6\text{-}14)$$

Since m_s is half-integral, a 2π rotation does not recover the original ket. In fact it requires a 4π rotation to obtain $|\chi'\rangle = |\chi\rangle$. For this reason, the spin kets are sometimes referred to as "half vectors" or more commonly as *spinors*. Under a 2π rotation an arbitrary spinor is multiplied by $e^{-i\pi} = -1$. However, since probabilities and expectation values involve inner products or "squares" of these spinors, observables remain invariant under 2π rotations.‡

Since $\exp(-i\hat{L}_z\bar{\phi}/\hbar)$ and $\exp(-i\hat{S}_z\bar{\phi}/\hbar)$ respectively transform the spatial and spin parts of the total state ket, it follows that

$$|\text{total}(\phi + \bar{\phi})\rangle = \exp(-i\hat{L}_z\bar{\phi}/\hbar)\exp(-i\hat{S}_z\bar{\phi}/\hbar)|\text{total}(\phi)\rangle$$
$$= \exp[-i(\hat{L}_z + \hat{S}_z)\bar{\phi}/\hbar]|\text{total}(\phi)\rangle. \qquad (6\text{-}15)$$

We define

$$\hat{\mathbf{J}} = \hat{\mathbf{L}} + \hat{\mathbf{S}} \qquad (6\text{-}16)$$

as the *total angular momentum operator*. We shall see in the next section that $\hat{\mathbf{J}}$ is of great importance in a variety of central force problems. From (6-15) it follows that $\exp(-i\hat{J}_z\bar{\phi}/\hbar)$ generates rotations about the z axis for both the spin and spatial characteristics simultaneously.

III   Spin and the Central Force Problem

To include the spin ($s = \frac{1}{2}$) of a particle in the central force problem we write the Hamiltonian as a product of a differential and a matrix operator,

---

‡ The effect of a 2π rotation in (6-14) is to produce a multiplicative phase factor $e^{i\pi} = -1$. Although there is no experimental evidence to suggest that this phase factor is observable, there is some debate on the issue.

that is,

$$\mathcal{H}_{cf} \to \left\{ -\frac{\hbar^2}{2m} \nabla^2 + V(r) \right\} \hat{1} = \begin{pmatrix} -\dfrac{\hbar^2}{2m} \nabla^2 + V)r) & 0 \\ 0 & -\dfrac{\hbar^2}{2m} \nabla^2 + V(r) \end{pmatrix} \qquad (6\text{-}17)$$

where $\hat{1} = \begin{pmatrix} 1 & 0 \\ 0 & 1 \end{pmatrix}$ is the unit or identity matrix. The differential and matrix parts operate respectively on the spatial and spin parts of the state kets. As was already pointed out, the set

$$|nlm_l m_s\rangle \to \psi_{nlm_l m_s} = \psi_{nlm_l}(r, \theta, \phi)|m_s\rangle \qquad (6\text{-}18)$$

contains the simultaneous eigenfunctions of $\mathcal{H}_{cf}$, $\hat{L}^2$, $\hat{L}_z$, $\hat{S}^2$, and $\hat{S}_z$. Since the energy depends at most on n and l, these states are degenerate and are not unique eigenstates of $\mathcal{H}_{cf}$. Any linear combination of the form

$$|nl\rangle = \sum_{m_l=-l}^{l} \sum_{m_s=-\frac{1}{2}}^{\frac{1}{2}} a_{nlm_l m_s} |nlm_l m_s\rangle \qquad (6\text{-}19)$$

is still a new and distinct energy eigenstate of $\mathcal{H}_{cf}$ with energy ε_{nl} (Lemma 1, Chapter 3).

We shall first examine whether or not simultaneous eigenfunctions of the operators $\mathcal{H}_{cf}$, $\hat{L}^2$, $\hat{S}^2$, $\hat{J}^2$, and $\hat{J}_z$ exist, where $\hat{\mathbf{J}} = \hat{\mathbf{L}} + \hat{\mathbf{S}}$ refers to the total angular momentum operator introduced above. To do this we merely observe that all five of the above operators mutually commute and are therefore compatible. For example, since $\mathcal{H}_{cf}$ commutes with the components of $\hat{\mathbf{L}}$ and $\hat{\mathbf{S}}$, we find

$$[\hat{J}^2, \mathcal{H}_{cf}] = [(\hat{L}^2 + \hat{S}^2 + 2\hat{L}_x \hat{S}_x + 2\hat{L}_y \hat{S}_y + 2\hat{L}_z \hat{S}_z), \mathcal{H}_{cf}] = 0$$

and

$$[\hat{J}_z, \mathcal{H}_{cf}] = [(\hat{L}_z + \hat{S}_z), \mathcal{H}_{cf}] = 0.$$

The components of $\hat{\mathbf{J}}$ satisfy the familiar commutation relations for angular momentum (see Problem 6-7),

$$[\hat{J}_i, \hat{J}_j] = i\hbar \hat{J}_k \qquad (i, j, k = x, y, z \text{ in cyclic order}). \qquad (6\text{-}20)$$

We have already shown at the end of Chapter 5, using ladder algebra, that vector operators whose components satisfy (6-20) must satisfy the following eigenvalue relations:

$$\hat{J}^2 |jm_j\rangle = j(j+1)\hbar^2 |jm_j\rangle \qquad (6\text{-}21\text{a})$$

$$\hat{J}_z |jm_j\rangle = m_j \hbar |jm_j\rangle \qquad (|m_j| \leqslant j) \qquad (6\text{-}21\text{b})$$

where $\hat{J}^2 = \hat{J}_x^2 + \hat{J}_y^2 + \hat{J}_z^2$ and where the quantum numbers may be either integral or half-integral.

The fact that the operators $\mathscr{H}_{cf}$, $\hat{L}^2$, $\hat{J}^2$, and $\hat{J}_z$ are compatible suggests that a set of common eigenvectors exists with the following properties:

$$\begin{aligned}
\mathscr{H}_{cf}|nljm_j\rangle &= \varepsilon_{nl}|nljm_j\rangle \\
\hat{L}^2|nljm_j\rangle &= l(l+1)\hbar^2|nljm_j\rangle \quad (l = 0, 1, 2, \ldots) \\
\hat{J}^2|nljm_j\rangle &= j(j+1)\hbar^2|nljm_j\rangle \\
\hat{J}_z|nljm_j\rangle &= m_j\hbar|nljm_j\rangle \quad (|m_j| \leqslant j).
\end{aligned} \qquad (6\text{-}22)$$

These eigenvectors can be constructed using (6-19), that is,

$$|nljm_j\rangle = \sum_{m_l=-l}^{l} \sum_{m_s=-\frac{1}{2}}^{\frac{1}{2}} C^{nl}_{jm_jm_lm_s}|nlm_l m_s\rangle. \qquad (6\text{-}23)$$

The coefficients in (6-23) that couple eigenstates of $\hat{L}_z$ and $\hat{S}_z$ into eigenstates of $\hat{J}^2$ and $\hat{J}_z$ are examples of so-called *vector addition* coefficients, Clebsch–Gordan coefficients, or Wigner coefficients. For a given value of l (and $s = \frac{1}{2}$), only certain eigenstates of $\hat{J}^2$ can be constructed using (6-23). To find the possible values of the quantum number j we shall use a semiclassical aid known as the vector model (Figure 6-3). This model is based on the

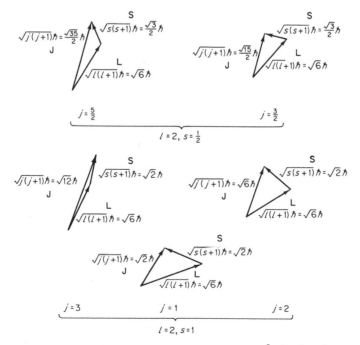

Figure 6-3   Using the vector model to generate states of $\hat{J}^2(\mathbf{\hat{J}} = \mathbf{\hat{L}} + \mathbf{\hat{S}})$ for the cases $l = 2$, $s = \frac{1}{2}$ and $l = 2$, $s = 1$.

triangle rule for vector addition and predicts that the values of j are restricted by $|l - s| \leqslant j \leqslant l + s$, or since $s = \frac{1}{2}$ in the case of an electron, $j = l \pm \frac{1}{2}$. We require that j be nonnegative and choose $j = \frac{1}{2}$ when $l = 0$. Since j is half-integral for the electron, m_j is also half-integral and subject to the restriction

$$m_j = \pm \tfrac{1}{2}, \pm \tfrac{3}{2}, \ldots, \pm j.$$

The Clebsch–Gordan coefficients in (6-23) are difficult to calculate although a knowledge of group theory simplifies the calculations. In our case for $s = \frac{1}{2}$ and arbitrary l the Clebsch–Gordan coefficients can be shown to give

$$\left| nljm_j \right\rangle = \left| n, l, l \pm \frac{1}{2}, m_j \right\rangle$$

$$= \frac{1}{(2l + 1)^{1/2}} \left\{ \pm \left(l \pm m_j + \frac{1}{2}\right)^{1/2} \left| n, l, m_j - \frac{1}{2}, \frac{1}{2} \right\rangle \right.$$

$$\left. + \left(l \mp m_j + \frac{1}{2}\right)^{1/2} \left| n, l, m_j + \frac{1}{2}, -\frac{1}{2} \right\rangle \right\} \tag{6-24}$$

or from (6-11) in wave mechanics, the eigenfunctions are

$$\psi_{nll \pm \frac{1}{2} m_j} = \frac{R_{nl}(r)}{(2l + 1)^{1/2}} \begin{pmatrix} \pm (l \pm m_j + \frac{1}{2})^{1/2} Y_l^{m_j - 1/2}(\theta, \phi) \\ (l \mp m_j + \frac{1}{2})^{1/2} Y_l^{m_j + 1/2}(\theta, \phi) \end{pmatrix}. \tag{6-25}$$

Thus, for example, the eigenfunction for $n = 2$, $l = 1$, $j = l + \frac{1}{2} = \frac{3}{2}$, and $m_j = \frac{1}{2}$ would be

$$\psi_{2,1,3/2,1/2}(r, \theta, \phi) = \frac{R_{21}(r)}{\sqrt{3}} \begin{pmatrix} \sqrt{2} \; Y_1^0(\theta, \phi) \\ \sqrt{1} \; Y_1^1(\theta, \phi) \end{pmatrix}. \tag{6-26}$$

We shall on occasion label a hydrogenic state of the type $\left| nljm_j \right\rangle$ by

$$^{2s+1}x_j = {}^2x_j \tag{6-27}$$

where x is the appropriate spectroscopic letter for the l value. For example, the state in (6-26) is a $^2p_{\frac{3}{2}}$ state. The (doubly degenerate) ground state of hydrogen ($n = 1, l = 0, j = \frac{1}{2}, m_j = \pm \frac{1}{2}$) is a $^2s_{\frac{1}{2}}$ state.

It is important to realize that the degeneracy of the level ε_{nl} is the same regardless of whether $\left| m_l m_s \right\rangle$ or $\left| jm_j \right\rangle$ is used. For example, the level ε_{21} is $2(2l + 1) =$ sixfold degenerate in the $\left| nlm_l m_s \right\rangle$ scheme. These states may be coupled to give $j = l + \frac{1}{2} = \frac{3}{2}$ and $j = l - \frac{1}{2} = \frac{1}{2}$. The former is $2j + 1 =$ four-fold degenerate (that is, $m_j = \frac{3}{2}, \frac{1}{2}, -\frac{1}{2}, -\frac{3}{2}$) while the latter is doubly degenerate. The total degeneracy is still sixfold as expected (Figure 6-4).

The spatial (m_l) and spin (m_s) degeneracies of a central force Hamiltonian

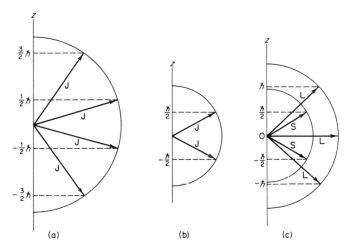

Figure 6-4 The six degenerate states for $l = 1$, $s = \frac{1}{2}$ in the $|jm_j\rangle$ and $|m_l m_s\rangle$ schemes. (a) $|l + s| = j = \frac{3}{2}$, $|\mathbf{J}| = (\sqrt{15}/2)\,\hbar$, $m_j = \pm\frac{1}{2}, \pm\frac{3}{2}$. (b) $|l - s| = j = \frac{1}{2}$, $|\mathbf{J}| = (\sqrt{3}/2)\,\hbar$, $m_j = \pm\frac{1}{2}$. (c) $m_l = 0, \pm 1$, $m_s = \pm\frac{1}{2}$.

lead to ambiguities in labeling the eigenstates. One set as we have seen is labeled $|nlm_l m_s\rangle$ and contains simultaneous eigenstates of $\mathscr{H}_{cf}$, $\hat{L}^2$, $\hat{L}_z$, and $\hat{S}_z$, while another set is $|nljm_j\rangle$ and contains simultaneous eigenkets of $\mathscr{H}_{cf}$, $\hat{L}^2$, $\hat{J}^2$, and $\hat{J}_z$. Since, for example, $\hat{J}^2$ and $\hat{L}_z$ are incompatible, the two sets are distinct. We must therefore be specific with regard to which set we are considering.

IV Spin Magnetism and the Spin–Orbit Interaction in Hydrogen

Any particle with intrinsic spin possesses a magnetic dipole moment and acquires energy in a magnetic field. This dipole moment cannot be calculated classically and must be *postulated* in analogy with (5-83) to be

$$\boldsymbol{\mu}_s = g_s \frac{q}{2mc}\, \mathbf{S} \tag{6-28}$$

where q, m, and $\mathbf{S}$ are respectively the charge, mass, and spin angular momentum of the particle and c is the speed of light. The proportionality factor g_s is called the spin "g" factor. Based on experimental data $g_s \simeq 2.0023 \simeq 2$ for the electron.

The energy acquired by this dipole in a magnetic field is $E_{mag} = -\boldsymbol{\mu}_s \cdot \mathbf{B}$.

Even in the absence of an external field, an orbiting electron in a central force potential experiences an internal field which we call $\mathbf{B}_{orbit}$, and the Hamiltonian for this problem becomes

$$\mathcal{H} = \mathcal{H}_{cf} - \boldsymbol{\mu}_s \cdot \mathbf{B}_{orbit}. \tag{6-29}$$

The last term is essentially due to an interaction between the particle's orbital motion and its intrinsic spin.

From classical electrodynamics we know that when a charged particle moves with a velocity $\mathbf{v}$ in an electric field $\mathbf{E}$, an observer traveling with the particle measures an induced magnetic field‡

$$\mathbf{B}_{orbit} = \mathbf{B}_{ind} \simeq -\frac{\mathbf{v}}{c} \times \mathbf{E}. \tag{6-30}$$

We shall assume that our particle moves in the field of an electrostatic central force potential $V(r) = q\Phi(r)$. We therefore set

$$\mathbf{E} = -\nabla\Phi(r) = -\frac{d\Phi}{dr}\frac{\mathbf{r}}{r}. \tag{6-31}$$

However, since the electron is accelerating, its frame is not inertial and corrections will have to be made if we are to accurately predict the spin–orbit magnetic energy. We state without proof that a relativistic transformation to the electron's frame introduces an additional motion known as Thomas precession.§ This motion introduces a factor of $\frac{1}{2}$ in our expression for the magnetic dipole energy. The energy associated with the orbital magnetism is now

$$-\boldsymbol{\mu}_s \cdot \mathbf{B}_{orbit} = +\boldsymbol{\mu}_s \cdot \frac{1}{2}\left(\frac{\mathbf{v}}{c} \times \mathbf{E}\right), \tag{6-32}$$

the factor $\frac{1}{2}$ being due to Thomas precession, which is a relativistic effect. Using (6-31) and (6-32), we find that the Hamiltonian in (6-29) becomes

$$\mathcal{H} = \mathcal{H}_{cf} + \tfrac{1}{2}\boldsymbol{\mu}_s \cdot \left(\frac{\mathbf{v}}{c} \times -\frac{d\Phi}{dr}\frac{\mathbf{r}}{r}\right)$$

$$= \mathcal{H}_{cf} - \frac{1}{2qmc}\frac{1}{r}\frac{dV}{dr}\boldsymbol{\mu}_s \cdot (m\mathbf{v} \times \mathbf{r}) \qquad (V = q\Phi)$$

$$= \mathcal{H}_{cf} + \frac{1}{2qmc}\frac{1}{r}\frac{dV}{dr}\boldsymbol{\mu}_s \cdot \mathbf{L}. \tag{6-33}$$

---

‡ See, for example, J. D. Jackson, "Classical Electrodynamics," pp. 380–383. Wiley, New York, 1962.

§ For an elementary discussion of Thomas precession, see R. M. Eisberg, "Fundamentals of Modern Physics," p. 340. Wiley, New York, 1960.

Next we consider the more general case where the central force system is inserted into an external uniform magnetic field, **B**. The interaction with this field involves both the orbital and spin magnetic dipoles. The Hamiltonian becomes

$$\mathcal{H} = \mathcal{H}_{cf} + \frac{1}{2qmc} \frac{1}{r} \frac{dV}{dr} \boldsymbol{\mu}_s \cdot \mathbf{L} - \boldsymbol{\mu}_l \cdot \mathbf{B} - \boldsymbol{\mu}_s \cdot \mathbf{B}$$

or using (5-83) and (6-28), the quantum-mechanical Hamiltonian takes the form

$$\mathcal{H} = \mathcal{H}_{cf} + \frac{g_s}{(2mc)^2} \frac{1}{r} \frac{dV}{dr} \hat{\mathbf{S}} \cdot \hat{\mathbf{L}} - \frac{qg_l}{2mc} \hat{\mathbf{L}} \cdot \mathbf{B} - \frac{qg_s}{2mc} \hat{\mathbf{S}} \cdot \mathbf{B}. \qquad (6\text{-}34)$$

The terms in (6-34) are respectively the central force energy, the "spin–orbit interaction" energy, and the energy of the orbital and spin magnetic moments in the external field. For the hydrogenic electron, we set $q = -e$, $V = -e^2/r$, $g_l = 1$, and $g_s = 2$. Eq. (6-34) now becomes

$$\mathcal{H} = \mathcal{H}_{cf} + \frac{e^2}{2m^2c^2} \frac{1}{r^3} \hat{\mathbf{S}} \cdot \hat{\mathbf{L}} + \frac{e}{2mc} \hat{\mathbf{L}} \cdot \mathbf{B} + \frac{e}{mc} \hat{\mathbf{S}} \cdot \mathbf{B} \qquad \text{(hydrogen)}$$

$$(6\text{-}35)$$

where

$$\mathcal{H}_{cf} = \frac{\hat{p}^2}{2m} - \frac{e^2}{r}.$$

V External Magnetic Fields—The Paschen–Back Effect

It is hopeless to seek exact eigensolutions of the completely general Hamiltonian in (6-35). However, there are two cases of practical importance worth considering. The first is the case in which the external field is sufficiently weak (compared with the $\mathbf{B}_{\text{orbit}}$) so that the spin–orbit energy is very much larger than the energy of the interaction with **B**. This effect is known as the *anomalous Zeeman effect* and occurs in hydrogen for external fields where

$$B \ll B_{\text{orbit}} \simeq \frac{e}{c} \frac{v}{r^2} \simeq 10^4 \quad \text{gauss.}$$

Even in this case only approximate solutions are possible and we defer a discussion of the anomalous Zeeman effect to the next chapter (Methods of Approximation).

The other case occurs for strong fields‡ $B \gg B_{\text{orbit}}$. The strong-field case is known as the *Paschen–Back effect*. The spin–orbit interaction is now completely ignored in a first approximation. The Paschen–Back Hamiltonian, whose eigenstates we seek, is

$$\hat{\mathscr{H}}_{\text{PB}} = \hat{\mathscr{H}}_{\text{cf}} + \frac{eB}{2mc}(\hat{L}_z + 2\hat{S}_z); \qquad (6\text{-}36)$$

we have aligned our z axis to coincide with the direction of **B**. We observe that the Paschen–Back and central force Hamiltonians commute, that is,

$$[\hat{\mathscr{H}}_{\text{PB}}, \hat{\mathscr{H}}_{\text{cf}}] = 0$$

which suggests that the two Hamiltonians have a set of common eigenstates. Assuming this set of states to be $|nlm_l m_s\rangle$, we write

$$\hat{\mathscr{H}}_{\text{PB}}|nlm_l m_s\rangle = \left\{\hat{\mathscr{H}}_{\text{cf}} + \frac{eB}{2mc}(\hat{L}_z + 2\hat{S}_z)\right\}|nlm_l m_s\rangle$$

$$= \left\{\varepsilon_{nl} + \frac{eB}{2mc}(m_l + 2m_s)\hbar\right\}|nlm_l m_s\rangle. \qquad (6\text{-}37)$$

Our assumption is therefore justified. While the $\psi_{nlm_l m_s}$ are exact eigenfunctions of $\hat{\mathscr{H}}_{\text{PB}}$,§ the energy eigenvalues are shifted and become

$$\varepsilon_{nlm_l m_s} = \varepsilon_{nl} + \mu_{\text{B}} B(m_l + 2m_s) \qquad \text{(Paschen–Back shift).} \qquad (6\text{-}38)$$

The splitting of magnetic sublevels in hydrogen is diagrammed in Figure 6-5.

A simple calculation would show that

$$[\hat{J}^2, \hat{\mathscr{H}}_{\text{PB}}] \neq 0$$

so that the kets $|nljm_j\rangle$, which are eigenkets of $\hat{J}^2$ and $\hat{J}_z$, would definitely not represent eigenstates of $\hat{\mathscr{H}}_{\text{PB}}$. For strong fields, only $\hat{L}^2$, $\hat{S}^2$, $\hat{L}_z$, and $\hat{S}_z$ are quantized (precise); the classical picture would have **L** and **S** "precessing" independently about **B** (Figure 6-6).

What we have done is to decrease the symmetry by imposing a strong magnetic field on the system. This removes the spatial degeneracy of a central force problem. The eigenfunctions are now unique as only the set $|nlm_l m_s\rangle$ (and not $|nljm_j\rangle$) constitutes an $\hat{\mathscr{H}}_{\text{PB}}$ eigenbasis.

The existence of intrinsic spin was made dramatically clear by the Stern–Gerlach experiment. The theory as suggested by Goudsmit and Uhlenbeck

---

‡ It is still assumed that **B** is small enough (that is, $|\boldsymbol{\mu} \cdot \mathbf{B}| \ll \mathscr{H}_{\text{cf}}$) to neglect quadratic effects in **B**.

§ Recall that this Hamiltonian is only approximate since the spin–orbit interaction is being ignored.

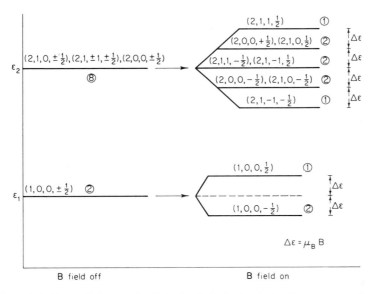

Figure 6-5 The splitting of the Bohr levels in hydrogen in the presence of a strong **B** field (Paschen–Back effect). The states are labeled by the scheme (n, l, m_l, m_s), with the degeneracy of the level circled.

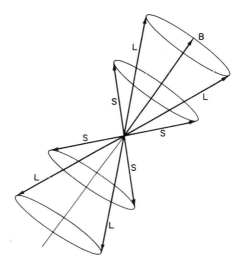

Figure 6-6 Independent statistical "precession" of **L** and **S** about **B** in the strong-field case (Paschen–Back effect).

and shortly thereafter formalized by Pauli explained many features in the emission spectrum of atoms. Yet, from a heuristic point of view, the postulational approach of spin is somewhat unsatisfying. We shall see in Chapter 11 that it is unnecessary to postulate the existence of intrinsic spin for an electron. In fact, it will be shown that spin evolves quite naturally as a consequence of quantum mechanics and Einstein's theory of relativity. The quantum relativistic equation for the electron analogous to the Schroedinger equation is known as the Dirac equation. In the limit $v \ll c$, the relativistic Dirac equation reduces to the nonrelativistic Schroedinger equation *with spin included.* Thus the existence of the intrinsic spin phenomenon may be regarded as a direct consequence of relativistic quantum mechanics.

In studying a variety of physical problems, we shall find that the exact solution is often beyond mathematical means. An important case already mentioned is that of hydrogen with the spin–orbit interaction included. Fortunately, in many cases the Hamiltonian of interest differs only slightly from one whose exact eigensolutions are known. The small difference between the two is called a *perturbation.* Obtaining the necessary corrections to the eigenfunctions and eigenvalues is the subject of perturbation theory, to be discussed in Chapter 7.

Suggested Reading

Bohm, D., "Quantum Theory," Chapter 17. Prentice-Hall, Englewood-Cliffs, New Jersey, 1951.

Eisberg, R. M., "Fundamentals of Modern Physics," Chapter 11. Wiley, New York, 1961.

Jackson, J. D., "Classical Electrodynamics," Chapter 11. Wiley, New York, 1962.

Merzbacher, E., "Quantum Mechanics," 2nd ed. Chapter 12. Wiley, New York, 1970.

Saxon, D., "Elementary Quantum Mechanics," Chapter 10. Holden-Day, San Francisco, 1968.

Stehle, P., "Quantum Mechanics," Chapter 4. Holden-Day, San Francisco, 1966.

Problems

6-1.   (a)   By using an analogy with electrostatics, show that the z component of the force on a magnetic dipole in a nonuniform magnetic field is

$$F_z = \mu_x \frac{\partial B_z}{\partial x} + \mu_y \frac{\partial B_z}{\partial y} + \mu_z \frac{\partial B_z}{\partial z}.$$

(b)   Assuming that $B_z \gg B_x$, B_y and that the dipole is "spatially" quantized along the z axis, show that

$$\langle F_z \rangle = \mu_z \frac{\partial B_z}{\partial z} = -\frac{eh}{2mc} m \frac{\partial B_z}{\partial z}$$

by showing that $\langle \mu_x \rangle = \langle \mu_y \rangle = 0$. (Why must B_x and B_y exist in order for the relation for $\langle F_z \rangle$ to be valid?)

6-2.   Verify the properties of the Pauli spin matrices as given by Eq. (6-8).

6-3.   (a)   Find the eigenvectors (in the $\hat{S}_z$ eigenbasis) of a particle whose spin projections are respectively $\pm \frac{1}{2}\hbar$ along the x axis.
(b)   If the particle is in a quantum state in which its projection is precisely $\frac{1}{2}\hbar$ along the x axis, then what is the probability of finding it with its projection $\frac{1}{2}\hbar$ along the z axis?

6-4.   Show directly using the appropriate matrices that $\langle S_x \rangle = \langle S_y \rangle = 0$ when the system is in an eigenstate of $\hat{S}_z$.

6-5.   (a)   Construct the matrices $\hat{S}_x$, $\hat{S}_y$, $\hat{S}_z$ (in the $\hat{S}_z$ eigenbasis) for a particle whose spin projections are quantized according to $+\hbar$, 0, $-\hbar$.
(b)   Construct the operator $\hat{S}^2 = \hat{S}_x^2 + \hat{S}_y^2 + \hat{S}_z^2$ and verify that its eigenvalues are $s(s+1)\hbar^2 = 1(1+1)\hbar^2 = 2\hbar^2$.

6-6.   Deduce the form of the unitary operator that rotates a *total* physical state by $\bar{\phi}$ radians about an *arbitrary* axis characterized by the unit vector **n**.

6-7.   Verify that the components of **J** satisfy

$$[\hat{J}_i, \hat{J}_j] = i\hbar \hat{J}_k.$$

6-8.   Verify that the function given in Eq. (6-25) is an eigenfunction of $\hat{J}_z = \hat{L}_z + \hat{S}_z$.

6-9.   Consider a particle of spin $s = \frac{1}{2}$ in an orbital state $l = 2$. Show that the degeneracy of the level is the same in the $|m_l m_s\rangle$ and $|j m_j\rangle$ schemes.

6-10.   The spin gyromagnetic factor for a proton ($s = \frac{1}{2}$) is $g_s^{(p)} = 5.59$.
(a)   Find the proton's magnetic dipole moment.
(b)   Find the projections of this moment along the direction of a uniform magnetic field B_z.
(c)   Assuming that a nucleus can be represented by a single proton moving in a central force potential, discuss qualitatively the nature of the splitting of the energy levels in a strong magnetic field.

7 | Methods of Approximation

I Perturbation Theory

In the previous chapters we have considered idealized systems for which it was possible to find the exact energy eigenfunctions. Unfortunately the complexities of nature lead to Hamiltonians for which exact solutions are not possible. In many physical situations the corrections to some idealized system are relatively minor. For example, we shall see that the spin–orbit interaction in hydrogen leads to fractional corrections of order $\sim 10^{-4}$ to the Bohr levels. In these cases approximate energy eigenfunctions can be obtained by a variety of approximation methods.

The Hamiltonian we shall consider is of the general form

$$\mathscr{H} = \mathscr{H}_0 + \lambda \hat{V}$$

where $\mathscr{H}_0$ is the unperturbed Hamiltonian and $\lambda \hat{V}$ is the perturbation. The latter is conveniently decomposed into a "smallness" parameter λ and a perturbation function $\hat{V}(\hat{\mathbf{r}}, \hat{\mathbf{p}})$. The perturbation may be considered using two fundamentally different approaches. The first, called *stationary state theory*, concerns itself with corrections to the unperturbed energy eigenfunctions and eigenvalues (of $\mathcal{H}_0$). This approach is used when the total (perturbed) Hamiltonian is of prime physical interest. The theory, for example, might be used to calculate the energy level corrections due to the spin–orbit interaction in hydrogen.

A second viewpoint, called *time-dependent perturbation theory*, is used when the *unperturbed* Hamiltonian is of fundamental importance. Since $\mathcal{H}_0$ is not the complete Hamiltonian (without $\lambda \hat{V}$), its eigenstates are not stationary. In particular, if the system is initially $(t = t_0)$ in an unperturbed eigenstate—let us say $|\varepsilon_i^0\rangle$—then at some later time it will evolve according to the rule

$$|\beta_i, t\rangle = \exp[-i\mathcal{H}(t - t_0)/\hbar]|\varepsilon_i^0\rangle = \exp[-i(\mathcal{H}_0 + \lambda \hat{V})(t - t_0)/\hbar]|\varepsilon_i^0\rangle$$

where we have assumed a time-independent perturbation, $\hat{V}$. According to the postulates of quantum mechanics, the probability of observing the system in another unperturbed eigenstate $|\varepsilon_j^0\rangle$ at a later time t is

$$\mathscr{P}_{ij}(t) = |\langle \varepsilon_j^0|\beta_i, t\rangle|^2.$$

The quantity $d\mathscr{P}_{ij}(t)/dt$ is the rate at which the system makes "transitions" from the ith to the jth state as a result of the perturbation. Thus the perturbation is regarded here as inducing transitions between unperturbed eigenstates of $\mathcal{H}_0$. This approach is often used in scattering problems, where the scattering potential is regarded as the perturbation inducing transitions between free-particle states. The transition rate is closely related to the scattering cross section. We shall first begin with a discussion of stationary perturbation theory.

II Nondegenerate–Bound–State–Stationary Perturbation Theory (Rayleigh–Schroedinger Method)‡

We shall assume that the unperturbed Hamiltonian is nondegenerate. This means that the eigenstates are unique and we may establish a one-to-one correspondence between the perturbed and unperturbed eigenstates $|\varepsilon_i\rangle$ and $|\varepsilon_i^0\rangle$. Thus we may write

$$\mathcal{H}_0|\varepsilon_i^0\rangle = \varepsilon_i^0|\varepsilon_i^0\rangle \qquad (7\text{-}1)$$

and

$$\mathcal{H}|\varepsilon_i\rangle = (\mathcal{H}_0 + \lambda \hat{V})|\varepsilon_i\rangle = \varepsilon_i|\varepsilon_i\rangle \qquad (7\text{-}2)$$

‡ This method of solution is originally due to Rayleigh, who applied it to the theory of sound. It was later applied by Schroedinger to quantum theory.

where $|\varepsilon_i\rangle \xrightarrow[\lambda \to 0]{} |\varepsilon_i^0\rangle$. We emphasize that this limiting condition is unique only in the case of nondegenerate Hamiltonians. We also restrict our discussion to the perturbation of bound (discrete) states.

It is possible to develop a solution to (7-2) in the form of a perturbation series‡ in powers of λ. The rate of the convergence of this series however depends on the smallness of λ.

It will be assumed that the perturbed eigenvectors and eigenvalues may be represented in a power series of the form

$$|\varepsilon_i\rangle = |\varepsilon_i^0\rangle + \lambda|\varepsilon_i^{(1)}\rangle + \lambda^2|\varepsilon_i^{(2)}\rangle + \cdots \qquad (7\text{-}3)$$

and

$$\varepsilon_i = \varepsilon_i^0 + \lambda\varepsilon_i^{(1)} + \lambda^2\varepsilon_i^{(2)} + \cdots. \qquad (7\text{-}4)$$

The kets $|\varepsilon_i^{(n)}\rangle$ and the parameters $\varepsilon_i^{(n)}$ represent the nth-order corrections to the eigenvectors and eigenvalues. We shall work in the Dirac representation but the wave-mechanical form follows the same pattern. For sufficiently small λ, we may assume that only the first few powers are necessary to approximate the exact solution. We shall limit our discussion to first- and second-order approximations.

Substituting (7-3) and (7-4) into (7-2) we obtain

$$(\mathcal{H}_0 + \lambda\hat{V})\{|\varepsilon_i^0\rangle + \lambda|\varepsilon_i^{(1)}\rangle + \lambda^2|\varepsilon_i^{(2)}\rangle + \cdots\}$$
$$= \{\varepsilon_i^0 + \lambda\varepsilon_i^{(1)} + \lambda^2\varepsilon_i^{(2)} + \cdots\}\{|\varepsilon_i^0\rangle + \lambda|\varepsilon_i^{(1)}\rangle + \lambda^2|\varepsilon_i^{(2)}\rangle + \cdots\}$$

and collecting terms of like powers of λ,

$$[\mathcal{H}_0|\varepsilon_i^0\rangle - \varepsilon_i^0|\varepsilon_i^0\rangle] + [\mathcal{H}_0|\varepsilon_i^{(1)}\rangle + \hat{V}|\varepsilon_i^0\rangle - \varepsilon_i^0|\varepsilon_i^{(1)}\rangle - \varepsilon_i^{(1)}|\varepsilon_i^0\rangle]\lambda$$
$$+ [\mathcal{H}_0|\varepsilon_i^{(2)}\rangle + \hat{V}|\varepsilon_i^{(1)}\rangle - \varepsilon_i^{(1)}|\varepsilon_i^{(1)}\rangle - \varepsilon_i^0|\varepsilon_i^{(2)}\rangle - \varepsilon_i^{(2)}|\varepsilon_i^0\rangle]\lambda^2 + \cdots$$
$$= 0. \qquad (7\text{-}5)$$

Since the equation is to be valid for any λ, every coefficient of λ^n must be equal to zero; we therefore have the following hierarchy of equations:

$$\mathcal{H}_0|\varepsilon_i^0\rangle = \varepsilon_i^0|\varepsilon_i^0\rangle \qquad (7\text{-}6)$$

$$\mathcal{H}_0|\varepsilon_i^{(1)}\rangle + \hat{V}|\varepsilon_i^0\rangle = \varepsilon_i^0|\varepsilon_i^{(1)}\rangle + \varepsilon_i^{(1)}|\varepsilon_i^0\rangle \qquad (7\text{-}7)$$

$$\mathcal{H}_0|\varepsilon_i^{(2)}\rangle + \hat{V}|\varepsilon_i^{(1)}\rangle = \varepsilon_i^{(1)}|\varepsilon_i^{(1)}\rangle + \varepsilon_i^0|\varepsilon_i^{(2)}\rangle + \varepsilon_i^{(2)}|\varepsilon_i^0\rangle \qquad (7\text{-}8)$$

$$\vdots$$

The zeroth-order equation is just the unperturbed eigenvalue equation (7-1); its eigenkets and eigenvalues are assumed known. The nth-order equation contains no higher than nth-order corrections. However, to obtain these corrections, all lower-order terms must have been previously calculated.

---

‡ This assumes, of course, that such a power series exists.

We begin by considering first-order theory and we make the expansion‡

$$|\varepsilon_i^{(1)}\rangle = \sum_{j=1} a_{ij}^{(1)}|\varepsilon_j^{0}\rangle \qquad (i = 1, \ldots, \infty). \tag{7-9}$$

This is allowed because the $|\varepsilon_j^{0}\rangle$ form a complete set in terms of which any arbitrary ket may be expanded. Once all the $a_{ij}^{(1)}$ have been evaluated, the first-order correction to the ith eigenket is complete. Substitution of (7-9) into (7-7) gives

$$\mathscr{H}_0 \sum_j a_{ij}^{(1)}|\varepsilon_j^{0}\rangle + \hat{V}|\varepsilon_i^{0}\rangle = \varepsilon_i^{0} \sum_j a_{ij}^{(1)}|\varepsilon_j^{0}\rangle + \varepsilon_i^{(1)}|\varepsilon_i^{0}\rangle$$

or

$$\sum_j (\varepsilon_j^{0} - \varepsilon_i^{0})a_{ij}^{(1)}|\varepsilon_j^{0}\rangle = (\varepsilon_i^{(1)} - \hat{V})|\varepsilon_i^{0}\rangle. \tag{7-10}$$

Multiplying (7-10) by $\langle\varepsilon_k^{0}|$ and using the orthonormality condition $\langle\varepsilon_k^{0}|\varepsilon_j^{0}\rangle = \delta_{kj}$, we obtain

$$\sum_{j=1} (\varepsilon_j^{0} - \varepsilon_i^{0})a_{ij}^{(1)}\delta_{kj} = \langle\varepsilon_k^{0}|(\varepsilon_i^{(1)} - \hat{V})|\varepsilon_i^{0}\rangle$$

or

$$(\varepsilon_k^{0} - \varepsilon_i^{0})a_{ik}^{(1)} = \varepsilon_i^{(1)}\delta_{ki} - \langle\varepsilon_k^{0}|\hat{V}|\varepsilon_i^{0}\rangle. \tag{7-11}$$

To obtain the first-order eigenvalue correction, we set $k = i$ and find

$$\varepsilon_i^{(1)} = \langle\varepsilon_i^{0}|\hat{V}|\varepsilon_i^{0}\rangle = V_{ii} = \langle V\rangle_i. \tag{7-12}$$

Using (7-4) the energy eigenvalue (to first order) becomes

$$\varepsilon_i = \varepsilon_i^{0} + \lambda V_{ii}. \tag{7-13}$$

To find the coefficients a_{ik}, we set $i \neq k$ in (7-11) and obtain

$$(\varepsilon_k^{0} - \varepsilon_i^{0})a_{ik}^{(1)} = -\langle\varepsilon_k^{0}|\hat{V}|\varepsilon_i^{0}\rangle \qquad (i \neq k)$$

or

$$a_{ik}^{(1)} = \frac{V_{ki}}{\varepsilon_i^{0} - \varepsilon_k^{0}}. \tag{7-14}$$

Since the ith state is assumed nondegenerate, then $\varepsilon_i^{0} \neq \varepsilon_k^{0}$ so that $a_{ik}^{(1)}$ is finite. From (7-9) and (7-3) the perturbed eigenket (to first order) becomes

$$|\varepsilon_i\rangle = |\varepsilon_i^{0}\rangle + \lambda a_{ii}^{(1)}|\varepsilon_i^{0}\rangle + \lambda \sum_{j\neq 1} \frac{V_{ji}}{\varepsilon_i^{0} - \varepsilon_j^{0}} |\varepsilon_j^{0}\rangle$$

$$= (1 + \lambda a_{ii}^{(1)})|\varepsilon_i^{0}\rangle + \lambda \sum_{j\neq i} \frac{V_{ji}}{\varepsilon_i^{0} - \varepsilon_j^{0}} |\varepsilon_j^{0}\rangle. \tag{7-15}$$

---

‡ The sum implies an integration over any continuum j states.

Only the coefficient $a_{ii}^{(1)}$ remains to be determined. We shall require that (7-15) be properly normalized to first order in λ. Setting $\langle \varepsilon_i | \varepsilon_i \rangle = 1$ and dropping terms of order λ^2, we find

$$|1 + \lambda a_{ii}^{(1)}|^2 \simeq 1 + 2\lambda \, \mathrm{Re} \, a_{ii}^{(1)} = 1 \qquad \text{or} \qquad \mathrm{Re} \, a_{ii}^{(1)} = 0.$$

Thus $a_{ii}^{(1)}$ is at worst imaginary. We may set $a_{ii}^{(1)} = 0$ without affecting the expectation values of operators to first order in λ.

Summarizing our first-order results, we have

$$\varepsilon_i = \varepsilon_i{}^0 + \lambda V_{ii} \tag{7-16}$$

and

$$|\varepsilon_i\rangle = |\varepsilon_i{}^0\rangle + \lambda \sum_{j \neq i} \frac{V_{ji}}{\varepsilon_i{}^0 - \varepsilon_j{}^0} |\varepsilon_j{}^0\rangle \tag{7-17}$$

(properly normalized to first order in λ). The wave-mechanical analog would be

$$\varepsilon_i = \varepsilon_i{}^0 + \lambda V_{ii} \tag{7-18}$$

$$\psi_i = \psi_i{}^0 + \lambda \sum_{j \neq i} \frac{V_{ji}}{\varepsilon_i{}^0 - \varepsilon_j{}^0} \psi_j{}^0 \tag{7-19}$$

where

$$V_{ji} = \int \psi_j^{0*} \, \hat{V}\!\left(\frac{\hbar}{i} \nabla, \mathbf{r}\right) \psi_i{}^0 \, d\mathbf{r}. \tag{7-20}$$

Inspection of (7-17) reveals why rapid convergence depends on the condition

$$\frac{|\lambda| \, |V_{ij}|}{|\varepsilon_j{}^0 - \varepsilon_i{}^0|} \ll 1.$$

III An Application of the First-Order Theory

We shall use first-order perturbation theory to find the eigenfunctions and eigenvalues of the modified one-dimensional oscillator Hamiltonian (Figure 7-1)

$$\mathcal{H} = \mathcal{H}_0 + \lambda x^4 \qquad \text{where} \quad \mathcal{H}_0 = \frac{\hat{p}^2}{2m} + \frac{1}{2} m\omega^2 x^2.$$

This is just the ordinary harmonic oscillator problem with a small anharmonic perturbation added. Using

$$\psi_n{}^0(x) = \left[\left(\frac{\alpha}{\pi}\right)^{1/2} \left(\frac{1}{2^n n!}\right)\right]^{1/2} e^{-\alpha x^2/2} H_n(\sqrt{\alpha} x) \qquad \left(\alpha = \frac{m\omega}{\hbar}\right)$$

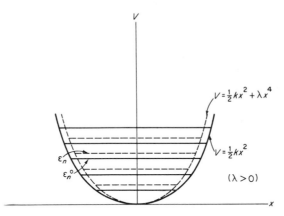

Figure 7-1 Energy shifts for the perturbed oscillator potential. Solid curves represent unperturbed quantities.

and $\varepsilon_n{}^0 = (n + \tfrac{1}{2})\hbar\omega$ as our unperturbed eigenfunctions and eigenvalues, (7-18) and (7-19) give the corresponding perturbed quantities as

$$\psi_n = \psi_n{}^0 + \lambda \sum_{m \neq n} \left(\frac{V_{mn}}{\varepsilon_n{}^0 - \varepsilon_m{}^0} \right) \psi_m{}^c \tag{7-21}$$

and $\varepsilon_n = \varepsilon_n{}^0 + \lambda V_{nn}$. The energy may be evaluated directly as

$$\varepsilon_n = \left(n + \frac{1}{2} \right)\hbar\omega + \lambda \left(\frac{\alpha}{\pi} \right)^{1/2} \frac{1}{2^n n!} \int_{-\infty}^{\infty} dx\, e^{-\alpha x^2} H_n{}^2(\sqrt{\alpha}x)x^4$$

or

$$\varepsilon_n = \left(n + \frac{1}{2} \right)\hbar\omega + \frac{3\lambda\hbar^2}{4m^2\omega^2}(2n^2 + 2n + 1). \tag{7-22}$$

We shall not evaluate the matrix elements in (7-21) but merely point out that they vanish if $|m - n| > 4$. This follows from the fact that $x^4 H_n$ may be expressed as a linear combination of the Hermite polynomials ranging from H_{n-4} to H_{n+4}. Using

$$x^4 \psi_n{}^0 = \sum_{l=n-4}^{n+4} a_l \psi_l{}^0$$

and the orthonormality of the $\psi_n{}^0$, we find

$$V_{mn} = \int_{-\infty}^{\infty} \psi_m^{0*}(x^4 \psi_n{}^0)\, dx = 0$$

when $|m - n| > 4$. It is therefore necessary to calculate at most eight matrix elements in (7-21) for the first-order corrections to the state ψ_n.

Experimentally the energy levels of a physical system are usually easier to observe than are its other characteristics. If the perturbation is small, the *zeroth*-order energy *eigenfunctions* are usually adequate for comparison with experiment. On the other hand the first-order energy eigenvalues can usually be directly observed. As we shall show below, the eigenfunctions corrected to $(p - 1)$th order are adequate to calculate the pth-order energy eigenvalues. Thus when we speak of first-order theory we shall be referring to the energies rather than the eigenfunctions, the latter being represented by their zeroth-order (unperturbed) counterparts.

IV Second-Order Theory

There are several reasons for pursuing second-order corrections even for small λ. First, these corrections may be experimentally observable (for example, spectroscopically) and therefore useful in confirming a particular theory. More important however, they may turn out to be the primary corrections should the first-order terms vanish. For example, suppose the perturbation to the oscillator problem above were λx^3 (instead of λx^4). Then the first-order corrections to the eigenvalues would be

$$V_{nn} = \int dx\ \psi_n^{0*}\, x^3 \psi_n^0 = 0$$

due to the oddness of the integrand. Here, second-order corrections become important.

The second-order terms are generated by solving (7-8) and using the established first-order result. The procedure is straightforward and quite similar to that used in first-order theory. The results are presented below:

$$\varepsilon_i = \varepsilon_i^0 + \lambda V_{ii} + \lambda^2 \sum_{j \neq i} \frac{V_{ji} V_{ij}}{\varepsilon_i^0 - \varepsilon_j^0} \tag{7-23}$$

and

$$|\varepsilon_i\rangle = \left[1 - \frac{1}{2}\lambda^2 \sum_{j \neq i} \frac{V_{ji} V_{ij}}{(\varepsilon_i^0 - \varepsilon_j^0)^2}|\varepsilon_i^0\rangle\right] + \lambda \left[\sum_{j \neq i} \frac{V_{ji}}{\varepsilon_i^0 - \varepsilon_j^0}|\varepsilon_j^0\rangle\right]$$
$$+ \lambda^2 \left[\sum_{j \neq i} \frac{1}{\varepsilon_i^0 - \varepsilon_j^0} \left\{\sum_{k \neq i} \frac{V_{ki} V_{jk}}{\varepsilon_i^0 - \varepsilon_k^0} - \frac{V_{ii} V_{ji}}{\varepsilon_i^0 - \varepsilon_j^0}\right\}|\varepsilon_j^0\rangle\right] \tag{7-24}$$

(properly normalized to second order in λ).

Inspection of $\varepsilon_i^{(1)}$ and $\varepsilon_i^{(2)}$ in (7-23) shows that

$$\varepsilon_i^{(1)} = V_{ii} = \langle \varepsilon_i^0 | \hat{V} | \varepsilon_i^0 \rangle$$

and

$$\varepsilon_i^{(2)} = \sum_{j \neq i} \frac{V_{ji} V_{ij}}{\varepsilon_i^0 - \varepsilon_j^0} = \langle \varepsilon_i^0 | \hat{V} | \varepsilon_i^{(1)} \rangle.$$

Generalizing, we see that the $(p-1)$th-order correction in the eigenvector leads directly to the pth-order correction in the eigenvalue, that is,

$$\varepsilon_i^{(p)} = \langle \varepsilon_i^0 | \hat{V} | \varepsilon_i^{(p-1)} \rangle. \tag{7-25}$$

This rule emphasizes the adequacy of the $(p-1)$th-order eigenfunction in the pth-order energy calculations.

The second-order theory can be used to calculate the approximate energy levels of an oscillator with an anharmonic perturbation, λx^3. Taking

$$\mathcal{H}_0 = \frac{\hat{p}^2}{2m} + \frac{1}{2} m\omega^2 x^2, \qquad \varepsilon_n^0 = \left(n + \frac{1}{2}\right) \hbar\omega$$

and

$$\psi_n^0 = \left[\left(\frac{\alpha}{\pi}\right)^{1/2} \left(\frac{1}{2^n n!}\right) \right]^{1/2} e^{-\frac{1}{2}\alpha x^2} H_n(\sqrt{\alpha} x)$$

(7-23) becomes

$$\varepsilon_n = \left(n + \frac{1}{2}\right) \hbar\omega + \lambda^2 \sum_{m \neq n} \frac{|V_{mn}|^2}{\hbar\omega(n-m)}$$

where

$$V_{mn} = \int_{-\infty}^{\infty} \psi_m^{*0} x^3 \psi_n^0 \, dx.$$

The evaluation of the integrals is left as an exercise (Problem 7-3).

V   Perturbation of a Degenerate Level

We observe that if the ith level is degenerate (for example, $\varepsilon_i^0 = \varepsilon_k^0$), then any term in (7-24) involving $V_{ki}/(\varepsilon_i^0 - \varepsilon_k^0)$ diverges. Of course the difficulty might be remedied if V_{ki} were also to vanish. However, this approach merely represents the cure, so to speak, rather than the cause of the pathological problem.

The difficulty with a degenerate level originates with the lack of uniqueness of the eigenstates. Let us suppose that $\mathcal{H}_0$ is degenerate but that the perturbed Hamiltonian $\mathcal{H} = \mathcal{H}_0 + \lambda \hat{V}$ is nondegenerate, that is, the perturbation removes the degeneracy. For example, $\mathcal{H}_0$ might be a central force Hamiltonian with a spatial (orientational) degeneracy; $\lambda \hat{V}$ could then be a uniform magnetic field removing this degeneracy. Now, the unperturbed eigenstates are not unique and any one set is generally different from any other. However, the eigenstates of $\mathscr{H}$, $|\varepsilon_i\rangle$, must be unique, and can only be approximately equal (for small λ) to one particular unperturbed set $|\widetilde{\varepsilon_i{}^0}\rangle$. We shall call this the "select" unperturbed set. Obviously, if we wish to perform a first-order calculation, we must use this select set as a zeroth-order approximation. Equivalently, we must establish a one-to-one correspondence between the perturbed states and the "select" unperturbed set in such a way that

$$|\varepsilon_i\rangle \xrightarrow[\lambda \to 0]{} |\widetilde{\varepsilon_i{}^0}\rangle. \tag{7-26}$$

The first task therefore is to find the correct zeroth-order eigenstates to make the correspondence above. Once these states have been found, as we shall see, no divergent terms will appear in (7-24). Furthermore, the first-order shifts in the energy will be given simply by the first two terms of (7-23) which are

$$\varepsilon_i = \varepsilon_i{}^0 + \lambda \langle \widetilde{\varepsilon_i{}^0} | \hat{V} | \widetilde{\varepsilon_i{}^0} \rangle. \tag{7-27}$$

We consider the perturbation of an s-fold degenerate level and set

$$\varepsilon_i{}^0 = \varepsilon^0 \qquad (i = 1, \ldots, s).$$

The kets $|\varepsilon_i{}^0\rangle$, $i = 1, \ldots, s$, are not unique energy eigenstates and are not necessarily orthogonal although we shall assume that we have constructed the latter.‡ The select set we are searching for can always be constructed using

$$|\widetilde{\varepsilon_i{}^0}\rangle = \sum_{j=1}^{s} b_{ij} |\varepsilon_j{}^0\rangle \qquad (i = 1, \ldots, s). \tag{7-28}$$

Since the select set (as well as the original set) is to be orthonormal, we find

$$\langle \widetilde{\varepsilon_i{}^0} | \widetilde{\varepsilon_i{}^0} \rangle = 1$$

$$\sum_{j,\,k=1}^{s} b_{ij}^* b_{ik} \langle \varepsilon_j{}^0 | \varepsilon_k{}^0 \rangle = \sum_{j,\,k=1}^{s} b_{ij}^* b_{ik} \delta_{jk}$$

$$= \sum_{j=1}^{s} |b_{ij}|^2 = 1. \tag{7-29}$$

The coefficients b_{ij} determine the "new" select set. Using $|\widetilde{\varepsilon_i{}^0}\rangle$ in the first-order equation (7-7), we find

$$\mathscr{H}_0 |\varepsilon_i^{(1)}\rangle + \hat{V} |\widetilde{\varepsilon_i{}^0}\rangle = \varepsilon_i{}^0 |\varepsilon_i^{(1)}\rangle + \varepsilon_i^{(1)} |\widetilde{\varepsilon_i{}^0}\rangle. \tag{7-30}$$

---

‡ Recall that it is always possible to construct such an orthonormal set.

We expand $|\varepsilon_i^{(1)}\rangle$ in the "old" basis as

$$|\varepsilon_i^{(1)}\rangle = \sum_{j=1} a_{ij}^{(1)}|\varepsilon_j^{\,0}\rangle. \qquad (7\text{-}31)$$

Using (7-31) and (7-6) in (7-30), we obtain

$$\sum_{j=1}^{\infty} a_{ij}^{(1)}\varepsilon_j^{\,0}|\varepsilon_j^{\,0}\rangle + \sum_{j=1}^{s} \hat{V}b_{ij}|\varepsilon_j^{\,0}\rangle = \varepsilon_i^{\,0} \sum_{j=1}^{\infty} a_{ij}^{(1)}|\varepsilon_j^{\,0}\rangle + \varepsilon_i^{(1)} \sum_{j=1}^{s} b_{ij}|\varepsilon_j^{\,0}\rangle. \quad (7\text{-}32)$$

Since $\varepsilon_j^{\,0} = \varepsilon_i^{\,0} = \varepsilon^0$ for $i \leqslant s$, the first s terms of the infinite series on each side cancel so that (7-32) becomes

$$\sum_{j>s}^{\infty} a_{ij}^{(1)}\varepsilon_j^{\,0}|\varepsilon_j^{\,0}\rangle + \sum_{j=1}^{s} \hat{V}b_{ij}|\varepsilon_j^{\,0}\rangle = \varepsilon^0 \sum_{j>s}^{\infty} a_{ij}^{(1)}|\varepsilon_j^{\,0}\rangle + \varepsilon_i^{(1)} \sum_{j=1}^{s} b_{ij}|\varepsilon_j^{\,0}\rangle. \quad (7\text{-}33)$$

Next, we multiply by $\langle\varepsilon_k^{\,0}|$ where $k = 1, \ldots, s$ and find, using orthonormality,

$$\sum_{j=1}^{s} b_{ij}\langle\varepsilon_k^{\,0}| \hat{V}|\varepsilon_j^{\,0}\rangle = \varepsilon_i^{(1)}b_{ik}$$

or

$$\sum_{j=1}^{s} V_{kj} b_{ij} = \varepsilon_i^{(1)}b_{ik}. \qquad (7\text{-}34)$$

This set of homogeneous equations for the coefficients b_{ij} has nontrivial solutions only for certain values of $\varepsilon_i^{(1)}$. In fact (7-34) can be written as an eigenvalue problem in matrix form

$$\mathbf{V}\mathbf{b}_i = \varepsilon_i^{(1)}\mathbf{b}_i$$

or

$$\begin{pmatrix} V_{11} & \cdots & V_{1s} \\ & \ddots & \\ V_{s1} & \cdots & V_{ss} \end{pmatrix} \begin{pmatrix} b_{i1} \\ \vdots \\ b_{is} \end{pmatrix} = \varepsilon_i^{(1)} \begin{pmatrix} b_{i1} \\ \vdots \\ b_{is} \end{pmatrix} \qquad (i = 1, \ldots, s) \qquad (7\text{-}35)$$

with the normalization restriction (7-29), that is,

$$\sum_{j=1}^{s} |b_{ij}|^2 = 1.$$

We are thus led to the following rule:

The coefficients which determine the ith select eigenket $|\widetilde{\varepsilon_i^{\,0}}\rangle$ are given by the components of the ith (normalized) eigenvector $\mathbf{b}_i$ of the $s \times s$ perturbation matrix $\mathbf{V}$. Furthermore, the first-order energy shift of the state is simply the corresponding eigenvalue $\varepsilon_i^{(1)}$ of $\mathbf{V}$.

The above rule suggests the following procedures for constructing the select eigenkets of a given s-fold degenerate level:

(a)   For each degenerate level, use the given eigenkets $|\varepsilon_i{}^0\rangle$ to construct the $s \times s$ matrix, that is,

$$\mathbf{V} = \begin{pmatrix} \langle \varepsilon_1{}^0 | \hat{V} | \varepsilon_1{}^0 \rangle & \cdots & \langle \varepsilon_1{}^0 | \hat{V} | \varepsilon_s{}^0 \rangle \\ \langle \varepsilon_s{}^0 | \hat{V} | \varepsilon_1{}^0 \rangle & \cdots & \langle \varepsilon_s{}^0 | \hat{V} | \varepsilon_s{}^0 \rangle \end{pmatrix}. \tag{7-36}$$

(b)   Find the s eigenvalues, $\varepsilon_i^{(1)}$, and the corresponding normalized eigenvectors (column matrices)

$$\mathbf{b}_i = \begin{pmatrix} b_{i1} \\ \vdots \\ b_{ij} \\ \vdots \\ b_{is} \end{pmatrix} \qquad (i = 1, \ldots, s). \tag{7-37}$$

Use the components to construct the new select set as

$$|\widetilde{\varepsilon_i{}^0}\rangle = \sum_{j=1}^{s} b_{ij} |\varepsilon_j{}^0\rangle. \tag{7-38}$$

(c)   The corresponding eigenvalues of $\mathbf{V}$, $\varepsilon_i^{(1)}$, give the first-order corrections to the energy of $|\varepsilon_i\rangle$, that is,

$$\varepsilon_i = \varepsilon^0 + \lambda \varepsilon_i^{(1)} \qquad (i = 1, \ldots, s). \tag{7-39}$$

In general, these eigenvalues are distinct ($\varepsilon_i^{(1)} \neq \varepsilon_k^{(1)}$, $i \neq k$) so that the degeneracy is usually removed in first order.

Had we originally used the select set instead of the given set as a zeroth-order approximation, we could have proceeded with the first-order nondegenerate theory as outlined in the previous section. In that case, the matrix elements would satisfy

$$V_{ij} = \langle \widetilde{\varepsilon_i{}^0} | \hat{V} | \widetilde{\varepsilon_j{}^0} \rangle = V_{ii} \delta_{ij} \tag{7-40}$$

and the divergent terms ($j \neq i$) in (7-24) would be absent. Furthermore, since the $\varepsilon_i^{(1)}$ represent the eigenvalues of $\hat{V}$, they must be equal to its diagonal elements, namely,

$$\varepsilon_i^{(1)} = \langle \widetilde{\varepsilon_i{}^0} | V | \widetilde{\varepsilon_i{}^0} \rangle.$$

Thus we have

$$\varepsilon_i = \varepsilon_i{}^0 + \lambda \langle \widetilde{\varepsilon_i{}^0} | \hat{V} | \widetilde{\varepsilon_i{}^0} \rangle$$

which is just the result of nondegenerate perturbation theory.

VI An Application of the Perturbation Theory to a Degenerate Level—The Stark Effect in Hydrogen‡

The effect of a uniform electric field $\mathbf{E}$ on the levels of hydrogen is known as the Stark effect. If this field is sufficiently strong, we may omit the spin–orbit interaction entirely as we did in the Paschen-Back effect. When an electron is inserted into a *uniform* electric field, it acquires a potential energy, $\int e\mathbf{E} \cdot d\mathbf{r} = e\mathbf{E} \cdot \mathbf{r}$. In hydrogen, this expression may also be written as $-\mathbf{p} \cdot \mathbf{E}$ where $\mathbf{p} = -e\mathbf{r}$ is the dipole moment of the proton–electron system. The appropriate Hamiltonian is

$$\mathscr{H}_{\text{Stark}} = \frac{\hat{p}^2}{2m} - \frac{e^2}{r} + e\mathbf{E} \cdot \mathbf{r}. \qquad (7\text{-}41)$$

Taking the direction of $\mathbf{E}$ to be along the z axis, the perturbation takes the form λV where $\lambda = +eE$ and $V = z = r\cos\theta$. In the scheme§ $|nlm_l\rangle$, the ground state $|100\rangle$ is nondegenerate. The correction to the energy for this level is, from (7-16), simply

$$\lambda\varepsilon^{(1)} = +eE\langle 100|z|100\rangle.$$

The matrix element is

$$\langle 100|z|100\rangle = \int \psi_{100}^* z\psi_{100}\, d\mathbf{r}.$$

However, since $|\psi_{100}|^2$ is an even function and z is an odd function, the integrand is odd. The integral therefore vanishes and there is no first-order Stark shift in the ground state of hydrogen.

The first excited state is fourfold degenerate, that is, $|200\rangle$, $|210\rangle$, and $|21\pm1\rangle$ all have energy $\varepsilon_2 = \varepsilon_1/2^2$. In principle, we have to construct the 4×4 V matrix. Fortunately this problem can be reduced to one involving a 2×2 matrix. To establish this, we first observe that the off-diagonal elements of the 4×4 matrix associated with the states $|21+1\rangle$ and $|21-1\rangle$ vanish. Since the perturbation is independent of the azimuthal angle ϕ, the matrix

---

‡ The Stark effect is actually not a proper example of a bound-state perturbation problem. Along the direction of the field, the potential varies as $V = -eEz$. As $r \to \infty$, the potential becomes sufficiently negative so as to be incapable of supporting bound states. Thus the probability of finding the particle at infinity is not zero and the field is capable of "ionizing" the initially bound system. The states are quasi-discrete and have certain properties associated with a continuous spectrum. Nevertheless, as long as the electric field is small the probability of ionization is small and the problem may be treated using bound-state perturbation techniques.

§ We shall suppress the spin indices.

elements under consideration involve inner products of the ϕ parts of the central force states. However, since these parts are orthogonal we find

$$\langle 211|z|21-1\rangle = \langle 211|z|210\rangle = \langle 211|z|200\rangle = 0$$

$$\langle 21-1|z|211\rangle = \langle 21-1|z|210\rangle = \langle 21-1|z|200\rangle = 0.$$

Thus the states $|21\pm1\rangle$ are already part of the select set. Their energy shifts are given by $\langle 21\pm1|z|21\pm1\rangle$. However it turns out that the integrals associated with these elements also vanish. This can be established by applying arguments of parity to the corresponding integrands. Thus there is no first-order Stark shift associated with the states $|21\pm1\rangle$.

The remaining select states are to be constructed from $|210\rangle$ and $|200\rangle$. The matrix elements for these states involve integrals which are evaluated using the hydrogenic functions in (5-74) as

$$\langle 210|z|210\rangle = \langle 200|z|200\rangle = 0$$

and

$$\langle 210|z|200\rangle = \int \psi_{210}^* r \cos\theta \psi_{200} \, d\mathbf{r} = \frac{-3\hbar^2}{me^2} = -3a.$$

The reduced matrix becomes

$$\mathbf{V} = \begin{pmatrix} \langle 200|r\cos\theta|200\rangle & \langle 200|r\cos\theta|210\rangle \\ \langle 200|r\cos\theta|210\rangle & \langle 210|r\cos\theta|210\rangle \end{pmatrix}$$

$$= \begin{pmatrix} 0 & -3a \\ -3a & 0 \end{pmatrix} \qquad (a = \hbar^2/me^2).$$

The eigenvalues of the matrix relation

$$\begin{pmatrix} 0 & -3a \\ -3a & 0 \end{pmatrix}\begin{pmatrix} b_{i1} \\ b_{i2} \end{pmatrix} = \varepsilon_i^{(1)}\begin{pmatrix} b_{i1} \\ b_{i2} \end{pmatrix} \qquad (i = 1, 2) \tag{7-42}$$

are determined by the roots of the (secular) equation

$$\begin{vmatrix} 0 - \varepsilon_i^{(1)} & -3a \\ -3a & 0 - \varepsilon_i^{(1)} \end{vmatrix} = 0$$

or

$$\varepsilon_i^{(1)} = \pm 3a. \tag{7-43}$$

Setting $\varepsilon_+^{(1)} = 3a$ and $\varepsilon_-^{(1)} = -3a$, the corresponding normalized eigenvectors become

$$\mathbf{b}_+ = \frac{1}{\sqrt{2}}\begin{pmatrix} 1 \\ -1 \end{pmatrix} \quad \text{and} \quad \mathbf{b}_- = \frac{1}{\sqrt{2}}\begin{pmatrix} 1 \\ +1 \end{pmatrix}. \tag{7-44}$$

The correct zeroth-order (select) states for the $n = 2$ level and their energy shifts are therefore (Figure 7-2)

$$|21 \pm 1\rangle \qquad\qquad\qquad \text{(no shift)}$$

$$|2^+\rangle = \frac{1}{\sqrt{2}} (|200\rangle - |210\rangle) \qquad [\text{shift} = \lambda \varepsilon_+^{(1)} = +eE(3a)]$$

$$(7\text{-}45)$$

$$|2^-\rangle = \frac{1}{\sqrt{2}} (|200\rangle + |210\rangle) \qquad [\text{shift} = \lambda \varepsilon_-^{(1)} = +eE(-3a)].$$

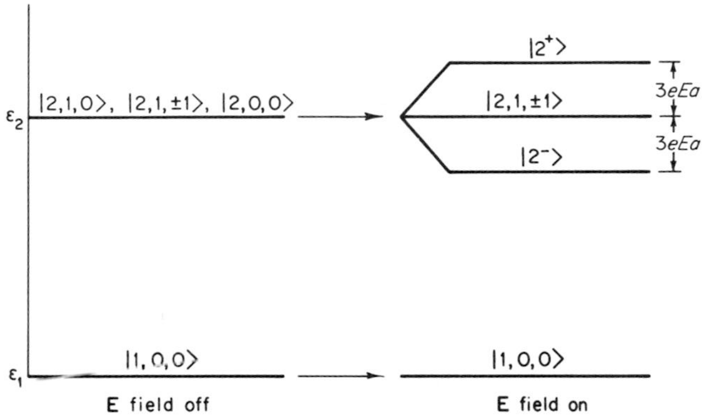

Figure 7-2 The first-order Stark shift in hydrogen.

Summarizing, if an unperturbed level is degenerate, we must first find a suitable "select eigenbasis" $|\widetilde{\varepsilon_i^0}\rangle$ with respect to which the perturbation matrix is diagonal. These kets have a one-to-one correspondence with the perturbed kets and serve as a suitable zeroth-order approximation. In most cases, it is necessary to find the select set by solving the **V** matrix equation (7-35).

In some cases however, physical intuition and symmetry assist in finding this set. We shall illustrate this in the case of the spin–orbit interaction in hydrogen.

VII   The Hydrogen Atom with Spin–Orbit Interaction

We have seen that even in an isolated situation an atomic electron does not see a purely central force potential. The small deviation, due to magnetic effects, is known as the spin–orbit interaction. The appropriate Hamiltonian is, from (6-34),

$$\mathcal{H}_{s-o} = \mathcal{H}_0 + \frac{e^2 f(r)}{2m^2 c^2} \hat{\mathbf{L}} \cdot \hat{\mathbf{S}}$$

$$(7\text{-}46)$$

where

$$\mathscr{H}_0 = \mathscr{H}_{cf} = \frac{\hat{p}^2}{2m} + V(r) \quad \text{and} \quad f(r) = \frac{1}{e^2 r}\frac{dV}{dr}.$$

We shall treat this problem using the perturbation methods introduced above. The unperturbed central force eigenstates are degenerate with respect to m_l and m_s and have energy ε_{nl}; two particularly interesting unperturbed sets already established are $|nlm_l m_s\rangle$ and $|nljm_j\rangle$. It is convenient to decompose the spin–orbit perturbation into

$$\lambda = \frac{e^2}{2m^2 c^2}, \quad \hat{V} = f(r)\hat{\mathbf{L}} \cdot \hat{\mathbf{S}}.$$

An unperturbed set of eigenvectors may serve as the zeroth-order approximation to a perturbed problem provided the set diagonalizes the $\mathbf{V}$ matrix. The set $|nlm_l m_s\rangle$ fails the test for the spin–orbit interaction since

$$\langle nlm_l m_s|f(r)\hat{\mathbf{L}} \cdot \hat{\mathbf{S}}|nlm_{l'} m_{s'}\rangle \neq C\delta_{m_l m_{l'}}\delta_{m_s m_{s'}}.$$

We could use the set $|nlm_l m_s\rangle$ to construct a suitable set by diagonalizing the $\mathbf{V}$ matrix. Instead we shall find this select set using some physical intuition.

The Hamiltonian in (7-46), $\mathscr{H}_{s-o}$, does not commute with $\hat{L}_z$ and $\hat{S}_z$. This incompatibility is due to the fact that $\hat{\mathbf{L}} \cdot \hat{\mathbf{S}}$ contains $\hat{L}_x$, $\hat{L}_y$, $\hat{S}_x$, and $\hat{S}_y$. It is not surprising then that the eigenstates of $\hat{L}_z$ and $\hat{S}_z$, $|nlm_l m_s\rangle$, are not eigenstates of $\mathscr{H}_{s-o}$. On the other hand, $\mathscr{H}_{s-o}$ is still compatible with $\hat{J}^2$ and $\hat{J}_z$. For example, using the identity

$$\hat{J}^2 = \hat{L}^2 + 2\hat{\mathbf{L}} \cdot \hat{\mathbf{S}} + \hat{S}^2$$

or

$$\hat{\mathbf{L}} \cdot \hat{\mathbf{S}} = \tfrac{1}{2}(\hat{J}^2 - \hat{L}^2 - \hat{S}^2) \tag{7-47}$$

we find

$$[\hat{\mathbf{L}} \cdot \hat{\mathbf{S}}, \hat{J}^2] = \tfrac{1}{2}\{[\hat{J}^2, \hat{J}^2] - [\hat{L}^2, \hat{J}^2] - [\hat{S}^2, \hat{J}^2]\} = 0.$$

A similar calculation would show that

$$[\hat{\mathbf{L}} \cdot \hat{\mathbf{S}}, \hat{J}_z] = 0.$$

Since $\mathscr{H}_{s-o}$, $\hat{J}^2$, and $\hat{J}_z$ mutually commute, these operators share at least one eigenbasis. In particular, while the central force eigenstates $|nljm_j\rangle$ may not be exact, it is likely that they constitute a select set and may be used as a zeroth-order approximation to the exact eigenbasis of $\mathscr{H}_{s-o}$.

If this physical argument is unconvincing, we need merely check to see whether the $\mathbf{V}$ matrix is diagonal in the $|nljm_j\rangle$ basis. Writing

$$V^{nl}_{jj'm_j m_{j'}} = \langle nlj'm_j'|f(r)\hat{\mathbf{L}} \cdot \hat{\mathbf{S}}|nljm_j\rangle$$

and using (7-47), we find

$$V^{nl}_{jj'm_jm_{j'}} = \langle nlj'm_j'|f(r)\tfrac{1}{2}(\hat{J}^2 - \hat{L}^2 - \hat{S}^2)|nljm_j\rangle$$

$$= \tfrac{1}{2}[j(j+1) - l(l+1) - s(s+1)]\hbar^2\langle nlj'm_j'|f(r)|nljm_j\rangle$$

$$= \tfrac{1}{2}[j(j+1) - l(l+1) - s(s+1)]\hbar^2\langle nl|f(r)|nl\rangle\,\delta_{jj'm_jm_{j'}} \qquad (s = \tfrac{1}{2})$$

$$(7\text{-}48)$$

that is, **V** is diagonal.

It is now clear that the set $|nljm_j\rangle$ represents the correct zeroth-order approximation to the exact eigenstates of the spin–orbit Hamiltonian. Setting $j = j'$ and $m_j = m_j'$ in (7-48), the energy corrected to first order becomes simply

$$\varepsilon_{nlj} = \varepsilon_{nl} + \lambda V^{nl}_{jjm_jm_j}$$

$$= \varepsilon_{nl} + \frac{e^2}{2m^2c^2}\left[\frac{j(j+1) - l(l+1) - s(s+1)}{2}\right]\hbar^2\langle n, l|f(r)|n, l\rangle. \quad (7\text{-}49)$$

The level splitting in (7-49) is known as *Lande's interval rule* and applies quite generally to any central force problem. For example, it may be applied to the valence electron of an alkali atom in which the Coulomb potential is modified by the screening of the inner electrons.

In hydrogen, the matrix element $\langle nl|f(r)|nl\rangle$ is

$$\langle nl|\frac{1}{r^3}|nl\rangle = \int_0^\infty R_{nl}(r)\frac{1}{r^3}R_{nl}(r)r^2\,dr$$

where we have set

$$f(r) = \frac{1}{e^2r}\frac{dV}{dr} = \frac{1}{r^3};$$

R_{nl} is the radial hydrogenic eigenfunction. The integral gives[‡]

$$\langle nl|\frac{1}{r^3}|nl\rangle = \frac{1}{a^3n^3(l+1)(l+\tfrac{1}{2})l} \qquad \left(a = \frac{\hbar^2}{me^2}\right)$$

so that Lande's rule for hydrogen becomes

$$\varepsilon_{nlj} = \varepsilon_n + \frac{\hbar^2}{m^2c^2a^2}\left[\frac{1}{2}\frac{j(j+1) - l(l+1) - \tfrac{3}{4}}{n(l+1)(l+\tfrac{1}{2})l}\right]\frac{e^2}{2n^2a}$$

$$= \varepsilon_n + \alpha^2|\varepsilon_n|\left[\frac{1}{2}\frac{j(j+1) - l(l+1) - \tfrac{3}{4}}{n(l+1)(l+\tfrac{1}{2})l}\right] \qquad (7\text{-}50)$$

‡ See H. A. Bethe and E. E. Salpeter, "Quantum Mechanics of One and Two Electron Atoms," p. 17. Springer-Verlag, Berlin, 1957.

where we have used the abbreviations

$$|\varepsilon_n| = \frac{e^2}{2n^2 a} = \frac{|\varepsilon_1|}{n^2} \qquad \text{and} \qquad \alpha = \frac{\hbar}{mca} = \frac{e^2}{\hbar c} \simeq \frac{1}{137}.$$

The term in brackets is of order unity so that the fractional correction due to the spin–orbit interaction is

$$\frac{\varepsilon_{s-o}}{|\varepsilon_n|} = \alpha^2 \simeq \left(\frac{1}{137}\right)^2 \simeq 10^{-4}.$$

For this reason, the dimensionless constant α is called the *fine-structure* constant. Note that the accidental degeneracy is reduced as the energy now depends on n, l, and j. The orientational degeneracy remains.

Experimentally it is found that the fine-structure corrections for the hydrogen atom are only in qualitative agreement with (7-50). This is due to the fact that these energy corrections are of the same order as the leading relativistic corrections that we have been neglecting throughout. Thus, in addition to the spin–orbit energy we should include the leading correction caused by the variation of mass with velocity. There is a third term to be included which appears quantumrelativistically (Darwin term) and has no classical analog. All three terms (s–o + mass variation + Darwin term) appear quite naturally in the nonrelativistic limit of the relativistic Dirac equation (Chapter 11) for the electron. Their combined effects give in place of (7-50)

$$\varepsilon_{nj} = \varepsilon_n + \alpha^2 \varepsilon_n \left[\frac{1}{n} \left(\frac{1}{j + \frac{1}{2}} - \frac{3}{4n} \right) \right] \qquad (j = l \pm \tfrac{1}{2}). \qquad (7\text{-}51)$$

This fine-structure formula will be derived in Chapter 11; it demonstrates a double degeneracy‡ (in addition to the m_j degeneracy) since for a given j there are generally two l values ($l = j \mp \frac{1}{2}$).

Summarizing, we see that the complete hydrogenic Hamiltonian with fine structure is

$$\mathscr{H} = \mathscr{H}_{cf} + (\text{fine-structure terms})$$

where

$$\mathscr{H}_{cf} = \frac{p^2}{2m} - \frac{e^2}{r}$$

and

fine-structure terms = (s–o interaction + mass variation + Darwin term).

Furthermore, the select (zeroth-order) eigenstates are $|nljm_j\rangle$ with the first-order energies given by (7-51).

---

‡ This degeneracy is actually slightly split due to the rather complex effects of quantum electrodynamics (Lamb shift).

VIII   The Anomalous Zeeman Effect in Hydrogen

When a hydrogen atom is inserted into a uniform magnetic field (along the z direction), the Hamiltonian becomes

$$\mathcal{H} = \frac{\hat{p}^2}{2m} - \frac{e^2}{r} + \text{(fine-structure terms)} + \frac{eB}{2mc}(\hat{L}_z + 2\hat{S}_z). \quad (7\text{-}52)$$

In practice, the corrections due to the fine structure and the external field are always small compared to the electrostatic central force energy. In principle, we should be required to choose an unperturbed set which diagonalizes both fine structure and the magnetic energy. However in the two important cases to be considered, a simpler approach is sufficient. If the last term in (7-52) is larger than the fine-structure energy (strong **B** field), we may neglect the latter altogether. This is the Paschen–Back effect already discussed in Chapter 6. If however **B** is weak enough, we use the select set which diagonalizes only the dominant perturbation, in this case the fine structure. Since we have found this set to be $|nljm_j\rangle$, the energy corrections due to the external **B** field become simply

$$\varepsilon_B = \frac{eB}{2mc} \langle nljm_j | \hat{L}_z + 2\hat{S}_z | nljm_j \rangle$$

$$= \frac{eB}{2mc} \langle nljm_j | \hat{J}_z + \hat{S}_z | nljm_j \rangle$$

$$= \frac{eB}{2mc} \{m_j \hbar + \langle nljm_j | \hat{S}_z | nljm_j \rangle\}. \quad (7\text{-}53)$$

The matrix element $\langle S_z \rangle$ may be evaluated using (6-24) to express the state $|nljm_j\rangle$ in terms of $|nlm_l m_s\rangle$. Since the latter are eigenkets of $\hat{S}_z$ the calculation is straightforward. A more interesting approach is to use the vector model. In the state $|nljm_j\rangle$, the **S** vector "precesses" about the **J** vector. Thus the average value of the component of **S** normal to **J** vanishes. As far as averages are concerned, we may write (Figure 7-3)

$$\langle \mathbf{S} \rangle = \langle \mathbf{S_J} \rangle = \langle (\hat{\mathbf{S}} \cdot \hat{\mathbf{J}}_u) \hat{\mathbf{J}}_u \rangle \quad (7\text{-}54)$$

where $\hat{\mathbf{J}}_u = \hat{\mathbf{J}}/\hat{J}$ is a unit vector along $\hat{\mathbf{J}}$. Using the identity

$$(\hat{\mathbf{J}} - \hat{\mathbf{S}})^2 = \hat{L}^2, \qquad \hat{J}^2 - 2\hat{\mathbf{J}} \cdot \hat{\mathbf{S}} + \hat{S}^2 = \hat{L}^2,$$

or

$$\hat{\mathbf{J}} \cdot \hat{\mathbf{S}} = \tfrac{1}{2}(\hat{J}^2 + \hat{S}^2 - \hat{L}^2)$$

(7-54) becomes

$$\langle \mathbf{S} \rangle = \left\langle \frac{(\hat{J}^2 + \hat{S}^2 - \hat{L}^2)\hat{\mathbf{J}}}{2\hat{J}^2} \right\rangle. \quad (7\text{-}55)$$

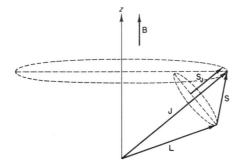

Figure 7-3   The vector

$$S_J = \frac{(S \cdot J)J}{J^2}.$$

Note that for weak **B** fields the "precession" of **S** and **L** about **J** is much more rapid than the "precession" of **J** about **B**.

For the z component we find

$$\langle S_z \rangle = \left\langle \left(\frac{\hat{J}^2 + \hat{S}^2 - \hat{L}^2}{2\hat{J}^2} \right) \hat{J}_z \right\rangle$$

or

$$\langle nlsjm_j | \hat{S}_z | nlsjm_j \rangle = \frac{j(j+1) + s(s+1) - l(l+1)}{2j(j+1)} m_j \hbar. \qquad (7\text{-}56)$$

Returning to (7-53), we find that the magnetic splitting becomes

$$\varepsilon_B = \frac{eB\hbar}{2mc} \left\{ 1 + \frac{j(j+1) - l(l+1) + s(s+1)}{2j(j+1)} \right\} m_j. \qquad (7\text{-}57)$$

It is convenient to define the *Lande "g" factor* by

$$g_L = 1 + \frac{j(j+1) - l(l+1) + s(s+1)}{2j(j+1)}$$

$$= 1 \pm \frac{1}{2l+1} \qquad \text{for} \quad s = \tfrac{1}{2}, \quad j = l \pm \tfrac{1}{2} \qquad (7\text{-}58)$$

in terms of which (7-57) may be written

$$\varepsilon_B = \mu_B B g_L m_j.$$

The parameter g_L is actually the effective gyromagnetic factor associated with **J** and depends on the state of coupling between **L** and **S**. If $l = 0$ is zero, then $g_L = g_s = 2$; if $s = 0$, then $g_L = g_l = 1$.

The energy levels for the hydrogenic electron in the presence of a weak **B** field become (Figure 7-4)

$$\varepsilon_{njlm_j} = \varepsilon_n + \alpha^2 \varepsilon_n \left\{ \frac{1}{n} \left(\frac{1}{j + \tfrac{1}{2}} - \frac{3}{4n} \right) \right\} + \mu_B B g_L m_j$$

(anomalous Zeeman effect).   (7-59)

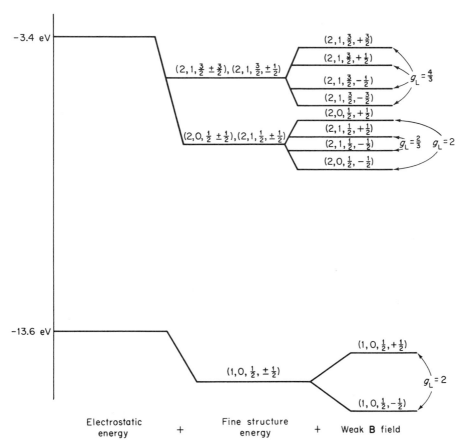

Figure 7-4 The anomalous Zeeman effect (hydrogen). The parentheses refer to $(nljm_j)$.

Equation (7-58) shows that there are two distinct values of g_L for any given j $(l = j \mp \frac{1}{2})$. This leads to unequal splitting in the magnetic sublevels m_j. For this reason, the effect is termed the *anomalous* Zeeman effect. As the field is increased from a weak field to a strong field, the anomalous Zeeman pattern gradually passes into the Paschen–Back pattern, (6-38). A treatment of the intermediate case where the magnetic and spin–orbit energy are of the same order of magnitude is beyond the level of this text; a discussion of this case may be found in any advanced treatise on atomic spectra.‡

---

‡ See for example, E. U. Condon and G. H. Shortley, "The Theory of Atomic Spectra," p. 152. Cambridge Univ. Press, London and New York, 1935.

IX   Time–Dependent Perturbation Theory

When the physical system of interest is represented by the unperturbed Hamiltonian $\mathscr{H}_0$, it is a common practice to regard the perturbation $\lambda \hat{V}$ as inducing transitions between unperturbed states $|\varepsilon_i^0\rangle$. From (3-66) the total evolution operator for our system satisfies the integral equation

$$\hat{U}(t, t_0) = 1 + \frac{1}{i\hbar} \int_{t_0}^{t} \mathscr{H}(\tau_1)\hat{U}(\tau_1, t_0)\, d\tau_1 \qquad (7\text{-}60\text{a})$$

where

$$\mathscr{H}(t) = \mathscr{H}_0 + \lambda \hat{V}(t) \qquad \text{and} \qquad t > t_0.$$

In the special case where the perturbation is independent of time, the solution to (7-60a) becomes

$$\hat{U}(t, t_0) = \exp\left[-\frac{i}{\hbar} \mathscr{H}(t - t_0)\right] \qquad (7\text{-}60\text{b})$$

where $\mathscr{H} = \mathscr{H}_0 + \lambda \hat{V}$. We shall be interested in the way that unperturbed states of $\mathscr{H}_0$ evolve under the influence of the operator given by (7-60). For this reason it is convenient to express $\hat{U}(t, t_0)$ in terms of the unperturbed evolution operator

$$\hat{U}^{(0)}(t, t_0] = \exp\left[-\frac{i}{\hbar} \mathscr{H}_0(t - t_0)\right].$$

It is tempting to do this by writing (7-60b) as

$$\hat{U}(t, t_0) = \exp\left\{-\frac{i}{\hbar}[\mathscr{H}_0 + \lambda \hat{V}](t - t_0)\right\} = \hat{U}^{(0)}(t - t_0)\exp\left\{-\frac{i}{\hbar}\lambda \hat{V}(t - t_0)\right\}.$$

$$(7\text{-}61)$$

We emphasize that (7-61) is totally *incorrect* since $\mathscr{H}_0$ and $\hat{V}$ *do not* commute and the identity $e^{\hat{A}+\hat{B}} = e^{\hat{A}}e^{\hat{B}}$ is valid only when $\hat{A}$ and $\hat{B}$ commute.

Relating $\hat{U}$ to $\hat{U}^{(0)}$ in a closed simple form is virtually an impossible task. However, if λ is small, then a perturbation series in powers of λ is usually possible. The mathematical derivation of such a series is too complicated to be developed here and so we present only the results.‡ The series takes the form

$$\hat{U}(t, t_0) = \hat{U}^{(0)}(t, t_0) + \sum_{n=1}^{\infty} \hat{U}^{(n)}(t, t_0) \qquad (7\text{-}62)$$

---

‡ For a derivation, see A. Messiah, "Quantum Mechanics," Volume II, pp. 722–724. Wiley, New York, 1962. See also Appendix B of the present book.

where

$$\hat{U}^{(n)} = \left(\frac{\lambda}{i\hbar}\right)^n \int_{t_0}^t d\tau_n \int_{t_0}^{\tau_n} d\tau_{n-1} \cdots \int_{t_0}^{\tau_2} d\tau_1$$

$$\times \hat{U}^{(0)}(t, \tau_n) \hat{V}(\tau_n) \hat{U}^{(0)}(\tau_n, \tau_{n-1}) \cdots \hat{U}^{(0)}(\tau_2, \tau_1) \hat{V}(\tau_1) \hat{U}^{(0)}(\tau_1, t_0).$$

Equation (7-62) is an exact representation for the solution to (7-60a) provided that the series converges. It is most useful when λ is small, so that the first few terms represent a good approximation. As $\lambda \to 0$, $\hat{U}$ reduces to $\hat{U}^{(0)}$.

Note the ordering of the time integrals for the nth-order term $\hat{U}^{(n)}$, that is, $\tau_n > \tau_{n-1}$. This is closely related to the fact that, nonrelativistically, evolution occurs *forward* in time only. In relativistic situations physical significance can be attributed to propagation backward in time.‡

X   Transitions Induced by a Constant Perturbation

We first consider the case in which $\hat{V}$ is independent of time. Furthermore, let the system be initially (at $t = t_0$) in one of the eigenstates of $\mathcal{H}_0$, let us say $|\varepsilon_i^0\rangle$. This state's evolution is given by

$$|\beta_i, t\rangle = \hat{U}(t, t_0)|\varepsilon_i^0\rangle.$$

It is obvious from (7-60) that when $\hat{V}$ is time independent, the evolution operator is a function of the time interval $\tau = t - t_0$, that is,

$$\hat{U}(t, t_0) = \hat{U}(\tau).$$

The probability of observing the system in another eigenstate of $\mathcal{H}_0$, let us say $|\varepsilon_j^0\rangle$, after a time τ has elapsed, is given by

$$\mathcal{P}_{ij}(\tau) = |\langle \varepsilon_j^0 | \beta_i t \rangle|^2 = |\langle \varepsilon_j^0 | \hat{U}(\tau) | \varepsilon_i^0 \rangle|^2. \tag{7-63}$$

Equivalently the transition probability from the state i to the state j is given by the square of the corresponding matrix element of the evolution operator. Note the symmetry with respect to an interchange of i and j. This demonstrates the *microscopic reversibility* of the process.

We next consider the behavior of this transition probability as the time interval tends to infinity. We shall *define* the transition *rate* between the states ε_i^0 and ε_j^0 as

$$R_{ij} = \lim_{\tau \to \infty} \frac{d\mathcal{P}_{ij}(\tau)}{d\tau} = \lim_{\tau \to \infty} \frac{d}{d\tau} |\langle \varepsilon_j^0 | \hat{U}(\tau) | \varepsilon_i^0 \rangle|^2. \tag{7-64}$$

‡ See, for example, R. P. Feynman, The Theory of Positrons, *Phys. Rev.* **76**, 749 (1949).

Using the series representation of $\hat{U}(\tau)$, (7-62), we obtain for the transition rate

$$R_{ij} = \lim_{\tau \to \infty} \frac{d}{d\tau} \left\{ \left| \langle \varepsilon_j{}^0 | \hat{U}^0 | \varepsilon_i{}^0 \rangle + \langle \varepsilon_j{}^0 | \sum_{n=1}^{\infty} \hat{U}^{(n)} | \varepsilon_i{}^0 \rangle \right|^2 \right\}. \qquad (7\text{-}65)$$

However, using

$$\langle \varepsilon_j{}^0 | \hat{U}^0 | \varepsilon_i{}^0 \rangle = \langle \varepsilon_j{}^0 | \exp\left[-\frac{i}{\hbar} \mathscr{H}(t - t_0) \right] | \varepsilon_i{}^0 \rangle$$

$$= \exp\left[-\frac{i}{\hbar} \varepsilon_i{}^0 (t - t_0) \right] \langle \varepsilon_j{}^0 | \varepsilon_i{}^0 \rangle$$

and recalling that $\langle \varepsilon_j{}^0 | \varepsilon_i{}^0 \rangle = 0$ when $j \neq i$, we find that

$$R_{ij} = \lim_{\tau \to \infty} \frac{d}{d\tau} \left\{ \left| \langle \varepsilon_j{}^0 | \sum_{n=1}^{\infty} \hat{U}^{(n)}(\tau) | \varepsilon_i{}^0 \rangle \right|^2 \right\}. \qquad (7\text{-}66)$$

The total transition rate from the state i to the state j involves the square of an infinite sum of terms, the nth of which is proportional to λ^n. As $\lambda \to 0$, we find $R_{ij} \to 0$ and the transitions cease. *The perturbation is therefore completely responsible for transitions.* We shall consider first the case where λ is sufficiently small so that only the term $n = 1$ need be retained. This leads directly to what is known as *Fermi's Golden Rule.*

XI First-Order Transitions—Fermi's Golden Rule

The first-order transition rate is obtained from (7-66) using only the term $n = 1$. We find this rate to be

$$R_{ij} = \lim_{\tau \to \infty} \frac{d}{d\tau} \left| \langle \varepsilon_j{}^0 | \hat{U}^{(1)}(\tau) | \varepsilon_i{}^0 \rangle \right|^2 \qquad (\tau = t - t_0). \qquad (7\text{-}67)$$

Using (7-62) we have (for $\hat{V}$ independent of time)

$$\hat{U}^{(1)} = \frac{\lambda}{i\hbar} \int_{t_0}^{t} d\tau_1 \, \hat{U}^{(0)}(t, \tau_1) \hat{V} \hat{U}^{(0)}(\tau_1, t_0)$$

$$= \frac{\lambda}{i\hbar} \int_{t_0}^{t} d\tau_1 \exp\left[-\frac{i}{\hbar} \mathscr{H}_0(t - \tau_1) \right] \hat{V} \exp\left[-\frac{i}{\hbar} \mathscr{H}_0(\tau_1 - t_0) \right]$$

so that the first-order transition rate in (7-67) can be written

$$R_{ij} = \lim_{\tau \to \infty} \frac{d}{d\tau} \left| \langle \varepsilon_j{}^0 | \frac{\lambda}{i\hbar} \int_{t_0}^{t} d\tau_1 \right.$$

$$\left. \times \exp\left[-\frac{i}{\hbar} \mathscr{H}_0(t - \tau_1) \right] \hat{V} \exp\left[-\frac{i}{\hbar} \mathscr{H}_0(\tau_1 - t_0) \right] | \varepsilon_i{}^0 \rangle \right|^2.$$

Using the relation

$$\exp\left[\frac{i}{\hbar}\mathscr{H}_0(\tau_1 - t_0)\right]|\varepsilon_i{}^0\rangle = \exp\left[\frac{i}{\hbar}\varepsilon_i{}^0(\tau_1 - t_0)\right]|\varepsilon_i{}^0\rangle$$

we have, after some rearrangement,

$$R_{ij} = \frac{\lambda^2}{\hbar^2} \lim_{\tau \to \infty} \frac{d}{d\tau} \left|\exp\left[-\frac{i}{\hbar}(\varepsilon_j{}^0 t - \varepsilon_i{}^0 t_0)\right]\langle\varepsilon_j{}^0|\hat{V}|\varepsilon_i{}^0\rangle\right.$$

$$\left. \times \int_{t_0}^{t} \exp\left[\frac{i}{\hbar}(\varepsilon_j{}^0 - \varepsilon_i{}^0)\tau_1\right] d\tau_1\right|^2.$$

The integral is straightforward and it is left as an exercise (see Problem 7-10) to show that this can be simplified to

$$R_{ij} = \frac{\lambda^2}{\hbar^2} |V_{ji}|^2 \lim_{\tau \to \infty} \frac{d}{d\tau} g(\omega_{ji}, \tau) \tag{7-68}$$

where

$$g(\omega, \tau) = \frac{2(1 - \cos \omega\tau)}{\omega^2} \quad \text{and} \quad \omega_{ji} = \frac{\varepsilon_j{}^0 - \varepsilon_i{}^0}{\hbar}.$$

We first study the behavior of the function $g(\omega, \tau)$ in order to understand its meaning. Introducing the auxiliary function $f(\omega, \tau) = g(\omega, \tau)/\tau$, we note that $f(\omega, \tau)$ is an oscillatory function of ω for fixed τ (Figure 7-5). Its peak is reached when $\omega = 0$, with the maximum value there being‡

$$f(0, \tau) = \lim_{\omega \to 0} \frac{2(1 - \cos \omega\tau)}{\tau\omega^2} \to \tau.$$

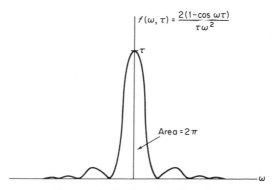

Figure 7-5 The function $f(\omega, \tau)$.

---

‡ This indeterminate form is easily evaluated using l'Hospital's rule.

The peak of the function $f(\omega, \tau)$ therefore tends to infinity with increasing τ. The area under the curve is evaluated as

$$\int_{-\infty}^{\infty} f(\omega, \tau) \, d\omega = \int_{-\infty}^{\infty} \frac{2(1 - \cos \omega\tau)}{\tau\omega^2} \, d\omega = \int_{-\infty}^{\infty} \frac{2(1 - \cos \theta)}{\theta^2} \, d\theta = 2\pi$$

where we have set $\theta = \omega\tau$. The important point is that *the area under the curve is independent of* τ. Thus as $\tau \to \infty$, the central peak rises and the curve in Figure 7-5 is drawn in about $\omega = 0$ in order to conserve area. But this is exactly the property of a delta function and we may make the identification

$$\lim_{\tau \to \infty} f(\omega, \tau) = 2\pi\delta(\omega) \qquad \text{or} \qquad \lim_{\tau \to \infty} g(\omega, \tau) = 2\pi\tau\delta(\omega).$$

The transition rate in (7-68) becomes finally‡

$$R_{ij} = \frac{2\pi\lambda^2}{\hbar^2} |V_{ji}|^2 \, \delta(\omega_{ji}) \qquad \left(\omega_{ji} = \frac{\varepsilon_j{}^0 - \varepsilon_i{}^0}{\hbar} \right)$$

$$= \frac{2\pi\lambda^2}{\hbar} |V_{ji}|^2 \, \delta(\varepsilon_j{}^0 - \varepsilon_i{}^0). \qquad (7\text{-}69)$$

The transition rate is therefore nonvanishing only between states of equal energy, that is, $\varepsilon_j{}^0 = \varepsilon_i{}^0$. We conclude that a static (time-independent) perturbation induces only energy-conserving transitions and neither supplies nor removes energy in the unperturbed system.

We focus our attention on transitions where the final state is part of the continuum. This is particularly important in scattering processes where a perturbation potential is responsible for transitions between free particle states, the latter, of course, being part of the continuous spectrum. In general, fundamental processes such as chemical and nuclear reactions and photoelectric interactions involve states in the continuum.

We relabel the states using $|n\rangle$ and $|m\rangle$ in place of $|\varepsilon_i{}^0\rangle$ and $|\varepsilon_j{}^0\rangle$ and use the conventional notation to rewrite (7-69) as

$$R_{nm} = \frac{2\pi\lambda^2}{\hbar} \int |\langle m| \hat{V} |n\rangle|^2 \delta(\varepsilon_m - \varepsilon_n) \, dm.$$

The integral is required because in the continuum we must consider the total transition rate to a group of final states $|m\rangle$, rather than to a single discrete state.

The integral may be transformed to one over energy (rather than one over final states) by using

$$dm = \rho(m) \, d\varepsilon_m \qquad (7\text{-}70)$$

---

‡ The delta function has the property $\delta(ax) = |a|^{-1} \delta(x)$.

where $\rho(\mathbf{m}) = d\mathbf{m}/d\varepsilon_\mathbf{m}$ represents the number of continuum states per energy interval, that is, the *density of states*. The transition rate becomes, finally,

$$R_{nm} = \frac{2\pi\lambda^2}{\hbar} \int |\langle \mathbf{m}| \hat{V} |\mathbf{n}\rangle|^2 \delta(\varepsilon_\mathbf{m} - \varepsilon_\mathbf{n})\rho(\mathbf{m})\, d\varepsilon_\mathbf{m}$$

or, using the sifting property of the delta function,

$$R_{nm} = \frac{2\pi\lambda^2}{\hbar} \left[|V_{mn}|^2 \rho(\mathbf{m})\right]_{\varepsilon_\mathbf{n} = \varepsilon_\mathbf{m}}. \tag{7-71}$$

Equation (7-71) is *Fermi's Golden Rule* and states that the transition rate R_{nm}

(a)   is nonzero only between continuum states of the same energy,
(b)   is proportional to the square of the matrix element of the perturbation connecting these states, and
(c)   is proportional to the density of final states.

In Chapter 8, we shall apply the Golden Rule to scattering problems where the scattering potential acts as a perturbation which induces a transition from an initial $|\mathbf{k}\rangle$ to some final $|\mathbf{k}'\rangle$ momentum eigenstate. The change in the momentum vector (its energy being unchanged) represents elastic scattering and the rate predicted by (7-71) is closely related to the scattering cross section.

XII   Higher-Order Corrections to the Golden Rule

Fermi's Golden Rule (7-71) was derived using only the first-order term in the expansion of the evolution operator. Very often first-order theory is inadequate. A derivation of the higher-order terms is too involved for our purpose; however the result is as follows‡:

$$R_{nm} = \frac{2\pi}{\hbar} \left[\left|\sum_{j=1}^{\infty} \lambda^j M_{nm}^{(j)}\right|^2 \rho(\mathbf{m})\right]_{\varepsilon_\mathbf{n} = \varepsilon_\mathbf{m}} \tag{7-72}$$

where

$$M_{nm}^{(1)} = V_{nm}$$

$$M_{nm}^{(2)} = \int \frac{V_{nl} V_{lm}\, d\mathbf{l}}{\varepsilon_\mathbf{m} - \varepsilon_\mathbf{l}}$$

$$M_{nm}^{(3)} = \int \frac{V_{nl} V_{ll'} V_{l'm}\, d\mathbf{l}\, d\mathbf{l'}}{(\varepsilon_\mathbf{m} - \varepsilon_\mathbf{l})(\varepsilon_\mathbf{m} - \varepsilon_{\mathbf{l'}})}$$

$$\vdots$$

‡ For a discussion of (7-72) see L. D. Landau and E. M. Lifshitz, "Quantum Mechanics," 2nd ed., Section 43. Addison-Wesley, Reading, Massachusetts, 1965.

The $M^{(j)}$ are referred to as transition amplitudes. Each amplitude is multiplied by λ^j and for sufficiently weak perturbations only the first few amplitudes need be retained in (7-72). If in fact $M_{nm}^{(1)}$ happens to vanish (this is sometimes the case), then the leading (second-order) term in the Golden Rule gives

$$R_{nm} = \frac{2\pi}{\hbar} [\lambda^4 | M_{nm}^{(2)} |^2 \rho(\mathbf{m})]_{\varepsilon_n = \varepsilon_m}. \tag{7-73}$$

Usually it will not be necessary to go beyond first order.

The importance of Fermi's Golden Rule, (7-71) and (7-72), rests with its complete generality. Note that nowhere has the coordinate representation (wave mechanics) been used in its derivation. There are a variety of fundamental scattering processes for which $\mathbf{r}$ is not a relevant observable. A very important example already considered is the spin-dependent interaction. In nuclear scattering, spin-dependent forces play an important role. A second example occurs in photoelectric interactions, that is, scattering of photons and electrons. As we shall see in Chapter 12, the quantization of the radiation field which leads directly to the concept of a photon does not involve the operator $\hat{\mathbf{r}}$. In the two examples just mentioned wave mechanics is inadequate to completely describe the scattering or transition process. However the Golden Rule is completely general and is probably the most commonly used formula for calculating transition rates (or cross sections) for fundamental processes.

XIII Transitions Induced by a Harmonic Perturbation

A most common way of inducing transitions between stationary states of quantum systems is by applying a harmonic perturbation. For example, the oscillating electric and magnetic fields associated with light can induce transitions between stationary states in atoms.

We shall apply (7-64) to transitions between discrete states $|\varepsilon_i\rangle$ and $|\varepsilon_j\rangle$ induced by a harmonic perturbation of the form

$$\lambda \hat{V} = \lambda \hat{V}(\hat{\mathbf{r}}) \cos \omega t = \lambda \hat{V}(\hat{\mathbf{r}}) \tfrac{1}{2}(e^{i\omega t} + e^{-i\omega t}).$$

The procedures are essentially similar to those used in the time-independent case. The transition rate in analogy with (7-69) is

$$R_{ij}(\omega) = \frac{2\pi\lambda^2}{\hbar^2} | V_{ij} |^2 \frac{1}{2} [\delta(\omega_{ji} - \omega) + \delta(\omega_{ji} + \omega)]. \tag{7-74}$$

The transition rate is frequency-dependent and vanishes unless $\omega = |\omega_{ji}|$. This is an example of a quantum-mechanical resonance in which the quantum system is unaffected unless the perturbing frequency is properly "tuned" to one of the excitation frequencies

$$\omega = |\omega_{ji}| = \frac{|\varepsilon_j - \varepsilon_i|}{\hbar}.$$

In practice it is impossible to generate a truly monochromatic wave. Rather the perturbation is usually of the form of a wave packet

$$\lambda \hat{V}(\hat{\mathbf{r}}, t) = \lambda \hat{V}(\hat{\mathbf{r}}) \int_0^\infty A(\omega) \cos \omega t \, d\omega$$

representing a superposition of different frequencies. The total transition rate must therefore be written as an integral over the incident frequencies, that is,

$$R_{ij} = \frac{2\pi\lambda^2}{\hbar^2} |V_{ij}|^2 \frac{1}{2} \int_0^\infty |A(\omega)|^2 [\delta(\omega_{ji} - \omega) + \delta(\omega_{ji} + \omega)] \, d\omega. \quad (7\text{-}75)$$

If $\varepsilon_j > \varepsilon_i$ ($\omega_{ji} > 0$), then the second delta function gives zero and the sifting property of the first gives

$$R_{ij} = \frac{2\pi\lambda^2}{\hbar^2} |V_{ij}|^2 \frac{1}{2} |A(\omega_{ji})|^2 \qquad (\varepsilon_j > \varepsilon_i). \qquad (7\text{-}76)$$

The rate in (7-76) describes a process in which the quantum system goes from a state of lower energy to one of higher energy. This is the process of *induced absorption* in which energy is transferred from the perturbing source to the system. The rate here is proportional to the intensity (square of the amplitude) of the perturbation at the resonant frequency ω_{ji} and to the square of the matrix element $|V_{ij}|^2$.

If $\varepsilon_j < \varepsilon_i$ ($\omega_{ji} < 0$), then the first delta function vanishes and the second gives

$$R_{ij} = \frac{2\pi\lambda^2}{\hbar^2} |V_{ij}|^2 \frac{1}{2} |A(\omega_{ji})|^2 \qquad (\varepsilon_j < \varepsilon_i) \qquad (7\text{-}77)$$

so that the *induced emission* rate is identical to the absorption rate between the two levels. This relationship is an example of the *principle of microscopic reversibility*.

Thus when a system is exposed to a harmonic perturbation containing the frequency $\omega = |\omega_{ji}|$, transitions between the levels ε_j and ε_i are induced. In the case of absorption, transitions are accompanied by an increase in the energy of the quantum system, the perturbation being the source of that energy. For emission, energy is delivered to the perturbing source. If the matrix element coupling the states vanishes, the transition does not occur (in first order) and is said to be *forbidden*.

XIV   Radiative Transitions in Hydrogen

We shall apply (7-76) and (7-77) to transitions in hydrogen induced by thermal (black-body) electromagnetic radiation. If the wavelength of the radiation is long when compared with the dimensions of the atomic system,‡ the spatial variation of the electric field may be ignored in a first approximation. The perturbation is then equivalent to a uniform oscillating electric field,§ $\mathbf{E} = \mathbf{E}_0 \cos \omega t$. The perturbing potential is similar to the one encountered in the Stark effect, that is, $\lambda \hat{V} = e\mathbf{E} \cdot \mathbf{r} = e\mathbf{E}_0 \cdot \mathbf{r} \cos \omega t$, where ω is the frequency of the incident radiation. Since $\mathbf{p} = -e\mathbf{r}$ represents the dipole moment of the hydrogen atom, we have therefore applied what is called the *electric dipole approximation*. Again, this approximation is valid only when the wavelength of the radiation is long compared to the dimensions of the atom.

The induced absorption and emission rates become

$$R_{ij} = \frac{2\pi e^2}{\hbar^2}\frac{1}{2}\,|\mathbf{E}_0(\omega_{ij})\cdot\langle i|\mathbf{r}|j\rangle|^2. \tag{7-78}$$

Since thermal (black-body) radiation is randomly polarized we must average over the *three* components of $\mathbf{E}$ and write

$$R_{ij} = \frac{2\pi e^2}{\hbar^2}\frac{1}{2}\frac{|E_0(\omega_{ij})|^2}{3}\,|\langle i|\mathbf{r}|j\rangle|^2 \tag{7-79}$$

where

$$|\langle i|\mathbf{r}|j\rangle|^2 = |x_{ij}|^2 + |y_{ij}|^2 + |z_{ij}|^2.$$

The radiation energy density is related to the electric field by $\rho_{\text{rad}} = (|E|^2/4\pi)$ so that the rate becomes finally

$$R_{ij} = \frac{4\pi^2 e^2}{3\hbar^2}\,\rho_{\text{rad}}(\omega_{ij})|\langle i|\mathbf{r}|j\rangle|^2. \tag{7-80}$$

Thus radiative transitions occur in first order only if

(a)   the radiation of the excitation frequency is present, and
(b)   the matrix element $\langle i|\mathbf{r}|j\rangle$ does not vanish.

Condition (b) leads to so-called *selection rules* which govern allowed and forbidden transitions. For example, in a central force problem in which the kets $|nlm_l m_s\rangle$ characterize the system, transitions will occur only between states in which the matrix element, $\langle n'l'm_l'm_s'|\mathbf{r}|nlm_l m_s\rangle$, does not vanish.

---

‡ The dimensions of an atom are $d \sim 10^{-8}$ cm while the radiation of interest usually has $\lambda \sim 10^{-5}$ cm.

§ The effects of the magnetic field are negligible in this *electric dipole approximation*.

The dipole matrix element z_{ij} is

$$z_{ij} = \int \psi^*_{n'l'm_l'm_s'} \, r \cos \theta \psi_{nlm_lm_s} \, d\mathbf{r}$$

$$= \int R_{n'l'} r R_{nl} r^2 \, dr \int_{-1}^{1} du \, P_{l'm_l'}(u) u P_{lm_l}(u) \int_0^{2\pi} e^{-im_l'\phi} e^{im_l\phi} \, d\phi \langle m_s' | m_s \rangle.$$

$$(7\text{-}81)$$

The last two products demonstrate that

$$z_{ij} \propto \delta_{m_l m_l'} \delta_{m_s m_s'}$$

so that z_{ij} exists only for transitions in which $\Delta m_l = 0$ and $\Delta m_s = 0$.

The θ (or μ) integral may be evaluated using an identity for the associated Legendre functions

$$\mu P_{lm_l} = \left[\frac{l - m_l + 1}{2l + 1}\right] P_{l+1, m_l} + \left[\frac{l + m_l}{2l + 1}\right] P_{l-1, m_l}.$$

Substituting in (7-81) and using the orthogonality of the associated Legendre functions, we find

$$z_{ij} \propto \delta_{m_l m_l'} \delta_{m_s m_s'} [\delta_{l, l'+1} + \delta_{l, l'-1}];$$

z_{ij} vanishes unless $\Delta l = \pm 1$. There may be additional restrictions derived from the radial integral but these vary with the central force problem being considered.

It is left as an exercise to show that the x_{ij} and y_{ij} elements lead to the selection rules $\Delta l = \pm 1, \Delta m_l = \pm 1,$ and $\Delta m_s = 0$. Summarizing, we see that in the electric dipole approximation, systems characterized by the states $|nlm_lm_s\rangle$ undergo radiative transitions subject to the selection rules $\Delta l = \pm 1$, $\Delta m_l = 0, \pm 1,$ and $\Delta m_s = 0$. All other transitions are forbidden in first order. It is also possible to verify the selection rules $\Delta l = \pm 1$, $\Delta j = 0, \pm 1,$ and $\Delta m_j = 0, \pm 1$ for the $|nljm_j\rangle$ scheme. The rate in (7-80) is valid provided both the wavelength of the radiation is long (dipole approximation) and the radiation intensity is low (first-order perturbation).

We will now show, using a statistical argument, that (7-80) is actually incomplete. In fact the same argument provides the missing terms in the transition process. The argument is as follows:

Consider a collection of quantum systems, let us say atoms, each of which can be in only one of two states $|\varepsilon_i\rangle$ or $|\varepsilon_j\rangle$, with $\varepsilon_i > \varepsilon_j$. Let us assume that the atoms are enclosed in a box along with radiation at the resonant frequency $\omega = \omega_{ij}$. Furthermore, assume that initially the numbers in each state are equal, that is, $N_i = N_j$. From the principle of microscopic reversibility, we find that

$$\frac{dN_{i \to j}}{dt} = \frac{dN_{j \to i}}{dt}$$

so that the levels remain evenly populated. Since no *net* energy is absorbed or emitted by the atoms, the radiation state remains unaltered. Unfortunately, these results are inconsistent with experimental observations. We have seen in Chapter 1 that the frequency distribution of radiation in a cavity will tend toward a black-body spectral density determined solely by the temperature of the walls of the enclosure and not by the composition of the box or its contents.

It may also be verified that the atoms in the box reach an equilibrium distribution consistent with *Boltzmann's law*. The ratio of the occupation numbers of the levels must tend to

$$\frac{N_j}{N_i} = \frac{e^{-\varepsilon_j/kT}}{e^{-\varepsilon_i/kT}}. \tag{7-82}$$

Since the levels of lower energy tend to be more populated, there must be a process that favors transitions from higher to lower energy. Furthermore, since it must always lead to a distribution as in (7-82) regardless of the initial radiation state, we conclude that this process must be independent of $\rho_{\text{rad}}(\omega)$ and can depend only on the atomic system. Thus radiative transitions must occur spontaneously from higher to lower energy states even in the absence of radiation. If no transition can occur from a given state spontaneously, it must be the ground or lowest energy state. *The origin of this spontaneous emission cannot be explained nor can a rate be calculated using the quantum mechanics of matter alone.*

There are however two methods commonly used for obtaining this rate. The rigorous one recognizes the fact that a time-dependent radiative perturbation is an entity in its own right and is subject to the laws of quantum theory. Quantization of this radiation field leads directly to the concept of light quanta or photons. In addition, such quantization leads quite naturally to the spontaneous as well as the induced rates. The entire system is therefore composed of the matter plus the radiation with energy being exchanged in the form of photons. We shall defer a detailed discussion of this approach to Chapter 12.

For our present purposes it is possible to deduce the spontaneous rate using a statistical method. While this approach appears to be lacking in rigor, it is nevertheless valid and leads to the same results as the method discussed above.

XV Einstein's Approach to Spontaneous Emission—
Detailed Balancing

Following Einstein's approach, it is convenient to write the induced (absorption and emission) rate in (7-80) as

$$R_{ij}^{(ind)} = B_{ij}\,\rho_{rad}(\omega_{ij})$$

where

$$B_{ij} = \frac{4\pi^2}{3}\frac{e^2}{\hbar^2}\,|\langle\varepsilon_i|\mathbf{r}|\varepsilon_j\rangle|^2 \qquad (7\text{-}83)$$

is called the Einstein " B " coefficient. Similarly we introduce the spontaneous rate for transitions $|\varepsilon_j\rangle$ to $|\varepsilon_i\rangle$ as

$$R_{ij}^{(spont)} = A_{ij} \qquad (\varepsilon_j > \varepsilon_i) \qquad (7\text{-}84)$$

where A_{ij} is the " A " coefficient. The number of atoms (per unit time) making the transitions from $|\varepsilon_j\rangle$ to $|\varepsilon_i\rangle$, where $\varepsilon_j > \varepsilon_i$, is

$$\frac{dN_{j\to i}}{dt} = N_j[R_{ij}^{(ind)} + R_{ij}^{(spont)}]$$

$$= N_j[B_{ij}\,\rho_{rad}(\omega_{ji}) + A_{ij}]. \qquad (7\text{-}85)$$

Similarly, the number going from lower to higher energy is $(\varepsilon_j > \varepsilon_i)$

$$\frac{dN_{i\to j}}{dt} = N_i R_{ij}^{(ind)} = N_i B_{ij}\,\rho_{rad}(\omega_{ji}). \qquad (7\text{-}86)$$

Once thermal equilibrium is reached, we have

$$\frac{dN_{i\to j}}{dt} = \frac{dN_{j\to i}}{dt}$$

or

$$N_j[B_{ij}\,\rho_{rad}(\omega_{ji}) + A_{ij}] = N_i B_{ij}\,\rho_{rad}(\omega_{ji}). \qquad (7\text{-}87)$$

Furthermore, in equilibrium the population ratio is given by Boltzmann's law

$$\frac{N_j}{N_i} = \frac{e^{-\varepsilon_j/kT}}{e^{-\varepsilon_i/kT}} = e^{-\hbar\omega_{ji}/kT} \qquad \left(\omega_{ji} = \frac{\varepsilon_j - \varepsilon_i}{\hbar}\right). \qquad (7\text{-}88)$$

Combining (7-87) and (7-88) and solving for ρ_{rad}, we obtain

$$\rho_{rad}(\omega_{ji}) = \frac{A_{ij}}{B_{ij}}\frac{1}{e^{\hbar\omega_{ji}/kT} - 1}. \qquad (7\text{-}89)$$

Finally we expect the equilibrium spectral density to conform to Planck's law

$$\rho(\omega, T) = \frac{\hbar}{\pi^2 c^3} \frac{\omega^3}{e^{+\hbar\omega/kT} - 1} . \tag{7-90}$$

Comparing (7-89) with (7-90) we observe that

$$\frac{A_{ij}}{B_{ij}} = \frac{\hbar\omega_{ij}^3}{\pi^2 c^3}$$

so that

$$A_{ij} = \frac{\hbar\omega_{ji}^3}{\pi^2 c^3} B_{ij} . \tag{7-91}$$

Note that statistical detailed balancing leads to a relationship between the A and B coefficients, but gives no information regarding their individual structures. Since B_{ij} has already been obtained in (7-83), we find the spontaneous rate as

$$R_{ij}^{(\text{spont})} = A_{ij} = \frac{\hbar\omega_{ji}^3}{\pi^2 c^3} \frac{4\pi^2}{3} \frac{e^2}{\hbar^2} |\langle \varepsilon_i | \mathbf{r} | \varepsilon_j \rangle|^2$$

$$= \frac{4\omega_{ji}^3 e^2}{3\hbar c^3} |\langle \varepsilon_i | \mathbf{r} | \varepsilon_j \rangle|^2 .$$

Summarizing, we observe that perturbation theory when applied to a harmonic perturbation leads to incomplete transition rates. This is because we have assumed that it is possible to "control" the time-dependence of a perturbation in the neighborhood of a quantum system. In classical mechanics, we rarely question the existence of such time-dependent constraints. In quantum theory, the perturbation itself is subject to restrictions imposed by the uncertainty principle. While statistical arguments may be used to deduce spontaneous rates, a rigorous approach requires quantization of the perturbing field.

XVI The Variational (Rayleigh–Ritz) Method

The approximation methods discussed above require that the perturbation constant λ be small. Furthermore, as $\lambda \to 0$ the first-order terms are usually adequate in approximating the energies as well as the eigenstates.

We shall now consider a method of approximation which is independent of the value of λ and applies even in the case where the perturbation is large.

However, the accuracy of this method is difficult to determine. Furthermore the method is usually more useful in estimating the energy eigenvalues rather than the eigenfunctions.

The variational method is particularly useful in estimating the ground-state energy of a Hamiltonian when this level is part of a discrete spectrum. We shall require the following lemma.

Lemma *Let $|\alpha\rangle$ be an arbitrary normalized ket and ε_1 the ground-state energy of $\mathscr{H}$. Then the following inequality always holds:*

$$\langle \alpha | \mathscr{H} | \alpha \rangle \geqslant \varepsilon_1. \tag{7-92}$$

Proof Expanding $|\alpha\rangle$ in the orthonormal eigenbasis of $\mathscr{H}$, that is, $|\alpha\rangle = \sum_i a_i |\varepsilon_i\rangle$, the left side of (7-92) becomes

$$\left\{ \sum_i^\infty \langle \varepsilon_i | a_i^* \right\} \mathscr{H} \left\{ \sum_j^\infty a_j | \varepsilon_j \rangle \right\} = \sum_{ij}^\infty a_i^* a_j \varepsilon_j \langle \varepsilon_i | \varepsilon_j \rangle$$

$$= \sum_{ij}^\infty a_i^* a_j \varepsilon_j \delta_{ij} = \sum_{i=1}^\infty \varepsilon_i |a_i|^2.$$

Since ε_1 is the ground-state energy, we have $\varepsilon_1 \leqslant \varepsilon_i$ and

$$\langle \alpha | \mathscr{H} | \alpha \rangle = \sum_{i=1}^\infty \varepsilon_i |a_i|^2 \geqslant \varepsilon_1 \sum_{i=1}^\infty |a_i|^2.$$

The ket $|\alpha\rangle$ is assumed normalized; thus the sum on the right is unity and the lemma is established.

In wave mechanics, the inequality takes the form

$$I = \int \psi_\alpha^* \mathscr{H} \psi_\alpha \, d\mathbf{r} \geqslant \varepsilon_1 \tag{7-93}$$

where

$$\int \psi_\alpha^* \psi_\alpha \, d\mathbf{r} = 1.$$

Note that the integral on the left represents an *upper bound* to the exact ground-state energy of $\mathscr{H}$ regardless of ψ_α.

We apply the variational method by choosing a "reasonable" trial function which depends on a set of N parameters, $\xi_1, \ldots, \xi_N$,

$$\psi_{\text{trial}} = \psi_{\text{trial}}(\mathbf{r}, \xi_1, \xi_2, \ldots, \xi_N).$$

Using the assumed form of ψ in (7-93), the matrix element $\langle \alpha | \mathscr{H} | \alpha \rangle$ becomes

$$\int \psi_{\text{trial}}^* \mathscr{H} \psi_{\text{trial}} \, d\mathbf{r} = I(\xi_1, \xi_2, \ldots, \xi_N). \tag{7-94}$$

From the lemma above it follows that $I \geqslant \varepsilon_1$ regardless of the numerical values of the ξ's. We may therefore find a *least upper bound* to the exact ground-state energy ε_1 by minimizing I. This is accomplished by finding the solutions to the N equations

$$\frac{\partial I}{\partial \xi}\bigg|_{\xi_i = \xi_{i_0}} = 0 \qquad (i = 1, \ldots, N).$$

The variational energy becomes

$$\varepsilon_{\text{var}} = I_{\min}(\xi_{1_0}, \xi_{2_0}, \ldots, \xi_{N_0}) \geqslant \varepsilon_1. \tag{7-95}$$

The method is repeated with different trial functions. The lowest variational energy is retained and represents the best approximation to ε_1. In practice, better results (lower energies) are obtained with trial functions having many parameters, since then ψ_{trial} will have greater flexibility with respect to variations.

If the trial function is such that it can be matched perfectly to the exact ground-state function by a variation of the parameters, then the minimum value of the variational integral will correspond to the exact ground-state energy. In fact, if $\psi_{\text{trial}} = \psi_1$, then

$$\langle \varepsilon_1 | \mathcal{H} | \varepsilon_1 \rangle = \int \psi_1^* \mathcal{H} \psi_1 \, d\mathbf{r} = \varepsilon_1 \tag{7-96}$$

Changing the form of the trial function to give a lower value of $I_{\min}$ and thus a value closer to that of ε_1 may actually result in a poorer approximation for ψ_1. Since ψ_1 is usually not known, it is difficult to determine the accuracy of $\psi_{\text{var}}(\mathbf{r}, \xi_{1_0}, \ldots, \xi_{N_0})$.

We shall illustrate the method in a calculation involving the ground state of hydrogen.‡ In this case, of course, we do know the exact solution to be

$$\psi_1 = \frac{1}{\sqrt{\pi}} \frac{1}{a^{3/2}} e^{-r/a} \quad \text{and} \quad \varepsilon_1 = -e^2/2a \quad (a = \hbar^2/me^2). \tag{7-97}$$

It will therefore be possible to evaluate the accuracy of the variational method for various trial functions.

If we use a normalized trial function of the form

$$\psi_{\text{trial}}(r, \xi) = \frac{1}{\sqrt{\pi}} \left(\frac{\xi}{a}\right)^{3/2} e^{-\xi r/a} \tag{7-98}$$

the variational integral takes the form

$$I(\xi) = \int \psi_{\text{trial}}^* \left(\frac{-\hbar^2}{2m} \nabla^2 - \frac{e^2}{r}\right) \psi_{\text{trial}} \, d\mathbf{r} = \frac{-e^2}{2a} (\xi^2 - 2\xi). \tag{7-99}$$

---

‡ See, for example, A. Messiah, "Quantum Mechanics," Volume II, p. 767. Wiley, New York, 1962.

This integral is a minimum when

$$\frac{\partial I}{\partial \xi}\bigg|_{\xi=\xi_0} = \frac{-e^2}{2a}(2\xi_0 - 2) = 0 \qquad \text{or} \qquad \xi_0 = 1.$$

The variational energy is

$$I_{min} = \frac{-e^2}{2a} = \varepsilon_1$$

which is identical to (7-97). Here is a case where the trial function is consistent with the exact eigenfunction. The variational method leads to a minimum value of energy when $\xi_0 = 1$. The variational function and energy represent the exact solution.

Had we chosen a normalized (one-parameter) trial function of the form

$$\psi_{trial}(r, \xi) = \left(\frac{\xi^5}{3\pi a^5}\right)^{1/2} r\, e^{-\xi r/a} \tag{7-100}$$

the variational integral analogous to (7-99) would have been

$$I_{var}(\xi) = \frac{-e^2}{2a}\left(\xi - \frac{\xi^2}{3}\right).$$

The minimum value occurs at $\xi_0 = \frac{3}{2}$, where I takes the value

$$\varepsilon_{var} = I_{min} = \frac{-e^2}{2a}\left(\frac{3}{2} - \frac{1}{3}\left(\frac{3}{2}\right)^2\right) = \frac{-3}{4}\frac{e^2}{2a} = 0.75\varepsilon_1 \tag{7-101}$$

generating a 25 percent error. Since ε_1 is negative, we find $\varepsilon_{var} > \varepsilon_1$ as required by the lemma above. The variational function in (7-100) becomes

$$\psi_{var}(r, \zeta) = \left[\frac{(\frac{3}{2})^5}{3\pi a^5}\right]^{1/2} r\, e^{-3r/2a}.$$

Finally, we consider the trial function

$$\psi_{trial}(r, \zeta) = \frac{1}{\pi a^{3/2}}\sqrt{\xi}\,\frac{1}{[\xi^2 + (r/a)^2]}. \tag{7-102}$$

The variational integral in this case gives

$$I_{var}(\xi) = \frac{-e^2}{2a}\left[\frac{8\xi - \pi}{2\pi\xi^2}\right]; \tag{7-103}$$

differentiation shows that $\xi_0 = \pi/4$ with

$$I_{min} = \frac{8\varepsilon_1}{\pi^2} \simeq 0.8\varepsilon_1. \tag{7-104}$$

This amounts to a 20 percent error. It is interesting that the apparently poorer trial function (7-102) actually leads to a lower and therefore a more accurate energy than the function in (7-100).

The variational method is particularly useful when accurate ground state energies are required. The accuracy of the corresponding trial function is difficult to estimate, and its reliability may rightly be questioned. With some modifications the method may also be applied to excited states, but we shall not do so here. We shall have occasion to use the variational method in our discussions of multielectron atoms (Chapter 10).

Time and space do not permit a description of all the methods of approximation. Among them are the numerical methods which have become more common in the age of the computer. An important analytical method omitted in our discussion is the "phase integral" or Wentzel–Kramers–Brillouin (WKB) approximation. A brief discussion of this method will be found in Appendix A.

Suggested Reading

Bethe, H. A., and Salpeter, E., "Quantum Mechanics of One and Two Electron Atoms." Springer-Verlag, Berlin, 1957.
Bohm, D., "Quantum Theory." Prentice-Hall, Englewood Cliffs, New Jersey, 1951.
Borowitz, S., "Fundamentals of Quantum Mechanics," Benjamin, New York, 1967.
Eisberg, R. M., "Fundamentals of Modern Physics." Wiley, New York, 1961.
Merzbacher, E., "Quantum Mechanics." 2nd ed. Wiley, New York, 1970.
Messiah, A. "Quantum Mechanics," Volume II. Wiley, New York, 1962.
Saxon, D., "Elementary Quantum Mechanics." Holden-Day, San Francisco, 1964.

Problems

7-1.   Find the ground-state energy, corrected to first order, for a particle in a perturbed one-dimensional well (Figure 7-6)

$$V = \frac{2\lambda}{a} x \qquad \text{for} \quad 0 \leqslant x \leqslant \tfrac{1}{2}a$$

$$V = 2\lambda\left(1 - \frac{x}{a}\right) \qquad \text{for} \quad \tfrac{1}{2}a < x \leqslant a$$

$$V = \infty \qquad \text{otherwise.}$$

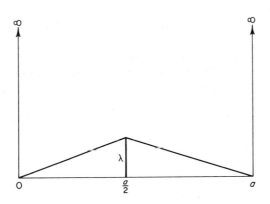

Figure 7-6

7-2. (a) Calculate the approximate energy eigenvalues (to second order) for a particle in a potential $V = \frac{1}{2}m\omega^2 x^2 + \lambda x$.

(b) Do this problem exactly by completing the square in the potential and transforming the independent variable. Compare with the result above.

7-3. Find the approximate ground state energy of a particle in a potential $V = \frac{1}{2}m\omega x^2 + \lambda x^3$.

7-4. Use (7-24) and (7-25) to derive a formula for the *third*-order corrections to the energy eigenvalues.

7-5. (a) Show that the electric dipole moment of hydrogen in the unperturbed ground state $|100\rangle$ is zero.

(b) Show that the dipole moment associated with the first excited states $|2^{\pm}\rangle$ [as given by (7-45)] is $p = \mp 3ea$ oriented along the z axis. Is this a permanent or an induced moment? Explain!

7-6. Set up the equation for finding the Stark shift in the first few energy levels of hydrogen for a *weak* **E** field. (Hint: Treat the fine-structure terms as the dominant perturbation as was done in the anomalous Zeeman effect. Next, assume ψ_{nljm_j} to be the unperturbed set with the unperturbed energy being a function of n and j only. Finally, construct and diagonalize the perturbation matrix associated with the **E** field with respect to states of different l and m_j.)

7-7. (a) Consider a spinless particle in a two-dimensional square well

$$V = 0 \qquad (0 \leqslant x, y \leqslant a)$$

$$V = \infty \qquad \text{(otherwise)}.$$

Find the energies and eigenfunctions for the ground and first excited states that evolve from a separation of variables in Cartesian coordinates.

(b) Next, assume that a perturbation is added of the form

$$V = \lambda xy \qquad \text{for} \quad 0 \leqslant x, y \leqslant a.$$

Find the unperturbed eigenfunctions and the first-order energy shifts for the ground and first excited states.

7-8. Verify the relation $[\hat{\mathbf{L}} \cdot \hat{\mathbf{S}}, \hat{J}_z] = 0$.

7-9. Show directly, using the results in (6-24), that (7-56) is correct.

7-10. Show by performing the integration and simplifying that (7-68) follows from the preceding equation.

7-11. On a level diagram for hydrogen (including the effects of fine structure) show the first few transitions permitted in the electric dipole approximation.

7-12. (a) Find the normalization constant for the trial function

$$\psi(\xi, r) = Ne^{-\xi(r/a)^2} \qquad (a = \hbar^2/me^2)$$

where ξ is a variational parameter.

(b) Use this function to find an upper bound to the ground-state energy of hydrogen.

(c) How does this energy compare with the exact value? (Note that this trial function is essentially the ground-state eigenfunction of the harmonic oscillator!)

7-13. Show that the matrix elements $\langle nlm_l m_s | \hat{x} | nl'm_l'm_s \rangle$ and

$$\langle nlm_l m_s | \hat{y} | nl'm_l'm_s' \rangle$$

are nonzero only if $\Delta l = l - l' = \pm 1$, $\Delta m_l = m_l - m_l' = \pm 1$, and $\Delta m_s = m_s' - m_s' = 0$.

7-14. Consider a hydrogenic electron (neglecting spin) in a uniform electric field directed along the z axis.

(a) Using (7-41), write the energy–eigenvalue equation for the electron in spherical polar coordinates. Set $\lambda = eE$.

(b) Verify that the function

$$\psi = \left(\frac{1}{\pi a^3}\right)^{1/2} e^{-r/a}\left[1 - \frac{\lambda \cos \theta}{e^2}\left(ar + \frac{r^2}{2}\right)\right]$$

(where $a = \hbar^2/me^2$) represents a *first*-order approximation to the ground-state eigenfunction of the hydrogen atom by directly substituting this result into the equation derived in part (a). Find the eigenvalue correct to first order. (Hint: Neglect terms of λ^2 and higher.)

(c)   Using (7-25), show that the *second*-order correction to the energy is

$$\lambda^2 \varepsilon^{(2)} = -\frac{9}{4} a^3 E^2.$$

(d)   The energy associated with an induced electric dipole moment of a system is expressed as

$$\varepsilon = -\frac{1}{2} \alpha E^2$$

where α is called the *polarizability* of the system. Show that the polarizability of the hydrogen atom in its ground state is

$$\alpha = \frac{9}{2} a^3.$$

8 | The Theory of Scattering

Throughout the development of modern physics, physicists have sought to understand the fundamental interactions of nature. Scattering experiments have provided much of our knowledge regarding these interactions. In this chapter we shall consider the quantum mechanics of the scattering process. We begin by briefly reviewing the classical theory of scattering.

I   The Classical Theory of Scattering

A typical scattering experiment consists of observing the initial and final trajectories of particles moving under the influence of a force—often a central force. In atomic and nuclear physics, a beam of particles is incident on a center of force and is subsequently scattered. Each particle of the beam is deflected through an angle which depends, in part, on the initial energy and momentum as well as on the nature of the center of force. We shall restrict ourselves to problems in which the projectile's energy is conserved in the scattering process (elastic scattering).

A central force potential is of the form $V(\mathbf{r}) = V(r)$; hence it is convenient to use spherical polar coordinates, taking the center of force to be at the origin and choosing the direction of incidence to coincide with the $\theta = 0$ direction, usually taken as the z axis. The direction of scattering is characterized by the polar and azimuthal angles θ and ϕ. The incident beam is characterized by its energy ε and its flux density or intensity $\mathbf{J}$. The number of particles scattered (per unit time) into a solid angle $d\Omega$ about a set of angles θ and ϕ is proportional to both J and $d\Omega$, and we may write $dN \propto J \, d\Omega$. The proportionality constant is a function of θ and ϕ and is called the *differential cross section*, namely,

$$dN = \sigma(\theta, \phi) J \, d\Omega$$

or

$$\sigma(\theta, \phi) = \frac{1}{J} \frac{dN}{d\Omega}. \tag{8-1}$$

For central force scatterers, symmetry suggests that the cross section is independent of the azimuthal angle. The scattering is therefore symmetric with respect to the direction of incidence (z axis).

It is convenient to calculate the total number of particles scattered (per unit time). This is accomplished by integrating (8-1) over a complete solid angle

$$N = J \int_{4\pi} \sigma(\theta, \phi) \, d\Omega = J\sigma_{\mathrm{T}}. \tag{8-2}$$

The integral generates the *total cross section*, which measures the effective scattering area as seen by the beam.

We shall restrict our discussion, in what follows, to central force scatterers. In order to calculate the cross section $\sigma(\theta)$, we must first know the scattering angle for each and every particle in the beam. The trajectory of a particle is depicted in Figure 8-1.

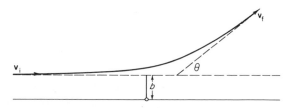

Figure 8-1 A typical scattering trajectory illustrating the initial and final velocities, the impact parameter, and the scattering angle.

For a given energy, the scattering angle is dependent on the impact parameter b. This parameter is defined as the perpendicular distance from the force center to the line of action of the initial velocity v_i (Figure 8-1). The equation $b = b(\theta)$, relating the impact parameter to the scattering angle, must first be calculated from the particle's trajectory before any predictions about the cross section can be made. For most scattering problems, the scattering angle decreases quite rapidly with increasing b.

The number of particles scattered into a solid angle between θ and $\theta + d\theta$ must equal the number situated in the incident beam in a ring about the z axis of radius b and thickness db (Figure 8-2). This number is

$$dN = J \, dA = J2\pi b \, db = J2\pi b \, \frac{db}{d\Omega} \, d\Omega.$$

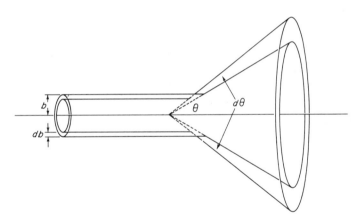

Figure 8-2 The scattering of particles from a differential ring into a differential solid angle.

Using (8-1), the cross section follows as

$$\sigma(\theta) = 2\pi b \, \frac{db}{d\Omega}.$$

This may be simplified, using‡

$$d\Omega = 2\pi \sin \theta \, d\theta$$

‡ Since b increases with decreasing θ, $db/d\theta$ is negative; a minus sign has therefore been introduced in (8-3) in order to keep the cross section positive.

to

$$\sigma(\theta) = \frac{-b(\theta)}{\sin \theta} \frac{db(\theta)}{d\theta}.$$ (8-3)

Once $b = b(\theta)$ is known, the cross section may be calculated directly.

We shall illustrate the technique in the case of a repulsive Coulomb force (Rutherford scattering). We shall assume that the projectile and target have respective charges $Z'e$ and Ze. The repulsive potential is

$$V = \frac{ZZ'e^2}{r}.$$

The trajectory of the projectile is always a hyperbola (Figure 8-3) with the

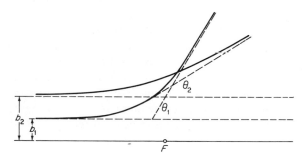

Figure 8-3 Two hyperbolic trajectories for different impact parameters observed in Coulomb scattering.

force center (target) at the far focus. We state without proof that the relation between b and θ is‡

$$b = \frac{ZZ'e^2}{2\varepsilon} \cot\left(\frac{\theta}{2}\right)$$

where ε is the energy of the projectile.

Using (8-3) the classical Rutherford cross section follows as

$$\sigma(\theta) = \left(\frac{ZZ'e^2}{4\varepsilon}\right)^2 \sin^{-4}\frac{\theta}{2}.$$ (8-4)

_____

‡ See, for example, J. B. Marion, "Classical Dynamics of Particles and Systems," 2nd ed. Academic Press, New York, 1970.

Upon integrating, we find that the total Rutherford cross section, in fact, diverges, namely,

$$\sigma_T = \left(\frac{ZZ'e^2}{4\varepsilon}\right)^2 \int \sin^{-4}\frac{\theta}{2} \, d\Omega = \left(\frac{ZZ'e^2}{4\varepsilon}\right)^2 2\pi \int_0^\pi \sin^{-4}\frac{\theta}{2} (\sin\theta \, d\theta) \to \infty.$$

The integral diverges at the lower limit, $\theta \to 0$, that is, $b \to \infty$. The divergence is related to the fact that the $1/r$ potential is very long ranged; consequently, even those particles with impact parameters tending toward infinity are scattered in such a way as to make σ_T diverge. For shorter-ranged potentials, σ_T will generally be finite.

The quantum-mechanical approach to scattering commonly takes one of two forms. The first, stationary scattering, obtains the cross section from the continuous eigenstates of the scattering potential. This method is similar to the one used in the one-dimensional case in Chapter 4 where the analogous scattering parameter is the reflection coefficient $\mathcal{R}$.

A second approach, dynamical scattering, analyzes the evolution of a momentum eigenstate in the presence of a scattering perturbation. As we shall show, the transition rate to other momentum eigenstates is directly related to the scattering cross section. While both approaches lead basically to the same cross sections, the dynamical method is regarded as a more general one. For one thing, it is independent of wave mechanics. We shall, however, begin our discussion with stationary scattering methods.

II The Stationary (Steady–State) Quantum Theory of Scattering

In the classical theory, we were able to establish a relationship between the unbound trajectories of a projectile and the corresponding scattering cross section. An analogous relationship exists between the continuous energy eigenstates of a potential and the corresponding quantum cross section. We shall therefore be concerned with what are called the scattering states of the Schroedinger equation

$$\left\{ -\frac{\hbar^2 \nabla^2}{2m} + V(\mathbf{r}) \right\} \psi = \varepsilon\psi. \tag{8-5}$$

It is assumed that $V(\mathbf{r})$ tends to zero as $|\mathbf{r}|$ becomes large. Since scattering eigenfunctions represent unbound states, we cannot require that they vanish at infinity. We can, however, impose certain asymptotic requirements consistent with experimental conditions.

In scattering experiments, the region of interest is located far from the target. In this asymptotic region, we expect an incident flux due to a plane wave traveling along the z axis, that is, e^{ikz}, and a scattered flux associated with an outgoing spherical wave, that is, $f(\theta, \phi)e^{ikr}/r$. These solutions correspond to free particles of energy $\varepsilon = \hbar^2 k^2/2m$. We shall therefore impose the asymptotic requirement

$$\psi \underset{r\to\infty}{\sim} e^{ikz} + f(\theta, \phi) \frac{e^{ikr}}{r} \qquad \left(k = \left(\frac{2m\varepsilon}{\hbar^2}\right)^{1/2}\right) \qquad (8\text{-}6)$$

on the solutions to the Schroedinger equation (8-5). It is not difficult to demonstrate that a solution with this asymptotic form exists provided the potential is not very long ranged. We shall show below that the Coulomb potential with its long range does not have a solution with the required asymptotic form. However it does have one which is close enough to be used in obtaining the Coulomb cross section.

While there are many solutions to (8-5) for fixed ε, we shall be concerned with finding the one that behaves asymptotically as in (8-6). This is not as simple as it may seem. However, once the desired ψ has been found and its asymptotic form generated, $f(\theta, \phi)$ can be directly related to the cross section. To do this, the incident flux is obtained from the incident wave e^{ikz} using the flux equation (3-96),

$$J_{\text{inc}} = -\frac{i\hbar}{2m}\left[(e^{ikz})^* \frac{\partial}{\partial z} e^{ikz} - e^{ikz}\frac{\partial}{\partial z}(e^{ikz})^*\right] = \frac{\hbar k}{m}.$$

Similarly, the scattered (radial) flux is obtained from $f(\theta, \phi)e^{ikr}/r$ using

$$J_{\text{scat}} = -\frac{i\hbar}{2m}|f(\theta, \phi)|^2\left[\left(\frac{e^{ikr}}{r}\right)^* \frac{\partial}{\partial r}\frac{e^{ikr}}{r} - \frac{e^{ikr}}{r}\frac{\partial}{\partial r}\left(\frac{e^{ikr}}{r}\right)^*\right] = \frac{\hbar k}{m}\frac{|f(\theta,\phi)|^2}{r^2}.$$

The number of particles scattered per solid angle is

$$\frac{dN}{d\Omega} = J_{\text{scat}}\,r^2 = \frac{\hbar k}{m}|f(\theta, \phi)|^2.$$

Dividing by $J_{\text{inc}} = \hbar k/m$, we find the cross section to be

$$\sigma(\theta, \phi) = \frac{1}{J_{\text{inc}}}\frac{dN}{d\Omega} = |f(\theta, \phi)|^2. \qquad (8\text{-}7)$$

This relation is exact. The angular modulation factor $f(\theta, \phi)$ is called the *scattering amplitude*. We have made a rough analogy between the classical and quantum approaches to scattering in Table 8-1.

Table 8-1

*A Comparison between the Methods of Classical and Quantum
Scattering Calculations*

Classical theory	Quantum theory
(a) Solve Newton's laws for the (unbound) trajectory of the particle for a given energy, impact parameter, and scattering potential.	(a) Solve Schroedinger's equation for the (continuous) eigenfunction with the desired asymptotic form for a given energy and scattering potential.
(b) Relate the scattering angle θ to the impact parameter b.	(b) From the asymptotic form of the solution in (a), find the scattering amplitude $f(\theta, \phi)$.
(c) Calculate the cross section using $$\sigma(\theta) = \frac{-b}{\sin\theta}\frac{db}{d\theta}.$$	(c) Calculate the cross section using $$\sigma(\theta, \phi) = |f(\theta, \phi)|^2.$$

We shall apply the theory above to the quantum-mechanical analysis of Rutherford scattering.

III   Rutherford Scattering (Quantum Case)

Like its classical analog, the quantum Rutherford problem can be solved exactly. We shall direct our attention to the appropriate solution of the Schroedinger equation

$$\left(-\frac{\hbar^2}{2m}\nabla^2 + \frac{K}{r}\right)\psi = \varepsilon\psi \tag{8-8}$$

or

$$\left(-\nabla^2 + \frac{2mK}{\hbar^2 r}\right)\psi = k^2\psi$$

where

$$K = ZZ'e^2 \quad \text{and} \quad k = \left(\frac{2m\varepsilon}{\hbar^2}\right)^{1/2}$$

This equation is, of course, separable in spherical coordinates. However, the resulting solutions do not have the required asymptotic form. We could, of course, take linear combinations of these degenerate spherical eigenfunctions to construct the proper asymptotic form. However, it is simpler to separate (8-8) in parabolic coordinates. The parabolic eigenfunctions naturally conform to the proper asymptotic requirement.

The transformation to parabolic coordinates is‡

$$\xi = r(1 + \cos\theta) = r + z$$
$$\eta = r(1 - \cos\theta) = r - z$$
$$\phi = \phi = \tan^{-1}\frac{y}{x}.$$

The metric coefficients, Laplacian, and volume element in parabolic coordinates can be written respectively as

$$h_\xi = \left[\frac{(\eta + \xi)}{4\xi}\right]^{1/2}, \qquad h_\eta = \left[\frac{(\eta + \xi)}{4\eta}\right]^{1/2}, \qquad h_\phi = (\xi\eta)^{1/2}$$

$$\nabla^2 = \left\{\frac{4}{\xi + \eta}\left[\frac{\partial}{\partial\xi}\left(\xi\frac{\partial}{\partial\xi}\right) + \frac{\partial}{\partial\eta}\left(\eta\frac{\partial}{\partial\eta}\right)\right] + \frac{1}{\xi\eta}\frac{\partial^2}{\partial\phi^2}\right\}$$

and

$$d\mathbf{r} = \tfrac{1}{4}(\xi + \eta)\,d\xi\,d\eta\,d\phi.$$

Using a solution of the form

$$\psi(\xi, \eta, \phi) = F(\xi)G(\eta)\Phi(\phi)$$

(8-8) separates into the following ordinary differential equations:

$$\frac{d^2\Phi}{d\phi^2} = -m_l^2\,\Phi, \qquad \frac{d}{d\xi}\left(\xi\frac{dF}{d\xi}\right) + \left(-\frac{m_l^2}{4\xi} + \frac{k^2}{4}\xi - \beta\right)F = 0$$

and   (8-9)

$$\frac{d}{d\eta}\left(\eta\frac{dG}{d\eta}\right) + \left(-\frac{m_l^2}{4\eta} + \frac{k^2\eta}{4} + \beta - \frac{Km}{\hbar^2}\right)G = 0$$

where m_l and β are separation constants.

The products of the solutions to these differential equations lead to solutions of Schroedinger's equation (8-8) regardless of the values of m_l and β. Our task is to choose β and m_l so that ψ will have the proper asymptotic form:

$$\psi \underset{r\to\infty}{\sim} e^{ikz} + f(\theta)\frac{e^{ikr}}{r}.$$

The absence of the variable ϕ in the scattering amplitude is a consequence of the central character of the Coulomb potential.

---

‡ The definition and notation of parabolic coordinates vary somewhat among authors.

The solution to the ϕ equation in (8-9) is $\Phi = e^{im_l\phi}$. However, since the asymptotic form is independent of azimuth, the exact solution cannot contain ϕ. We therefore choose the separation constant $m_l = 0$ so that $\Phi \equiv 1$. The solution now takes the form

$$\psi(\xi, \eta) = F(\xi)G(\eta)$$

where F and G are the solutions to

$$\frac{d}{d\xi}\left(\xi \frac{dF}{d\xi}\right) + \left(\frac{k^2}{4}\xi - \beta\right)F = 0 \tag{8-10a}$$

and

$$\frac{d}{d\eta}\left(\eta \frac{dG}{d\eta}\right) + \left(\frac{k^2}{4}\eta + \beta - \frac{Km}{\hbar^2}\right)G = 0 \qquad \left(k = \left(\frac{2m\varepsilon}{\hbar^2}\right)^{1/2}\right). \tag{8-10b}$$

The required asymptotic form can be expressed in parabolic coordinates as‡

$$\psi(\xi, \eta)\underset{r-z=\eta\to\infty}{\sim} e^{ikz} + f\frac{e^{ikr}}{r} = e^{ikz}\left(1 + \frac{fe^{ik(r-z)}}{r}\right)$$

$$= e^{ik(\xi-\eta)/2}\left(1 + \frac{fe^{ik\eta}}{r}\right). \tag{8-11}$$

Equation (8-11) suggests that the exact solution must be of the form

$$\psi(\xi, \eta) = e^{ik(\xi-\eta)/2}W(\eta) = e^{ik\xi/2}e^{-ik\eta/2}W(\eta)$$

where

$$W(\eta)\underset{\eta\to\infty}{\sim}\left(1 + \frac{fe^{ik\eta}}{r}\right).$$

Setting $F(\xi) = e^{ik\xi/2}$ in (8-10a), we find that $\beta = \frac{1}{2}ik$. Using this choice for the separation constant β and setting $G(\eta) = e^{-ik\eta/2}W(\eta)$ in (8-10b), we obtain

$$\eta \frac{d^2 W(\eta)}{d\eta^2} + (1 - ik\eta)\frac{dW(\eta)}{d\eta} - \gamma W(\eta) = 0 \tag{8-12}$$

where $\gamma = Km/\hbar^2 k$.

Equation (8-12) will be recognized as the confluent hypergeometric equation already discussed in Chapter 5. The solution that is well-behaved at the origin ($\eta = 0$) is called the confluent series and is labeled $F(-i\gamma|1|ik\eta)$. Finally, our total solution may be written

$$\psi(\xi, \eta) = Ne^{ik(\xi-\eta)/2}F(-i\gamma|1|ik\eta); \tag{8-13}$$

---

‡ Since, by definition, the scattered beam is observed off the incident (z) axis, (that is, $r > z$), the asymptotic limit may be written $(r - z) \to \infty$ rather than $r \to \infty$.

for convenience, we shall choose the multiplicative constant‡

$$N = \frac{\Gamma(1 + i\gamma)}{e^{\pi\gamma/2}}.$$

Equation (8-13) is an *exact* solution to the Schroedinger equation which hopefully has the proper asymptotic form. Using somewhat advanced mathematical techniques, it is possible to show that the asymptotic form of (8-13) is

$$\psi \underset{\eta\to\infty}{\sim} \exp i[kz + \gamma \ln kr(1 - \cos\theta)] + f(\theta)\frac{\exp[i(kr - \gamma \ln 2kr)]}{r}$$

where

$$f(\theta) = \frac{-\Gamma(1 + i\gamma)}{\Gamma(1 - i\gamma)}\frac{\gamma}{k}\frac{\exp[-i\gamma \ln \sin^2\theta/2]}{2\sin^2\theta/2}. \tag{8-14}$$

It appears that with all our effort, (8-14) is as close as we can come to the proper asymptotic form. The logarithmic phase terms in the incident and scattered wave exponents represent distortions which are characteristic of the long-ranged Coulomb potential. These distortions vanish in asymptotic solutions for potentials of shorter range. Fortunately, the logarithmic terms do not present any difficulty since they tend to zero more rapidly than the linear terms in the exponent of (8-14). Their contribution to the asymptotic fluxes can be ignored, and we calculate the cross section using

$$\sigma(\theta) = |f(\theta)|^2 = \frac{\gamma^2}{4k^2 \sin^4\theta/2}\left|\frac{\Gamma(1 + i\gamma)}{\Gamma(1 - i\gamma)}\right|^2.$$

Since $|\Gamma(1 + i\gamma)| = |\Gamma(1 - i\gamma)|$, we have finally

$$\sigma(\theta) = \left(\frac{\gamma^2}{4k^2}\right)\sin^{-4}\frac{\theta}{2} = \left(\frac{K}{4\varepsilon}\right)^2 \sin^{-4}\frac{\theta}{2} = \left(\frac{ZZ'e^2}{4\varepsilon}\right)^2 \sin^{-4}\frac{\theta}{2}. \tag{8-15}$$

Comparing this result to (8-4), we observe that the classical and quantum Rutherford cross sections are identical. This must be regarded as coincidental. Quantum-mechanical cross sections for other "well-behaved" potentials usually contain $\hbar$, and therefore differ from their classical counterparts.

In practice, the Rutherford formula is rarely observed since the Coulomb potential is modified by auxiliary factors. In Rutherford's original experiment involving scattering of α particles from gold nuclei, deviations from the pure Coulomb potential occurred at $r \sim 0$ (nuclear structure) and $r \sim \infty$ (screening by the electrons of the gold atoms).

---

‡ The *gamma function* is defined by $\Gamma(z) = \int_0^\infty t^{z-1}e^{-t}\,dt.$

As we have just seen, finding a solution to the Schroedinger equation with the proper asymptotic form is no simple matter. For the more compli- cated scattering potentials associated with nuclear forces, alternative methods and possibly methods of approximation are necessary. We shall consider two methods commonly used in obtaining cross sections. The first method, *the Born expansion*, is applicable when the strength λ of the scatterer is much smaller than the incident energy. The scattering amplitude is then expressed as a perturbation series involving powers of λ. This method is the analog of the Rayleigh–Schroedinger method used for bound states.

The second method, *partial wave analysis*, leads to a sum of a different sort. It is useful provided the potential is short ranged. This method is quite independent of the strength of the scatterer. Both Born's approach and the method of partial waves are useful; the particular approach taken usually depends on the energy of the incident beam and on the nature of the scatter- ing potential, for example, its strength and range.

IV   A Perturbation Treatment of Stationary Scattering— The Born Series

We direct our attention to the Schroedinger equation for a scattering potential $\lambda \hat{V}$, where the parameter λ characterizes the strength of the scatterer. The equation to be solved is

$$\left(-\frac{\hbar^2 \nabla^2}{2m} + \lambda V \right) \psi = \varepsilon \psi \tag{8-16}$$

subject to the asymptotic condition

$$\psi \underset{r \to \infty}{\sim} e^{i\mathbf{k} \cdot \mathbf{r}} + f(\theta, \phi)\frac{e^{ikr}}{r} \qquad \left(|\mathbf{k}| = k = \left(\frac{2m\varepsilon}{\hbar^2} \right)^{1/2} \right).$$

The incident momentum vector $\mathbf{k}$ is along the z axis (that is, $e^{i\mathbf{k} \cdot \mathbf{r}} = e^{ikz}$). This asymptotic form is only possible for potentials falling off faster than $1/r$, and so we shall require that the potential satisfy

$$\lim_{r \to \infty} rV(r) \to 0.$$

The difficulty with solving the differential equation (8-16) is that for any given energy many degenerate solutions can exist. We seek only the one with the proper asymptotic form. To single this solution out, it becomes con- venient to transform (8-16) into an *integral equation* which has only the desired eigenfunction as its solution.

Transposing in (8-16) and setting

$$u(\mathbf{r}) = \frac{2m}{\hbar^2} \lambda V(\mathbf{r})\psi(\mathbf{r})$$

the Schroedinger equation becomes

$$(\nabla^2 + k^2)\psi = u. \tag{8-17}$$

Actually, the original equation in (8-16) is a homogeneous differential equation in that each term contains ψ. However, (8-17) will now be regarded as an inhomogeneous equation with $u(\mathbf{r})$ as the "source" or inhomogeneity. The general solution to (8-17) is composed of the general solution to the homogeneous equation $(\nabla^2 + k^2)\psi_1 = 0$ plus any particular solution ψ_s of the complete equation $(\nabla^2 + k^2)\psi_s = u$. The general solution is $\psi = \psi_1 + \psi_s$.

Since ψ_1 is to represent the case where $u = 0$ (or $V = 0$), it should represent the incident plane wave. We therefore set $\psi_1 = e^{i\mathbf{k}\cdot\mathbf{r}}$. We next obtain the particular solution and make it subject to the asymptotic condition

$$\psi_s \underset{r\to\infty}{\sim} f(\theta, \phi)\frac{e^{ikr}}{r} \qquad (= \text{outgoing spherical wave}). \tag{8-18}$$

In other words, we expect that the particular solution will represent the scattered wave. The desired solution takes the form

$$\psi = e^{i\mathbf{k}\cdot\mathbf{r}} + \psi_s \underset{r\to\infty}{\sim} e^{i\mathbf{k}\cdot\mathbf{r}} + f(\theta, \phi)\frac{e^{ikr}}{r}.$$

We have reduced the problem to finding a particular solution to (8-17) which asymptotically represents an outgoing spherical wave.

Suppose we could find a function $G(\mathbf{r}, \mathbf{r}')$ (Green's function) with the property

$$(\nabla^2 + k^2)G(\mathbf{r}, \mathbf{r}') = \delta(\mathbf{r} - \mathbf{r}'); \tag{8-19}$$

then the function

$$\psi_s(\mathbf{r}) = \int d\mathbf{r}' \, G(\mathbf{r}, \mathbf{r}')u(\mathbf{r}') \tag{8-20}$$

is automatically a solution to (8-17). Direct substitution verifies that

$$(\nabla^2 + k^2)\psi_s = \int d\mathbf{r}'(\nabla^2 + k^2)G(\mathbf{r}, \mathbf{r}')u(\mathbf{r}')$$

$$= \int d\mathbf{r}' \, \delta(\mathbf{r} - \mathbf{r}')u(\mathbf{r}') = u(\mathbf{r});$$

thus ψ_s is indeed a solution.

While there are many possible Green's functions with the property in
(8-19), we are interested in the one which guarantees that

$$\psi_s = \int d\mathbf{r}' G(\mathbf{r}, \mathbf{r}') u(\mathbf{r},) \underset{r \to \infty}{\sim} f(\theta, \phi) \frac{e^{ikr}}{r}.$$

It can be verified that the particular Green's function

$$G(\mathbf{r}, \mathbf{r}') = \frac{-1}{4\pi} \frac{e^{ik|\mathbf{r}-\mathbf{r}'|}}{|\mathbf{r} - \mathbf{r}'|} \tag{8-21}$$

has all the desired properties. It satisfies (8-19), as a straightforward calculation
(see Problem 8-6) shows. But its very structure for large $\mathbf{r}$ ($r \gg r'$) assures the
proper asymptotic form required by (8-18). Using (8-20) and (8-21), we finally
have *a fundamental integral equation for scattering*:

$$\psi(\mathbf{r}) = e^{i\mathbf{k}\cdot\mathbf{r}} - \frac{\lambda 2m}{4\pi\hbar^2} \int d\mathbf{r}' \frac{e^{ik|\mathbf{r}-\mathbf{r}'|}}{|\mathbf{r} - \mathbf{r}'|} \psi(\mathbf{r}') V(\mathbf{r}'). \tag{8-22}$$

While (8-22) satisfies (8-16), it is a solution only in a formal sense. Equation (8-22) is termed an integral equation‡ because the unknown function ψ
is situated inside the integral on the right. We have succeeded in transforming
the differential equation into an integral equation. The advantage of this
transformation is that while the differential equation has many solutions for
fixed ε, the integral equation has only the solution with the proper asymptotic
form. Unfortunately, solving (8-22) is usually much more difficult than its
differential counterpart.

One technique that may be used in solving (8-22) is called the method of
iteration. It leads to a series solution in powers of λ (called the Born series)
and is valid provided λ is small enough for the series to converge. The zeroth-
order solution ($\lambda = 0$) is obviously $\psi^{(0)} = e^{i\mathbf{k}\cdot\mathbf{r}}$ and represents a state of zero
scattering. Inserting this solution in the integral on the right of (8-22), we
generate the first-order solution $\psi^{(1)}$. The nth-order solution is obtained
using $\psi^{(n-1)}$ in a similar manner, namely,

$$\psi^{(n)}(\mathbf{r}) = e^{i\mathbf{k}\cdot\mathbf{r}} - \frac{2m\lambda}{4\pi\hbar^2} \int d\mathbf{r}' \frac{e^{ik|\mathbf{r}-\mathbf{r}'|}}{|\mathbf{r} - \mathbf{r}'|} \psi^{(n-1)}(\mathbf{r}') V(\mathbf{r}'). \tag{8-23}$$

The nth-order approximation leads to a power series in λ with λ^n as the highest
power. Once $\psi^{(n)}$ is found, it must have the asymptotic form

$$\psi^{(n)} \underset{r \to \infty}{\sim} e^{i\mathbf{k}\cdot\mathbf{r}} + f(\theta, \phi) \frac{e^{ikr}}{r}$$

---

‡ More precisely, it is a Fredholm equation of the second kind.

where $f(\theta, \phi) = \sum_{j=1}^{n} f^{(j)}(\theta, \phi)$. Each term in the Born series for $\psi^{(n)}$ contributes a term known as the *jth-order Born amplitude*. The cross section to nth order is given by

$$\sigma^{(n)} = \left| \sum_{j=1}^{n} f^{(j)} \right|^2 .$$

The exact cross section is obtained as $n \to \infty$, provided the series converges, and we can write

$$\sigma_{\text{exact}} = \left| \sum_{j=1}^{\infty} f^{(j)} \right|^2 . \tag{8-24}$$

V   The First Born Approximation

If λ is sufficiently small, we need only retain the terms obtained from a single iteration of (8-22), namely,

$$\psi^{(1)}(\mathbf{r}) = e^{i\mathbf{k} \cdot \mathbf{r}} - \frac{m\lambda}{2\pi\hbar^2} \int d\mathbf{r}' \, \frac{e^{ik|\mathbf{r}-\mathbf{r}'|}}{|\mathbf{r}-\mathbf{r}'|} \, V(\mathbf{r}')e^{i\mathbf{k}\cdot\mathbf{r}'} . \tag{8-25}$$

This solution is valid everywhere in space (for small λ). However, we are interested only in the scattering amplitude which is obtained from the asymptotic form. For large $\mathbf{r}$, we can make the expansion

$$\frac{e^{ik|\mathbf{r}-\mathbf{r}'|}}{|\mathbf{r}-\mathbf{r}'|} \underset{r \gg r'}{\sim} \frac{\exp[i(kr - k\mathbf{r}\cdot\mathbf{r}'/r)]}{r} = (e^{-i\mathbf{k}'\cdot\mathbf{r}'}) \frac{e^{ikr}}{r} . \tag{8-26}$$

The term $\mathbf{k}' = k\mathbf{r}/r$ is a vector of magnitude k oriented along the radial (scattered) direction. The asymptotic form of (8-25) becomes

$$\psi^{(1)}(\mathbf{r}) \underset{r \to \infty}{\sim} e^{i\mathbf{k}\cdot\mathbf{r}} - \frac{m\lambda}{2\pi\hbar^2} \left[\int d\mathbf{r}' e^{i(\mathbf{k}-\mathbf{k}')\cdot\mathbf{r}'} V(\mathbf{r}') \right] \frac{e^{ikr}}{r} .$$

We can make the identification

$$f^{(1)}(\theta, \phi) = \frac{-m\lambda}{2\pi\hbar^2} \int d\mathbf{r}' e^{i(\mathbf{k}-\mathbf{k}')\cdot\mathbf{r}'} V(\mathbf{r}') \tag{8-27}$$

where $f^{(1)}(\theta, \phi)$ is called the *first Born amplitude*. The integral can be written as

$$V(\mathbf{K}) = \int d\mathbf{r}' e^{i\mathbf{K}\cdot\mathbf{r}'} V(\mathbf{r}') \tag{8-28}$$

where $\mathbf{K} = \mathbf{k} - \mathbf{k}'$ is the *momentum transfer* vector. Its geometrical significance is illustrated in Figure 8-4. The function $V(\mathbf{K})$ is the *Fourier transform* of $V(\mathbf{r})$. The cross section in the first Born approximation can now be written

$$\sigma^{(1)}(\theta, \phi) = |f^{(1)}(\theta, \phi)|^2 = \left(\frac{-m\lambda}{2\pi\hbar^2}\right)^2 |V(\mathbf{K})|^2. \qquad (8\text{-}29)$$

Figure 8-4   The relationships between the vectors $\mathbf{k}$, $\mathbf{k}'$, $\mathbf{K}$ and the scattering angle θ. Note that for elastic scattering, $k = k'$, the triangle is isosceles.

Note that $\sigma^{(1)}$ is independent of the sign of λ and is therefore the same whether the scatterer is attractive or repulsive. In higher orders the sign of λ is significant.

Equation (8-29) represents a good approximation to the exact cross section provided $f^{(1)}(\theta, \phi)/r$ is small for all r, that is,

$$\frac{|f^{(1)}|}{r} = \left| \frac{m\lambda}{2\pi\hbar^2} \frac{V(\mathbf{K})}{r} \right| \ll 1 \qquad \text{(Born's criterion).}$$

This condition is necessary in order that the Born series (8-24) converge. Quite generally this criterion is satisfied for short-ranged potentials when the energy of the incident beam is large compared to the strength of the scattering potential.‡ However, the degree to which the criterion is satisfied is quite difficult to determine in general and depends on the specific nature of the scattering potential function.

The scattering angle θ (that is, the angle between $\mathbf{k}$ and $\mathbf{k}'$) is implicit in the magnitude of $\mathbf{K}$ as can be seen from the relation

$$K^2 = |\mathbf{k} - \mathbf{k}'|^2 = k^2 + k'^2 - 2kk' \cos\theta.$$

---

‡ For high energies, the Born criterion can be expressed as

$$\frac{m\lambda^2 V_0^2 R^2}{2\hbar^2 \varepsilon} \ll 1$$

where R is the *effective range* and λV_0 is the strength of the potential; thus, the Born approximation is valid when the scattering energy is large and the scattering potential weak and short ranged.

Since the scattering is elastic (that is, the energy is conserved), we set $k = k'$ and obtain

$$K^2 = 2k^2(1 - \cos \theta) = 4k^2 \sin^2 \frac{\theta}{2} \quad \text{or} \quad K = 2k \sin \frac{\theta}{2}. \quad (8\text{-}30)$$

The Fourier transform for a central force potential, $V(r)$, can be evaluated using spherical polar coordinates. Orienting $\mathbf{K}$ along the z' axis (for the purpose of integration), we have

$$V(\mathbf{K}) = \int dr' e^{i\mathbf{K} \cdot \mathbf{r}'} V(r')$$

$$= 2\pi \int_0^\infty V(r') r'^2 \, dr' \int_{-1}^1 du' e^{iKr'u'} \quad (u' = \cos \theta')$$

$$= \frac{4\pi}{K} \int_0^\infty \frac{V(r') \sin Kr'}{r'} r'^2 \, dr'. \quad (8\text{-}31)$$

We observe that for a central force potential, $V(\mathbf{K})$ is only a function of the magnitude of $\mathbf{K}$, that is, $V(\mathbf{K}) = V(K)$. Since $|\mathbf{K}| = K$ involves only the polar angle θ, it follows that $f^{(1)}$ and $\sigma^{(1)}$ are independent of ϕ as expected. The first Born cross section for $V(r)$ becomes

$$\sigma^{(1)}(\theta) = \left(-\frac{m\lambda}{2\pi\hbar^2} \right)^2 \left| \frac{4\pi}{K} \int_0^\infty \frac{V(r') \sin Kr'}{r'} r'^2 \, dr' \right|^2. \quad (8\text{-}32)$$

Since the Coulomb potential does not fall off rapidly enough with increasing r, the integral equation (8-22) and the Born series are not relevant to Rutherford scattering. The results of the Born approximation in this case are nevertheless interesting. The Fourier transform of the Coulomb potential $1/r$ is,‡ using (8-31),

$$V(K) = \frac{4\pi}{K} \int_0^\infty \frac{1}{r'} \frac{\sin Kr'}{r'} r'^2 \, dr' = \frac{4\pi}{K^2}.$$

The cross section becomes

$$\sigma^{(1)}(\theta) = \left(-\frac{m\lambda}{2\pi\hbar^2} \right)^2 \left| \frac{4\pi}{K^2} \right|^2 = \frac{4m^2\lambda^2}{\hbar^4 16k^4 \sin^4 \theta/2}.$$

Setting $\varepsilon = \hbar^2 k^2/2m$ and $\lambda = ZZ'e^2$, the first Born approximation gives

$$\sigma^{(1)}(\theta) = \left(\frac{ZZ'e^2}{4\varepsilon} \right)^2 \sin^{-4} \frac{\theta}{2}$$

---

‡ The integral is usually performed using a convergence factor,

$$\int_0^\infty \sin Kr \, dr = \lim_{\alpha \to 0} \int_0^\infty \sin Kr \, e^{-\alpha r} \, dr = \frac{1}{K}.$$

which is exactly Rutherford's classical cross section. This is just another "accidental" coincidence with the Coulomb potential.

The Born approximation can be properly applied to a shorter-ranged central potential. For example, in the case of the Gaussian potential (Figure 8-5a)

$$V = \lambda \exp\left[-\left(\frac{r}{R}\right)^2\right]$$

the Fourier transform is

$$V(K) = \int d\mathbf{r}' e^{i\mathbf{K}\cdot\mathbf{r}'} e^{-(r/R)^2} = (2\pi)^{3/2} R^3 e^{-K^2 R^2/2}.$$

The Born cross section becomes (Figure 8-5b)

$$\sigma^{(1)}(\theta) = \left(=\frac{m\lambda}{2\pi\hbar^2}\right)^2 |V(K)|^2 = \frac{m^2\lambda^2}{4\pi^2\hbar^4}(2\pi)^3 R^6 e^{-K^2 R^2}$$

$$= \frac{2\pi m^2\lambda^2}{\hbar^4} R^6 \exp\left(-4k^2 R^2 \sin^2\frac{\theta}{2}\right)$$

$$= \frac{2\pi m^2\lambda^2}{\hbar^4} R^6 \exp\left[-\left(\frac{8m\varepsilon}{\hbar^2}\right)R^2 \sin^2\frac{\theta}{2}\right].$$

Note that this cross section is expected to be accurate only when λ is small and when ε is large (Born's criterion).

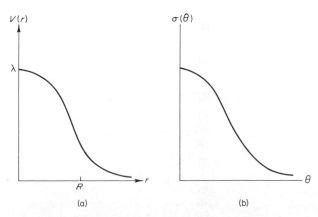

(a)   (b)

Figure 8-5   A plot of (a) a repulsive Gaussian potential and (b) its quantum-mechanical cross section in the Born approximation.

VI Higher Born Approximations

If we continue the iteration procedure of the integral equation (8-22), we generate the Born series. We are led to a set of Born amplitudes $f^{(n)}$ whose sum gives the total scattering amplitude, that is,

$$f(\theta, \phi) = \sum_{j=1}^{\infty} f^{(j)}(\theta, \phi).$$

The first few amplitudes turn out to be‡

$$f^{(1)} = -\frac{2m\lambda}{\hbar^2} \frac{1}{4\pi} V(\mathbf{k} - \mathbf{k}') = -\frac{m\lambda}{2\pi\hbar^2} V(\mathbf{K})$$

$$f^{(2)} = \left(\frac{2m\lambda}{\hbar^2}\right)^2 \frac{1}{4\pi} \frac{1}{(2\pi)^3} \int \frac{V(\mathbf{k} - \mathbf{k}'')V(\mathbf{k}'' - \mathbf{k}')\,d\mathbf{k}''}{(k''^2 - k^2)} \qquad (8\text{-}33)$$

$$f^{(3)} = -\left(\frac{2m\lambda}{\hbar^2}\right)^3 \frac{1}{4\pi} \frac{1}{(2\pi)^6} \int \frac{V(\mathbf{k} - \mathbf{k}'')V(\mathbf{k}'' - \mathbf{k}''')V(\mathbf{k}''' - \mathbf{k}')\,d\mathbf{k}''\,d\mathbf{k}'''}{(k'''^2 - k^2)(k''^2 - k^2)}.$$

As $\lambda \to 0$, the higher Born amplitudes will make a progressively smaller contribution to the cross section. For any short-ranged potential, the Born series generally converges provided that both λ is small and ε is sufficiently large. In the case of electron scattering from atoms, the scattering energies involved usually are well within the Born criterion. However, for nuclear scattering, energies must be $\gtrsim 100$ MeV. Since nuclear collisions often involve lower energies, Born's criterion is violated and other analytical methods are required to calculate nuclear scattering cross sections. One such method is referred to as partial wave analysis. This method is generally independent of the scattering strength and can be applied even to strong scatterers provided the range of the potential is short. As we shall see, the method is most useful for *low* incident energies.

VII The Method of Partial Waves

This method was originally developed by Rayleigh in his analysis of the scattering of sound from obstacles. It was later applied by Faxen and Holts-mark to quantum-mechanical scattering problems.

---

‡ The amplitudes in (8-33) involve integrations through singularities corresponding to those points at which the denominators vanish. Some vagueness is therefore introduced in the calculation of $f^{(2)}$ and $f^{(3)}$. This minor difficulty can be remedied by making the replacements $(k^2 - k''^2) \to (k^2 - k''^2 + i\Delta)$, $(k^2 - k'''^2) \to (k^2 - k'''^2 + i\Delta)$, and so forth, in the denominators and letting $\Delta \to 0$ *after* the (contour) integral has been evaluated.

The cross section for a spinless particle in a central force potential $V(r)$ can be readily calculated from an energy eigenfunction ψ_ε provided:

(a) ψ_ε satisfies the Schroedinger equation

$$\left\{-\frac{\hbar^2\nabla^2}{2m} + V(r)\right\}\psi_\varepsilon = \varepsilon\psi_\varepsilon \qquad \left(rV(r) \underset{r\to\infty}{\to} 0\right) \qquad (8\text{-}34a)$$

and

(b) ψ_ε has the asymptotic form

$$\psi_\varepsilon \underset{r\to\infty}{\sim} e^{ikz} + f(\theta)\frac{e^{ikr}}{r} \qquad \left(k = \left(\frac{2m\varepsilon}{\hbar^2}\right)^{1/2}\right). \qquad (8\text{-}34b)$$

Since we shall work in spherical coordinates, it is convenient to use the identity [see (5-59)]

$$e^{ikz} = e^{ikr\cos\theta} \equiv \sum_{l=0}^{\infty}(2l+1)i^l P_l(\cos\theta)j_l(kr) \qquad (8\text{-}35)$$

in (8-34b). Separation of (8-34a) in spherical coordinates gives

$$\psi = \sum_{l=0}^{\infty}\sum_{m_l=-l}^{l} a_{lm_l}Y_{lm_l}(\theta,\phi)R_{kl}(r)$$

where R_{kl} is the well-behaved solution to the radial equation

$$\left\{\frac{1}{r^2}\frac{d}{dr}\left(r^2\frac{d}{dr}\right) + \left[k^2 - \frac{2m}{\hbar^2}V(r) - \frac{l(l+1)}{r^2}\right]\right\}R_{kl}(r) = 0 \qquad \left(k = \left(\frac{2m\varepsilon}{\hbar^2}\right)^{1/2}\right).$$

$$(8\text{-}36)$$

Since the asymptotic form of (8-34) is to be independent of the azimuthal angle ϕ, we need only retain those terms which have $m_l = 0$. We must therefore find those coefficients a_l in the exact solution which lead to the proper asymptotic form, namely,

$$\sum_{l=0}^{\infty} a_l P_l(\cos\theta)R_{kl}(r) \underset{r\to\infty}{\sim} \sum_{l=0}^{\infty}(2l+1)i^l P_l(\cos\theta)j_l(kr) + f(\theta)\frac{e^{ikr}}{r}. \qquad (8\text{-}37)$$

The solution to the radial equation (8-36) when $V = 0$ is simply $j_l(kr)$; this solution has the asymptotic form

$$j_l(kr) \underset{r\to\infty}{\sim} \frac{\sin(kr - \tfrac{1}{2}l\pi)}{kr}.$$

Next we find the asymptotic form of $R_{kl}(r)$ in (8-36) for the general case where $V \neq 0$. Note that V tends to zero at infinity so that (8-36) tends to the free-particle equation. This suggests that since the projectile is almost free at infinity, R_{kl} and $j_l(kr)$ must have similar asymptotic forms. In fact, as long as $V \to 0$ faster than a $1/r$ potential,‡ the asymptotic form of the radial solution will be

$$R_{kl}(r) \underset{r \to \infty}{\sim} \frac{\sin(kr - \frac{1}{2}l\pi + \delta_l)}{kr}.$$

The δ_l is termed the lth partial phase shift and is a measure of the degree to which $R_{kl}(r)$ and $j_l(kr)$ differ at infinity. Since the scattering potential is responsible for this difference, we expect that the δ_l determine, in part, the form of the scattering cross section. Substituting the asymptotic forms of $R_{kl}(r)$ and $j_l(kr)$ in (8-37), we obtain

$$\sum_{l=0}^{\infty} a_l P_l(\cos\theta) \frac{\sin(kr - \frac{1}{2}l\pi + \delta_l)}{kr}$$

$$= \sum_{l=0}^{\infty} (2l + 1)i^l P_l(\cos\theta) \frac{\sin(kr - \frac{1}{2}l\pi)}{kr} + f(\theta) \frac{e^{ikr}}{r}. \qquad (8\text{-}38)$$

Transposing, we find

$$\sum_{l=0}^{\infty} \frac{P_l(u)}{kr} \left[a_l \sin\left(kr - \frac{1}{2}l\pi + \delta_l\right) - (2l + 1)i^l \sin\left(kr - \frac{1}{2}l\pi\right) \right] = f(\theta) \frac{e^{ikr}}{r}$$

or

$$\sum_{l=0}^{\infty} \frac{P_l(u)}{2ik} \left[e^{-\frac{1}{2}i\pi l}(a_l e^{i\delta_l} - (2l + 1)i^l) \frac{e^{ikr}}{r} - e^{\frac{1}{2}i l\pi}(a_l e^{-i\delta_l} - (2l + 1)i^l) \frac{e^{-ikr}}{r} \right]$$

$$= f(\theta) \frac{e^{ikr}}{r} \qquad (8\text{-}39)$$

where $u = \cos\theta$. The left and right sides of (8-39) can be equal if and only if the coefficient of e^{-ikr}/r vanishes, that is,

$$a_l = (2l + 1)i^l e^{i\delta_l}.$$

This determines the coefficients in (8-37). Finally, we substitute this result in (8-39) and find

$$\sum_{l=0}^{\infty} \frac{P_l(u)}{2ik} e^{-il\pi/2}(2l + 1)i^l(e^{i2\delta_l} - 1) \frac{e^{ikr}}{r} = f(\theta) \frac{e^{ikr}}{r}; \qquad (8\text{-}40)$$

---

‡ We have already seen that a $1/r$ potential produces deviations from the required asymptotic form at infinity.

we make the identification

$$f(\theta) = \sum_{l=0}^{\infty} \frac{P_l(u)}{2ik} e^{-il\pi/2}(2l + 1)i^l(e^{i2\delta_l} - 1)$$

or

$$f(\theta) = \frac{1}{k} \sum_{l=0}^{\infty} (2l + 1)e^{i\delta_l} \sin \delta_l \, P_l(\cos \theta). \tag{8-41}$$

This relation for the scattering amplitude, although not in closed form, *is exact*. It is not a perturbation series, and it is independent of the strength of the scatterer. The only restriction for its validity is that the potential fall off faster than a $1/r$ potential. To calculate the cross section, we first solve the radial equation (8-36) for each and every l value. Having obtained $R_{kl}(r)$, we then study its asymptotic form and deduce the δ_l, each of which depends on k (energy). Equation (8-41) immediately gives the scattering amplitude. Finally, the cross section is obtained using

$$\sigma(\theta) = |f(\theta)|^2$$

$$= \frac{1}{k^2} \sum_{l=0}^{\infty} \sum_{l'=0}^{\infty} (2l + 1)(2l' + 1)e^{i(\delta_l - \delta_{l'})} \sin \delta_l \sin \delta_{l'} \, P_l(u)P_{l'}(u). \tag{8-42}$$

Note that if $V = 0$, all the δ_l vanish and $\sigma(\theta) = 0$ as expected.
   The total cross section follows as

$$\sigma_T = \int \sigma \, d\Omega = 2\pi \int_{-1}^{1} \sigma(u) \, du = \frac{4\pi}{k^2} \sum_{l=0}^{\infty} (2l + 1) \sin^2 \delta_l. \tag{8-43}$$

In establishing (8-43), we have used the orthogonality relation

$$\int_{-1}^{1} du \, P_l(u)P_{l'}(u) = \frac{2}{2l + 1} \delta_{ll'}.$$

Note that

$$\text{Im} f(\theta) = \frac{1}{k} \sum_{l=0}^{\infty} (2l + 1) \sin^2 \delta_l P_l(u).$$

Since for $\theta = 0$, $P_l(1) = 1$, we find

$$\frac{4\pi}{k} \text{Im} f(0) = \frac{4\pi}{k^2} \sum_{l=0}^{\infty} (2l + 1) \sin^2 \delta_l = \sigma_T. \tag{8-44}$$

Equation (8-44) is exact and expresses the *optical theorem* of scattering theory. Actually, the imaginary part of $f(0)$ is proportional to the number of particles removed from the direction of incidence. The optical theorem relates this number to the total number appearing in the scattered beam and is therefore a statement of conservation of probability. The theorem is analogous to the relation $\mathscr{R} + \mathscr{T} = 1$ in one-dimensional scattering.

VIII   The Partial Phase Shift Approximation

The method of partial waves gives a cross section in the form of an infinite sum; it is therefore most useful when the first few terms represent an accurate approximation. The method actually decomposes the incident beam into partial waves according to their l (angular momentum) values. Each wave has a characteristic scattering amplitude determined by its phase shift δ_l. The total amplitude is the sum of the partial-wave amplitudes.

Those waves with large l correspond to particles with large impact parameters. As we noted in the classical case, these particles are not appreciably scattered. They do not contribute significantly to the cross section and can be neglected. The resulting approximate cross section will be accurate provided the potential is short ranged.

Classically, for a hard-sphere potential of radius R those particles with impact parameters in the range $b > R$ will not be scattered. Multiplying and dividing by the linear momentum, this condition can be expressed as $bp/p > R$. Setting $bp = L = \hbar(l(l + 1))^{1/2} \simeq \hbar l =$ angular momentum and $p = (2m\varepsilon)^{1/2} = \hbar k$, we are led to the condition

$$\frac{\hbar l}{\hbar k} > R \qquad \text{or} \qquad l > kR.$$

More generally, if R is the range of the potential, then those waves for which $l \gg l_{max} = kR$ are not appreciably scattered quantum-mechanically. For a given range and k value (energy), the approximate cross section in (8-42) can be written

$$\sigma \simeq \left| \sum_{l=0}^{l_{max}} \frac{1}{k} (2l + 1)e^{i\delta_l} \sin \delta_l P_l(\cos \theta) \right|^2. \tag{8-45}$$

For a fixed range, l_{max} decreases with decreasing k; consequently, fewer terms are necessary to approximate the cross section. Equivalently, when a fixed number of terms are retained in (8-45), the accuracy of the cross section increases with decreasing k. This approximation is therefore valid for *low* incident energies in contrast to the Born approximation.

The lth partial phase shift δ_l depends on the incident energy (or k) and on the nature of the scattering potential. For certain potentials, usually attractive, the scattered wave is drawn in toward the scatterer, producing a positive δ_l. The reverse is generally true for repulsive potentials (Figure 8-6). This can be understood by observing that for attractive potentials

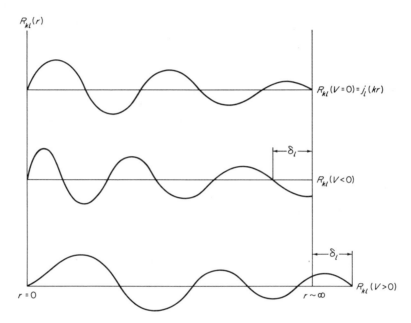

Figure 8-6 The effect of attractive and repulsive central force potentials on the asymptotic form of the radial eigenfunctions.

$V < 0$, the kinetic energy is large in the neighborhood of the scatterer. The parameter $k = (2m(\varepsilon - V)/\hbar^2)^{1/2}$ is also large, producing a contracted de Broglie wavelength $\lambda = 2\pi/k$. The wave function is thus drawn inward toward the scatterer. For repulsive potentials, the kinetic energy and the parameter k is smallest near the scatterer; hence the wavelength is largest and the wave shifts away from the scatterer.

If we assume that the dominant contribution to σ in (8-45) comes from s waves ($l = 0$), then this cross section depends critically on δ_0. If for a given energy δ_0 becomes $n\pi$, then $\sin \delta_l = 0$ and (8-45) predicts a marked drop in the scattering cross section. This effect has been observed experimentally in the scattering of low-energy electrons (~ 1 eV) from noble gas atoms. The notable absence of scattering at this energy is known as the *Ramsauer–Townsend effect*. This effect is observed in the scattering of electrons from the noble gas atoms where at certain energies the electrons experience a marked drop in scattering. At other energies, the phase shifts can become $\delta_0 = (n + \frac{1}{2})\pi$, in which case the scattering is enhanced. At these energies, we have s-wave *resonances*. Similar resonances can occur for higher l values.

IX s-Wave Scattering

We shall apply the partial-wave approximation to the calculation of the cross section associated with the core potential

$$V = V_0 \qquad \text{for} \quad 0 < r \leqslant R$$

and

$$V = 0 \qquad \text{for} \quad r > R.$$

The phase shifts are found by solving the corresponding radial equation

$$\frac{d^2}{dr^2} u_{kl}(r) + \left(K^2 - \frac{l(l+1)}{r^2} \right) u_{kl}(r) = 0 \qquad (0 < r \leqslant R)$$

$$\frac{d^2}{dr^2} u_{kl}(r) + \left(k^2 - \frac{l(l+1)}{r^2} \right) u_{kl}(r) = 0 \qquad (r > R)$$

(8-46)

where we have set

$$R_{kl} = \frac{u_{kl}}{r}, \qquad k = \left(\frac{2m\varepsilon}{\hbar^2} \right)^{1/2}, \qquad \text{and}$$

$$K = \left(\frac{2m(\varepsilon - V_0)}{\hbar^2} \right)^{1/2} \qquad \text{or} \qquad K = [k^2 - (2mV_0/\hbar^2)]^{1/2}.$$

The acceptable solutions for R_{kl} must be everywhere finite, continuous, and differentiable. We shall direct our attention to s-wave ($l = 0$) solutions. The interior solution ($r \leq R$) for $l = 0$ in (8-46) is

$$u_{k0} = A \sin Kr + B \cos Kr.$$

However, the radial eigenfunctions

$$R_{kl} = \frac{u_{kl}(r)}{r}$$

can remain finite at the origin if and only if $u_{kl}(0) = 0$. This requires that we set $B = 0$ and write

$$u_{k0}(r \leqslant R) = A \sin Kr. \qquad (8-47)$$

The exterior solution can be written

$$u_{k0}(r > R) = D \sin(kr + \delta_0). \qquad (8-48)$$

The continuity of R_{k0} and its derivative at $r = R$ requires that‡

$$\frac{u'_{k0}\ (r \leqslant R)}{u_{k0}\ (r \leqslant R)}\bigg|_{r=R} = \frac{u'_{k0}\ (r > R)}{u_{k0}\ (r > R)}\bigg|_{r=R}$$

which becomes, using (8-47) and (8-48),

$$K \cot KR = k \cot(kR + \delta_0)$$

or

$$\left(\frac{2m(\varepsilon - V_0)}{\hbar^2}\right)^{1/2} \cot\left(\frac{2m(\varepsilon - V_0)}{\hbar^2}\right)^{1/2} R = \left(\frac{2m\varepsilon}{\hbar^2}\right)^{1/2} \cot\left(\left(\frac{2m\varepsilon}{\hbar^2}\right)^{1/2} R + \delta_0\right).$$

$$(8\text{-}49)$$

The s-wave phase shift is obtained by solving (8-49) for δ_0 in terms of ε, R, and V_0. Using (8-45), the s-wave cross section becomes

$$\sigma^{(s)} = \frac{1}{k^2} \sin^2 \delta_0.$$

s-Wave resonances occur when $\delta_0 = (n + \frac{1}{2})\pi$, in which case $\sin^2 \delta_0$ is unity. The resonant incident energies ε_n, at which maximum scattering of s waves occurs (for fixed R and V_0), are obtained from the roots of (8-49), setting $\delta_0 = (n + \frac{1}{2})\pi$.

A Ramsauer-type effect occurs at those energies in (8-49) for which $\delta_0 = n\pi$. In that case, $\sin^2 \delta_0 = 0$ and no s-wave scattering occurs.

The s-wave cross section accurately describes the true cross section provided $kR \ll 1$ or $\varepsilon \ll \hbar^2/2mR$, that is, for low incident energies. On the other hand, when the energies are large, that is, $\varepsilon \gg V_0$, the Born cross section usually represents a better approximation.

It is interesting to note that in the limiting case $V_0 \to \infty$ (impenetrable sphere), the Born approximation is totally inapplicable. However, the method of partial waves can be applied directly. The inner solution for $V_0 = \infty$ becomes

$$u_{k0}(r \leqslant R) \equiv 0.$$

---

‡ The continuity requirement on the function and its derivative can also be satisfied by requiring the continuity of the logarithmic derivative. Mathematically

$$\frac{u_1'}{u_1} = \frac{u_2'}{u_2}$$

is equivalent to

$$(\ln u_1)' = (\ln u_2)'.$$

The continuity condition at $r = R$ for s waves becomes

$$u_{k0}(\leqslant R)|_{r=R} = B \sin(kR + \delta_0) = 0$$

from which it follows that $\delta_0 = -kR$. The differential s-wave cross section for an impenetrable sphere becomes

$$\sigma^{(s)} = \frac{1}{k^2} \sin^2 \delta_0 = \frac{1}{k^2} \sin^2 (-kR).$$

In the low-energy limit, $kR \ll 1$, we have

$$\sigma \simeq \sigma^{(s)}(kR \ll 1) = \frac{1}{k^2} \sin^2(-kR) \simeq \frac{1}{k^2}(-kR)^2 = \frac{1}{k^2} k^2 R^2.$$

The total s-wave cross section follows as

$$\sigma_T \simeq \sigma_T^{(s)} = 4\pi R^2. \tag{8-50}$$

Note that (8-50) does not contain $\hbar$. Yet, this cannot be a classical result since the classical limit is expected at high energies.‡ Furthermore, the classical cross section for the hard sphere is logically only πR^2. The factor 4 in (8-50) originates with the interference effects of quantum mechanics. The fact that $\hbar$ is absent must be attributed to the unphysical discontinuity of the scattering potential at $r = R$.

X Dynamical Quantum Scattering and Transitions

The scattering methods discussed above are quite general for elastic scattering of particles from a fixed scatterer; however, they cannot be generalized to non-wave-mechanical problems. One such problem, for example, is Compton scattering of photons from free electrons, which involves scattering of the quantized electromagnetic field.

We next develop a formalism which can be used for all of the fundamental quantum processes of nature. We shall apply it to scattering of particles from fixed potentials in order to demonstrate its equivalence to the stationary methods discussed above.

The dynamical method of scattering is based on Fermi's Golden Rule and treats the scattering phenomenon as a transition process from a given initial

‡ Actually it can be shown that as $\varepsilon \to \infty$, the quantum cross section approaches $\sim 2\pi R^2$ which is *twice* the classical result. This apparent anomaly is again related to the fact that the hard-sphere potential changes abruptly at $r = R$.

momentum state $|\mathbf{k}\rangle$ to some other final state $|\mathbf{k}'\rangle$. The transition is induced by the scattering perturbation $\lambda\hat{V}$ (Figure 8-7). The probability rate at which

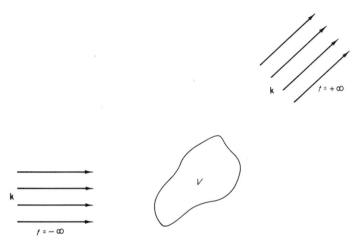

Figure 8-7   The scattering process pictured as a transition from an initial (at $t = -\infty$) to a final (at $t = +\infty$) momentum state. The transition is induced by the scattering potential.

particles go from $|\mathbf{k}\rangle$ to $|\mathbf{k}'\rangle$ is, from (7-72),

$$R_{\mathbf{k}\mathbf{k}'} = \left[\frac{2\pi}{\hbar}\left|\sum_{j=1}^{\infty}\lambda^j M_{\mathbf{k}\mathbf{k}'}^{(j)}\right|^2 \rho(\mathbf{k}')\right]_{\varepsilon_\mathbf{k} = \varepsilon_\mathbf{k}'} \tag{8-51}$$

The term $M_{\mathbf{k}\mathbf{k}'}^{(j)}$, represents the jth-order transition amplitude. The first few amplitudes are obtained from (7-72) by setting‡ $d\mathbf{n} = [\mathscr{V}/(2\pi)^3]\,d\mathbf{k}$, thus giving

$$M_{\mathbf{k}\mathbf{k}'}^{(1)} = \langle\mathbf{k}'|\hat{V}|\mathbf{k}\rangle$$

$$M_{\mathbf{k}\mathbf{k}'}^{(2)} = \frac{\mathscr{V}}{(2\pi)^3}\int\frac{\langle\mathbf{k}'|\hat{V}|\mathbf{k}''\rangle\langle\mathbf{k}''|\hat{V}|\mathbf{k}\rangle\,d\mathbf{k}''}{\varepsilon_\mathbf{k} - \varepsilon_{\mathbf{k}''}} \tag{8-52}$$

$$M_{\mathbf{k}\mathbf{k}'}^{(3)} = \left(\frac{\mathscr{V}}{(2\pi)^3}\right)^2\int\frac{\langle\mathbf{k}'|\hat{V}|\mathbf{k}''\rangle\langle\mathbf{k}''|\hat{V}|\mathbf{k}'''\rangle\langle\mathbf{k}'''|\hat{V}|\mathbf{k}\rangle\,d\mathbf{k}''\,d\mathbf{k}'''}{(\varepsilon_\mathbf{k} - \varepsilon_{\mathbf{k}''})(\varepsilon_\mathbf{k} - \varepsilon_{\mathbf{k}'''})}.$$

For our scattering problem, the box-normalized momentum eigenfunctions

$$|\mathbf{k}\rangle \to \psi_\mathbf{k} = \frac{1}{\sqrt{\mathscr{V}}}\,e^{i\mathbf{k}\cdot\mathbf{r}} \tag{8-53}$$

---

‡ The scattering process is assumed to be taking place in a large box of volume $\mathscr{V}$ (see Eq. (5-16)).

will represent free-particle eigenstates. The corresponding energies and fluxes are

$$\varepsilon_\mathbf{k} = \frac{\hbar^2 k^2}{2m} \quad \text{and} \quad \mathbf{J} = \frac{1}{\mathscr{V}} \frac{\hbar \mathbf{k}}{m}. \tag{8-54}$$

The first transition amplitude in (8-52) takes the form

$$M_{\mathbf{kk'}}^{(1)} = \langle \mathbf{k'} | \hat{V} | \mathbf{k} \rangle = \frac{1}{\mathscr{V}} \int V(\mathbf{r'}) e^{i(\mathbf{k}-\mathbf{k'})\cdot\mathbf{r'}} \, d\mathbf{r'}$$

$$= \frac{1}{\mathscr{V}} V(\mathbf{K} = \mathbf{k} - \mathbf{k'}). \tag{8-55}$$

The density of states (ignoring spin) associated with these eigenfunctions is obtained using (5-16), which is

$$d\mathbf{n} = \frac{\mathscr{V}}{(2\pi)^3} \, d\mathbf{k} = \rho(\mathbf{k}) \, d\varepsilon.$$

Expressing this equation in spherical polar coordinates and using (8-54), we find

$$\frac{\mathscr{V}}{(2\pi)^3} \, d\Omega_\mathbf{k} k^2 \, dk = \rho(\mathbf{k}) \, d\varepsilon = \rho(\mathbf{k}) \frac{\hbar^2 k}{m} \, dk$$

or

$$\rho(\mathbf{k}) = \frac{\mathscr{V}}{(2\pi)^3} \frac{mk}{\hbar^2} \, d\Omega_\mathbf{k}. \tag{8-56}$$

Equation (8-56) represents the number of states per unit energy whose $\mathbf{k}$ vector lies in a solid angle $d\Omega_\mathbf{k}$. The total number of particles entering this solid angle per unit time is, from (8-51),

$$R_{\mathbf{kk'}} = dN_{\mathbf{kk'}} = \left[\frac{2\pi}{\hbar} \left| \sum_{j=1}^{\infty} \lambda^j M_{\mathbf{kk'}}^{(j)} \right|^2 \frac{\mathscr{V}}{(2\pi)^3} \frac{mk'}{\hbar^2} \, d\Omega_{\mathbf{k'}} \right]_{k=k'}$$

or

$$dN_{\mathbf{kk'}} = \frac{\mathscr{V} m k}{(2\pi)^2 \hbar^3} \left| \sum_{j=1}^{\infty} \lambda^j M_{\mathbf{kk'}}^{(j)} \right|^2_{k=k'} d\Omega_{\mathbf{k'}}.$$

Dividing by the incident flux $J_{\text{inc}} = (1/\mathscr{V})(\hbar k/m)$, the cross section finally becomes

$$\sigma = \frac{1}{J_{\text{inc}}} \frac{dN_{\mathbf{kk'}}}{d\Omega_{\mathbf{k'}}} = \frac{\mathscr{V}^2 m^2}{(2\pi \hbar^2)^2} \left| \sum_{j=1}^{\infty} \lambda^j M_{\mathbf{kk'}}^{(j)} \right|^2_{k=k'}. \tag{8-57}$$

It appears strange, at first sight, that the volume of the box used in the normalization of $\psi_\mathbf{k}$ should enter into the cross section. However, each term involving the $M^{(j)}$ contains an appropriate factor which cancels any possible volume dependence in σ.

If λ is small, we need retain only the first amplitude,

$$M_{kk'}^{(1)} = \frac{1}{\mathcal{V}} V(\mathbf{k})$$

and the first-order cross section becomes

$$\sigma^{(1)} = \frac{\mathcal{V}^2 m^2}{(2\pi\hbar^2)^2} \left| \frac{\lambda}{\mathcal{V}} V(\mathbf{K}) \right|^2$$

$$= \left(\frac{\lambda m}{2\pi\hbar^2} \right)^2 |V(\mathbf{K})|^2. \tag{8-58}$$

This first-order result is identical to the first Born approximation (8-29). In fact, it can be verified (see Problem 8-10) that higher amplitudes also agree term-by-term with those of the Born series.

Each transition amplitude in (8-57) is said to represent a *process*. In the case of scattering from a fixed potential, we can assign a diagram to each $M_{kk'}^{(j)}$ in (8-57). The first-order process involves an entering arrow for $\mathbf{k}$, a vertex for the potential $\hat{V}$, and an exiting arrow for $\mathbf{k}'$ (Figure 8-8). The second-order process involves an entering arrow for $\mathbf{k}$, a vertex for $\hat{V}$, an intermediate arrow for $\mathbf{k}''$, another vertex for $\hat{V}$, and an exiting arrow for $\mathbf{k}'$. Diagrams of this type were introduced by R. P. Feynman, after whom they are named. They are particularly useful in the more complicated problems associated with the scattering of quantum fields where many-particle scattering occurs. Note the correspondence between the integrals for $M_{kk'}^{(j)}$ in (8-52) and the Feynman diagrams in Figure 8-8. We shall have occasion to use the Golden Rule in our discussion of quantum electrodynamics in Chapter 12.

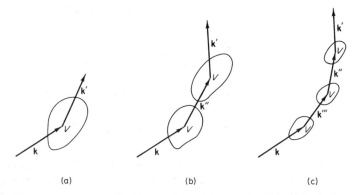

(a)   (b)   (c)

Figure 8-8 Feynman diagrams for scattering from a fixed potential in (a) first order, (b) second order, and (c) third order.

XI Inelastic Scattering and Absorption

In the scattering problems considered thus far, the static potential of the fixed target played a passive role. As a result, the following two conservation principles were reflected in the cross section:

(1) *Conservation of energy*, that is, the incident and scattered beams had equal energies (elastic scattering).

(2) *Conservation of particles*, that is, the number of particles removed from incidence appeared in the scattered beam (optical theorem).

In many situations, the target plays an active role in the scattering process. For example, in the actual Rutherford scattering process, the recoiling gold nuclei carry away energy as well as momentum. The scattered α particles necessarily lose some energy on impact. The total energy (projectile + target) is conserved in this situation. In the scattering of electrons from atoms, the projectiles may lose energy by internal excitation of the targets. The projectiles may therefore scatter inelastically.

In other scattering processes, it is possible for the projectile to be captured or absorbed by the target. This is quite common in nuclear reactions. The target is thus able to remove particles; the number removed from incidence need not appear in the scattered beam.

Each mode of scattering (elastic, inelastic, absorption) is said to represent a "channel," and each has a cross section which can be calculated using the Golden Rule. However, the detailed nature of the scatterer must be included in such a calculation. In general, the analysis of scattering processes requires understanding the many-body problem.

In Chapters 9 and 10, we shall discuss the quantum mechanics of the many-body problem as it pertains to bound and scattering states.

Suggested Reading

Bohm, D., "Quantum Theory," Chapter 21. Prentice-Hall, Englewood Cliff, New Jersey, 1951.

Landau, L. D., and Lifshitz, E. M., "Quantum Mechanics," 2nd ed., Chapter XVII. Addison-Wesley, Reading, Massachusetts, 1965.

Marion, J. B., "Classical Dynamics of Particles and Systems," 2nd ed. Academic Press, New York, 1970.

Merzbacher, E., "Quantum Mechanics," 2nd ed., Chapter 21. New York, Wiley, 1970.

Newton, R., "Scattering Theory of Waves and Particles." McGraw-Hill, New York, 1966.
(This advanced text is devoted entirely to scattering phenomena. It is aimed at the more advanced student and considerable mathematical background is required on the part of the student.)

Rodberg, L. S., and Thaler, R. M., "Introduction to the Quantum Theory of Scattering."
    Academic Press, New York, 1967.
    (This text on scattering is somewhat more introductory and less specialized in its dis-
    cussions of quantum mechanics than is R. Newton's book above.)

Problems

8-1. (a) Using classical mechanics, derive a general expression for the
hyperbolic trajectory of a particle repelled from the origin by a potential
$V = ZZ'e^2/r$.
(b) Show that the relation between the impact parameter b and the
scattering angle θ is

$$b = \left(\frac{ZZ'e^2}{2\varepsilon}\right) \cot\frac{\theta}{2}.$$

(c) Using (8-3), show that the Rutherford cross section is given by

$$\sigma(\theta) = \left(\frac{ZZ'e^2}{4\varepsilon}\right)^2 \sin^{-4}\left(\frac{\theta}{2}\right).$$

8-2. (a) Using classical mechanics, derive the relationship $b = b(\theta)$ for an
impenetrable-sphere potential $V = \infty$ for $r \leqslant R$ and $V = 0$ otherwise.
(b) Find the differential cross section for this potential.
(c) Integrate the result in (b) and find the total cross section. (Note
that the cross section is independent of ε.)

8-3. Derive the expressions for the metric coefficients, the Laplacian, and
the volume element in parabolic coordinates.

8-4. (a) Calculate the differential cross section classically for an attractive
hard-sphere potential $V = -V_0$ for $r \leqslant R$ and $V = 0$ otherwise.
(b) What is the quantum cross section in the Born approximation?
(c) Find the quantum cross section using partial-wave analysis
including s and p waves.

8-5. (a) Using classical mechanics, show that the relation between the
impact parameter and the scattering angle for a potential $V = K/r^2$ is

$$b^2 = \frac{K}{\varepsilon} \frac{1}{\theta} \frac{(\pi - \theta)^2}{(2\pi - \theta)}$$

where ε is the energy of the projectiles.
(b) Calculate the classical cross section for this potential.
(c) Find the quantum cross section for a $1/r^2$ potential in the Born
approximation.

8-6. Show that the Green's function

$$G(\mathbf{r}, \mathbf{r}') = -\frac{1}{4\pi} \frac{e^{ik|\mathbf{r}-\mathbf{r}'|}}{|\mathbf{r} - \mathbf{r}'|}$$

has the property

$$(\nabla^2 + k^2)G(\mathbf{r},\mathbf{r}') = \delta(\mathbf{r} - \mathbf{r}').$$

8-7. (a) Show that the second Born amplitude is given by

$$f^{(2)} = \left(\frac{2m\lambda}{\hbar^2}\right)^2 \frac{1}{4\pi} \frac{1}{(2\pi)^3} \int \frac{V(\mathbf{k} - \mathbf{k}'')V(\mathbf{k}'' - \mathbf{k}')\, d\mathbf{k}''}{(k''^2 - k^2)}.$$

(b) What is the expression for the cross section in the second Born approximation?

8-8. Express the general form of the angular dependence of the cross section when the $l = 2$ partial wave is included.

8-9. (a) Show that for low energies ($k \to 0$) the s-wave phase shift for a hard-sphere potential ($V = V_0$ for $r \le R$) follows from (8-49) as

$$\delta_0 \simeq kR\left[\frac{\tanh[k_0 R]}{k_0 R} - 1\right] \qquad \text{where} \quad k_0 = \left(\frac{2mV_0}{\hbar^2}\right)^{1/2}.$$

(b) Show that the total s-wave cross section is

$$\sigma_0 \simeq 4\pi R^2 \left[\frac{\tanh[k_0 R]}{k_0 R} - 1\right]^2.$$

(c) What happens to this expression as $V_0 \to \infty$?

8-10. Show that the first two terms in (8-57) give the same results as the first two terms in (8-24) (the second Born approximation).

II | MANY-PARTICLE SYSTEMS

9 | Noninteracting Particles

Many-particle systems are of great interest and importance in both classical and quantum mechanics. Even hydrogen, the simplest of atoms, consists of two particles—the proton and the electron. Most problems in atomic and nuclear physics involve intermediate numbers $(2 \leqslant N \lesssim 100)$ of particles. In the physics of solid, liquids, and gases, the situation is truly a many-body $(N \sim 10^{23})$ problem. Before formulating the quantum theory, we shall briefly discuss some of the classical concepts of many-particle systems.

I   Classical Mechanics

A system composed of N point particles is kinematically represented by a set of N position vectors, where $\mathbf{r}^{(q)}$ locates the qth particle $(q = 1, \ldots, N)$. Once the time dependence of each of these vectors has been established (subject to prescribed initial conditions), the classical problem is said to have been solved.

We begin by constructing the Lagrangian, $\mathcal{L} = T - V$, in terms of the $3N$

Cartesian coordinates and velocities, $x_i^{(q)}$ and $\dot{x}_i^{(q)}$ $(i = 1, \ldots, 3, q = 1, \ldots, N)$, where the kinetic energy is

$$T = \sum_{q=1}^{N} \tfrac{1}{2} m_q |\dot{\mathbf{r}}^{(q)}|^2 = \sum_{q=1}^{N} \sum_{i=1}^{3} \tfrac{1}{2} m_q \dot{x}_i^{(q)2}$$

$$= \sum_{q=1}^{N} \tfrac{1}{2} m_q (\dot{x}^{(q)2} + \dot{y}^{(q)2} + \dot{z}^{(q)2}). \qquad (9\text{-}1)$$

The potential energy will be decomposed into‡

$$V = \sum_{q}^{N} V_q + \tfrac{1}{2} \sum_{q}^{N} \sum_{q \neq t}^{N} V_{qt}$$

where V_q is the potential energy of the qth particle due to external forces and V_{qt} represents the interaction energy between the qth and tth particles. We shall assume that V_{qt} is independent of velocity and satisfies Newton's third law, that is, $V_{qt} = V_{tq}$. In many situations, the V_{qt} is simply a function of the separation of the two particles involved.

Applying Lagrange's equations (2-26) to (9-1), we obtain $3N$ differential equations for the $x_i^{(q)}$

$$m_q \ddot{x}_i^{(q)} - \frac{\partial V_q}{\partial x_i^{(q)}} - \frac{\partial}{\partial x_i^{(q)}} \sum_{t \neq q}^{N} V_{qt} = 0 \qquad (i = 1, 2, 3, \quad q = 1, \ldots, N)$$

or, in vector form,

$$m_q \ddot{\mathbf{r}}^{(q)} - \mathbf{\nabla}^{(q)} V(\mathbf{r}^{(q)}) - \mathbf{\nabla}^{(q)} \sum_{t \neq q}^{N} V(\mathbf{r}^{(q)} - \mathbf{r}^{(t)}) = 0 \qquad (q = 1, \ldots, N). \quad (9\text{-}2)$$

Equations (9-2) are to be solved for $\mathbf{r}^{(q)}(t)$, subject to the initial conditions $\mathbf{r}^{(q)}(0) = \mathbf{r}_0^{(q)}$ and $\dot{\mathbf{r}}^{(q)}(0) = \mathbf{v}_0^{(q)}$. The difficulty in solving (9-2), even for relatively few particles, rests with the fact that these equations are coupled, that is, the qth equation contains $\mathbf{r}^{(t)}$. Note that this coupling originates with the interaction potential energy and not with the external forces which act on each particle individually. Unfortunately, no general method exists which can separate (9-2) and render it soluble. For very special interactions, it

---

‡ The interaction term $q = t$ is omitted since a particle cannot exert a force on itself. The factor $\tfrac{1}{2}$ compensates for double counting. Some authors use the notation,

$$\tfrac{1}{2} \sum_{q} \sum_{q \neq t} V_{qt} = \sum_{q} \sum_{t < q} V_{qt}$$

which avoids double counting.

is sometimes possible to decouple the equations of motion by transforming to another set of canonical coordinates. For example, when only two interacting particles are involved, with no external forces, (9-2) can be decoupled using the *center of mass transformation*. Unfortunately, the method cannot even be generalized to three particles. In the special case of N particles interacting via elastic forces (that is, when the potential energy is a quadratic form in the coordinates) another transformation (the *normal coordinate transformation*) exists which renders the problem soluble.‡ We shall return to this problem in Chapter 10.

When the equations of motion are coupled, the motions of the particles are said to be *dynamically correlated*. The motion of each particle then influences the trajectories of all the others. In the rather rare situation when the interactions are negligible or perhaps nonexistent, the equations of motion essentially decouple and the many-body problem reduces to one involving N one-body problems. The motion of an "ideal" system, for example, the ideal gas, is dynamically uncorrelated and each particle moves as though the others did not exist. We can therefore conclude that the classically "ideal" or noninteracting N-body problem is no more difficult to treat than each of its constituent one-body problems. In practice, when the mean separation of the particles becomes very much larger than the range of the interaction potential, the system becomes approximately ideal. For this reason classical gases with short-ranged intermolecular forces begin to exhibit ideal gas behavior when sufficiently dilute.

The quantum-mechanical many-body problem has much in common with its classical counterpart. The solution to the quantum problem requires a solution of the many-body (coupled) Schroedinger equation. Decoupling this equation presents similar difficulties to those encountered in the classical Lagrange equations. In the case of an *ideal* (noninteracting) system, the decoupling is straightforward. However, there is one striking difference between classical and quantum ideal systems. This occurs when the particles are indistinguishable. In the quantum case, the motions of the particles are still correlated, that is, they mutually influence each other even though there are no forces between them. These correlations are termed *statistical* (as opposed to dynamical) and have no analog in the classical domain. They generally result in attractions for bosons (particles with integral spins) and repulsions for fermions (particles with half-integral spins). As a result of these correlations, the former are said to obey Bose–Einstein statistics while the latter are said to obey Fermi–Dirac statistics (see Section XI). Classical particles always obey Maxwell–Boltzmann statistics. We shall return to these points shortly.

_____

‡ See, for example, H. Goldstein, "Classical Mechanics." Addison-Wesley, Reading, Massachusetts, 1950.

As we have already seen, the transition to quantum theory requires the Hamiltonian formalism. The Hamiltonian in this case is just the total energy expressed in terms of $\mathbf{p}^{(q)}$ and $\mathbf{r}^{(q)}$, namely,

$$\mathcal{H} = \sum_{q=1}^{N} \frac{|\mathbf{p}^{(q)}|^2}{2m_q} + \sum_{q=1}^{N} V(\mathbf{r}^{(q)}) + \tfrac{1}{2} \sum_{q}^{N} \sum_{t \neq q}^{N} V(\mathbf{r}^{(q)} - \mathbf{r}^{(t)}). \tag{9-3}$$

The $6N$ Hamilton equations of motion are

$$-\frac{\partial \mathcal{H}}{\partial x_i^{(q)}} = \dot{p}_i^{(q)}; \qquad \frac{\partial \mathcal{H}}{\partial p_i^{(q)}} = \dot{x}_i^{(q)}. \tag{9-4}$$

In analogy with (2-76), any canonical function of the variables $x_i^{(q)}$ and $p_i^{(q)}$ satisfies the equation of motion

$$\frac{dA}{dt} = \{A, \mathcal{H}\} + \frac{\partial A}{\partial t}. \tag{9-5}$$

The Poisson bracket of two canonical functions is

$$\{A, B\} = \sum_{q=1}^{N} \sum_{i=1}^{3} \left(\frac{\partial A}{\partial x_i^{(q)}} \frac{\partial B}{\partial p_i^{(q)}} - \frac{\partial B}{\partial x_i^{(q)}} \frac{\partial A}{\partial p_i^{(q)}} \right). \tag{9-6}$$

The following Poisson bracket relations are simple to verify:

$$\{x_i^{(q)}, x_j^{(t)}\} = \{p_i^{(q)}, p_j^{(t)}\} = 0$$

and

$$\{x_i^{(q)}, p_j^{(t)}\} = \delta_{ij} \delta_{qt}. \tag{9-7}$$

In fact, if $A^{(q)}$ and $B^{(t)}$ are respectively canonical functions associated with the qth and tth particles, then

$$\{A^{(q)}, B^{(t)}\} = 0 \qquad (\text{for} \quad t \neq q).$$

II   The Transition to Quantum Mechanics

Suppose we consider a collection of N particles having $\mathbf{r}^{(q)}$ and $\mathbf{p}^{(q)}$ as canonical coordinates and momenta. We then make the associations:

$$\begin{aligned} \mathbf{r}^{(q)} &\to \hat{\mathbf{r}}^{(q)} \\ \mathbf{p}^{(q)} &\to \hat{\mathbf{p}}^{(q)} \end{aligned} \qquad (q = 1, \ldots, N)$$

and for a canonical function,

$$A(\mathbf{r}^{(1)}, \mathbf{p}^{(1)}, \ldots, \mathbf{r}^{(N)}, \mathbf{p}^{(N)}) \to \hat{A}(\hat{\mathbf{r}}^{(1)}, \hat{\mathbf{p}}^{(1)}, \ldots, \hat{\mathbf{r}}^{(N)}, \hat{\mathbf{p}}^{(N)}).$$

We require all the physically observable operators to be Hermitian. Using the Poisson bracket relations (9-7), we obtain the following fundamental commutation relations

$$[\hat{x}_i^{(q)}, \hat{x}_j^{(t)}] = [\hat{p}_i^{(q)}, \hat{p}_j^{(t)}] = 0$$

and

$$[\hat{x}_i^{(q)}, \hat{p}_j^{(t)}] = \delta_{ij}\delta_{qt}\,\hat{1}i\hbar. \tag{9-8}$$

We shall direct our attention to the stationary states, that is, the energy eigenstates of the many-body system. These are represented by kets satisfying‡

$$\mathcal{H}\,|E_\alpha\rangle = E_\alpha\,|E_\alpha\rangle \tag{9-9}$$

where $\mathcal{H}$ is the Hamiltonian derived from (9-3). We emphasize that the ket $|E_\alpha\rangle$ characterizes the *entire* many-particle system and not the component particles. Similarly, E_α reflects the quantized energy of the total system. The index α represents a set of $3N$ quantum numbers as there are this many degrees of freedom. When the particles of the system possess an intrinsic spin, each acquires one internal degree of freedom. For systems with intrinsic spin, there are therefore $4N$ quantum numbers.

III The Coordinate Representation and Wave Mechanics

While (9-9) represents a relation that is completely basis independent, it is convenient to deduce its wave-mechanical analog, that is, its form in the coordinate representation.

Since the operators $\hat{\mathbf{r}}^{(q)}$ mutually commute and are assumed nondegenerate, there must be a common eigenket which has the property

$$\hat{\mathbf{r}}^{(q)}|\mathbf{r}^{(1)}\cdots\mathbf{r}^{(q)}\cdots\mathbf{r}^{(N)}\rangle = \mathbf{r}^{(q)}|\mathbf{r}^{(1)}\cdots\mathbf{r}^{(q)}\cdots\mathbf{r}^{(N)}\rangle \qquad (q=1,\ldots,N). \tag{9-10}$$

This ket represents a state in which the first particle is precisely at $\mathbf{r}^{(1)}$, the second at $\mathbf{r}^{(2)}$, etc. It is possible to represent this ket as a direct product

$$|\mathbf{r}^{(1)}\cdots\mathbf{r}^{(q)}\cdots\mathbf{r}^{(N)}\rangle = |\mathbf{r}^{(1)}\rangle \otimes \cdots |\mathbf{r}^{(q)}\rangle \otimes \cdots |\mathbf{r}^{(N)}\rangle$$

$$= |\mathbf{r}^{(1)}\rangle \cdots |\mathbf{r}^{(q)}\rangle \cdots |\mathbf{r}^{(N)}\rangle$$

as was done in the case of intrinsic spin in Chapter 6. The operator $\hat{\mathbf{r}}^{(q)}$ acts only on the ket $|\mathbf{r}^{(q)}\rangle$ and bypasses the others. Since they are part of a

---

‡ We shall use Greek subscripts to identify many-particle states.

continuous spectrum, these kets have the obvious orthonormality property

$$\langle \mathbf{r}^{(1)\prime} \cdots \mathbf{r}^{(N)\prime} | \mathbf{r}^{(1)} \cdots \mathbf{r}^{(N)} \rangle = \delta(\mathbf{r}^{(1)\prime} - \mathbf{r}^{(1)}) \cdots \delta(\mathbf{r}^{(N)\prime} - \mathbf{r}^{(N)}).$$

Furthermore, they constitute a complete basis and any many-body state ket can be expressed as

$$|\beta\rangle = \int_N \cdots \int_1 d\mathbf{r}^{(1)} \cdots d\mathbf{r}^{(N)} \Psi_\beta(\mathbf{r}^{(1)} \cdots \mathbf{r}^{(N)}) | \mathbf{r}^{(1)} \cdots \mathbf{r}^{(N)} \rangle. \qquad (9\text{-}11)$$

The "coefficient"

$$\Psi_\beta(\mathbf{r}^{(1)} \cdots \mathbf{r}^{(N)}) = \langle \mathbf{r}^{(1)} \cdots \mathbf{r}^{(N)} | \beta \rangle \qquad (9\text{-}12)$$

is called the many-particle state function. According to the postulates of quantum mechanics, the quantity

$$d\mathscr{P}_\beta = |\Psi_\beta|^2 \, d\mathbf{r}_1 \cdots d\mathbf{r}_N \qquad (9\text{-}13)$$

represents the probability of finding the first particle in a volume $d\mathbf{r}_1$ about $\mathbf{r}_1$, the second particle in $d\mathbf{r}_2$ about $\mathbf{r}_2$, etc. If the state is bound, we must have

$$\lim_{\text{all } \mathbf{r}^{(q)} \to \infty} \Psi_\beta \to 0.$$

We can therefore normalize a bound state using

$$\langle \beta | \beta \rangle = 1$$

or

$$\int_N \cdots \int_1 |\Psi_\beta|^2 \, d\mathbf{r}^{(1)} \cdots d\mathbf{r}^{(N)} = 1.$$

For continuum states, the choice of the normalization scheme depends on the particular problem. For an unbounded ideal gas, box normalization is commonly used (see Section X).

A generalization of the proof developed in Chapter 4 for one-particle systems implies that the many-particle operators in wave mechanics are obtained using the replacements

$$\hat{\mathbf{r}}^{(q)} \to \mathbf{r}^{(q)} \qquad \text{and} \qquad \hat{\mathbf{p}}^{(q)} \to \frac{\hbar}{i} \nabla^{(q)}. \qquad (9\text{-}14)$$

The gradient operator acts only on the coordinate $\mathbf{r}^{(q)}$, treating the remaining coordinates as constants. The many-particle Schroedinger energy-eigenvalue equation (9-9) can be written in wave mechanics as

$$\mathscr{H}\Psi_\alpha = E_\alpha \Psi_\alpha \qquad (9\text{-}15\text{a})$$

where

$$\mathscr{H} = \sum_{q=1}^{N} -\frac{\hbar^2}{2m_q} \nabla^{(q)2} + \sum_{q=1}^{N} V(\mathbf{r}^{(q)}) + \tfrac{1}{2} \sum_{q}^{N} \sum_{t \neq q}^{N} V(\mathbf{r}^{(q)} - \mathbf{r}^{(t)}). \qquad (9\text{-}15\text{b})$$

Equation (9-15) is a partial differential equation for the many-particle energy eigenfunctions $\Psi_\alpha(\mathbf{r}^{(1)} \cdots \mathbf{r}^{(N)})$ and eigenvalues E_α. It is generally not possible to separate (9-15) into one-body equations. The difficulty in solving this equation rests with the two-body interaction terms. We shall see that when these terms are absent, (9-15) trivially separates into N one-particle Schroedinger energy-eigenvalue equations.

Once the spectrum and the eigenstates have been determined, expectation values of many-particle operators can be calculated in the usual way, namely,

$$\langle A \rangle_\alpha = \langle E_\alpha | \hat{A}(\hat{\mathbf{r}}^{(1)}, \hat{\mathbf{p}}^{(1)}, \cdots \hat{\mathbf{r}}^{(N)}, \hat{\mathbf{p}}^{(N)} | E_\alpha \rangle$$

$$= \int \Psi_\alpha^*(\mathbf{r}^{(1)} \cdots \mathbf{r}^{(N)}) \, \hat{A}\left(\mathbf{r}^{(1)}, \frac{\hbar}{i} \nabla^{(1)} \cdots \mathbf{r}^{(N)}, \frac{\hbar}{i} \nabla^{(N)}\right)$$

$$\times \Psi_\alpha(\mathbf{r}^{(1)} \cdots \mathbf{r}^{(N)}) \, d\mathbf{r}^{(1)} \cdots d\mathbf{r}^{(N)}. \tag{9-16}$$

The expectation values $\langle A \rangle_\alpha$ as well as the energy eigenvalues E_α refer to the system as a whole rather than to the component particles.

IV   The Permutation Operator

Although an exact solution to (9-15) is usually impossible, it is nevertheless possible to deduce various properties of the eigenstates by applying symmetry principles. Many-body operators generally involve indices associated with each of the particles. It is therefore interesting to investigate the behavior of the physical system under a permutation of these indices. We introduce the Hermitian *permutation operator*‡ $\hat{P}^{(qt)}$, with the following property

$$\hat{P}^{(qt)}\hat{A}(1, 2, \ldots, q, \ldots, t, \ldots, N) = \hat{A}(1, 2, \ldots, t, \ldots, q, \ldots, N)\hat{P}^{(qt)}$$

$$(t, q = 1, \ldots, N; \quad t \neq q) \tag{9-17}$$

where $\hat{A}$ is an arbitrary many-body operator. Moving the permutation operator from the left to the right of $\hat{A}$ results in an interchange of the indices q and t.

Since two consecutive operations of the permutation operator must leave any operator invariant, we expect that $[\hat{P}^{(qt)}]^2 = \hat{1}$; consequently the eigenvalues of $\hat{P}^{(qt)}$ are ± 1. In this respect, the permutation operator is similar to the parity operator. The eigenfunctions of $\hat{P}^{(qt)}$ are said to be of *definite symmetry* (instead of definite parity). Those corresponding to the $+1$ eigenvalue are said to be permutation *symmetric* and have the property

$$\Psi^S(1, \ldots, q, \ldots, t, \ldots, N) = \Psi^S(1, \ldots, t, \ldots, q, \ldots, N)$$

$$(t, q = 1, \ldots, N). \tag{9-18a}$$

‡ This operator is sometimes called the *exchange* operator.

The eigenfunctions associated with the -1 eigenvalue are *antisymmetric* and have the property

$$\Psi^A(1, \ldots, q, \ldots, t, \ldots, N) = -\Psi^A(1, \ldots, t, \ldots, q, \ldots, N). \quad (9\text{-}18b)$$

Note that in any state of definite symmetry, the probability density $|\Psi|^2$ is always invariant with respect to an interchange of indices. Equivalently, in such states, the particles are completely *indistinguishable*.

If an operator is symmetric with respect to permutations, then

$$\hat{P}^{(qt)}\hat{A}(1, \ldots, q, \ldots, t, \ldots, N) = \hat{A}(1, \ldots, t, \ldots, q, \ldots, N)\hat{P}^{(qt)}$$

$$= \hat{A}(1, \ldots, q, \ldots, t, \ldots, N)\hat{P}^{(qt)}$$

$$(q, t = 1, \ldots, N)$$

or

$$[\hat{A}, \hat{P}^{(qt)}] = 0. \quad (9\text{-}19)$$

Thus any (permutation) symmetric operator always commutes with the permutation operator. Note however, that the permutation operators do not generally commute among themselves. Consider for example the three-particle function

$$\Psi(x_1, x_2, x_3) = \frac{x_1 x_2}{x_3}.$$

Note that the two permutation operations

$$\hat{P}_{12}\hat{P}_{23}\Psi = \frac{x_2 x_3}{x_1}, \qquad \hat{P}_{23}\hat{P}_{12}\Psi = \frac{x_3 x_1}{x_2}$$

are not equivalent so that $\hat{P}_{12}$ and $\hat{P}_{23}$ do not in general commute. It turns out fortunately that nature has arranged matters such that for indistinguishable particles, the only physically realizable states are represented by state functions of definite symmetry with respect to permutations. It can be shown that it is possible to construct a *physically* complete set of eigenfunctions of definite symmetry for any symmetric operator for the purpose of expanding an arbitrary physical state function associated with a collection of quantum-mechanically indistinguishable particles. We stress that this symmetry (or antisymmetry) is always considered with regard to an interchange of *both* the spin *and* the spatial coordinates of each pair of particles. Thus the laws of nature have maintained the validity of Theorem VI (Chapter 3). We can therefore assume the existence of a complete set of energy eigenfunctions of definite symmetry for every symmetric Hamiltonian associated with indistinguishable particles. On the other hand, if the operator is not symmetric, none of its eigenstates can be of definite symmetry (Theorem IV).

If a Hamiltonian for a system is not permutation symmetric, then the particles are evidently different and distinguishable. From our discussion above, the energy eigenstates cannot be of definite symmetry. On the other hand, if the particles are indistinguishable the Hamiltonian as well as all other relevant observable operators must be permutation symmetric, and according to nature we must require that the energy eigenstates (as well as all other state functions) be of definite symmetry (that is, either symmetric or antisymmetric) with respect to permutations. Experiments have shown that the particular state of symmetry is related to the intrinsic spin of the indistinguishable particles. The correspondence is by the following postulates:

Postulate (a)   Particles with integral spins are termed *bosons*. All state functions for systems of indistinguishable bosons are completely symmetric and have the property

$$\Psi^S(1, 2, \ldots, t, \ldots, q, \ldots, N) = + \Psi^S(1, 2, \ldots, q, \ldots, t, \ldots, N)$$

$$\text{(bosons)}. \qquad (9\text{-}20a)$$

Postulate (b)   Particles with half-integral spins are termed *fermions*. All state functions for indistinguishable fermions are completely antisymmetric and have the property

$$\Psi^A(1, 2, \ldots, t, \ldots, q, \ldots, N) = - \Psi^A(1, 2, \ldots, q, \ldots, t, \ldots, N)$$

$$\text{(fermions)}. \qquad (9\text{-}20b)$$

The symmetry requirements for indistinguishable bosons and fermions lead to effects that are observable even in their macroscopic thermodynamic behavior. For fermions, we note that when we set $\mathbf{r}^{(q)} = \mathbf{r}^{(t)} = \mathbf{R}$, (9-20b) becomes

$$\Psi^A(1, 2, \mathbf{R}, \ldots, \mathbf{R}, \ldots, N) = - \Psi^A(1, 2, \mathbf{R}, \ldots, \mathbf{R}, \ldots, N).$$

However, since only a function that is identically zero can equal its own negative, it follows that

$$\Psi^A(1, 2, \ldots, \mathbf{R}, \mathbf{R}, \ldots, N) = 0.$$

We are thus led to the following principle for fermions:

The Pauli Exclusion Principle (general form)   The probability of finding any two indistinguishable fermions at the same point in space is zero.‡

---

‡ This exclusion effect applies to fermions with the same spin coordinates, that is, with the same values of m_s.

The Pauli principle implies that if one fermion is situated at a particular point, it excludes all others in the same spin state from that point. *This statistical repulsion is purely quantum-mechanical and is quite independent of the dynamical forces between the particles.* It originates with the indistinguishable nature of the particles.

Bosons in the same spin state have quite the opposite effect on each other. Their statistical interactions lead to attractions which favor clustering of the particles. We shall show this effect in Section VII.

It should be pointed out that not all identical particles are indistinguishable. For example, consider two noninteracting identical masses bound by springs to points a and b on the x axis (Figure 9-1). The one-dimensional Hamiltonian for this system is

$$\mathcal{H} = \frac{\hat{p}^{(1)2}}{2m} + \frac{\hat{p}^{(2)2}}{2m} + \frac{1}{2} k[(x^{(1)} - a)^2 + (x^{(2)} - b)^2].$$

Figure 9-1   Two identical but distinguishable systems.

Although the particles themselves are identical, the fact that they are bound to different points in space destroys the permutation invariance $(1 \leftrightarrow 2)$ of $\mathcal{H}$. These particles cannot be regarded as indistinguishable.

We emphasize that for two particles to be indistinguishable it is necessary that *all* physically measurable operator observables be permutation symmetric. For example, a positronium atom is very much like a hydrogen atom except that the proton is replaced by a positron, the latter having equal mass but opposite charge to that of the electron. The electrostatic Hamiltonian for an isolated positronium atom is

$$\mathcal{H} = \frac{\hat{p}^{(1)2}}{2m} + \frac{\hat{p}^{(2)2}}{2m} - \frac{e^2}{|\mathbf{r}^{(1)} - \mathbf{r}^{(2)}|}$$

where 1 and 2 refer, respectively, to the positron and electron. Although this Hamiltonian is permutation symmetric, the two particles are not truly indistinguishable. This can be seen by applying a uniform electric field whereby the atom acquires an additional energy $V = -eE \cdot (\mathbf{r}^{(1)} - \mathbf{r}^{(2)})$ where $e = 4.8 \times 10^{-10}$ esu. The Hamiltonian of the positronium atom in an electric field is no longer permutation symmetric; the particles are therefore not indistinguishable and states of definite symmetry do not exist.

Solving the Schroedinger equation for a system of interacting particles is very difficult. However for "ideal" systems where the two-body interactions are neglected, it is always possible to separate (9-15) into N one-particle equations. The ideal many-body problem is therefore soluble in principle. Ideal systems are particularly useful when they closely approximate real systems. Usually, when a real gas is sufficiently dilute, the interparticle separation is large. The interactions become negligible and the gas can be regarded as ideal. However, the quantum effects of indistinguishability can still persist.

V Distinguishable Ideal Systems

We consider a system of N noninteracting distinguishable particles each with a different mass $m^{(q)}$ and each experiencing a different external potential $V^{(q)} = V^{(q)}(\mathbf{r}^{(q)})$. The Schroedinger equation for this system is

$$\mathscr{H}\Psi_\alpha = \left\{ \sum_{q=1}^{N} \frac{-\hbar^2}{2m^{(q)}} \nabla^{(q)2} + V^{(q)}(\mathbf{r}^{(q)}) \right\} \Psi_\alpha(\mathbf{r}^{(1)}, \ldots, \mathbf{r}^{(N)})$$

$$= E_\alpha \Psi_\alpha(\mathbf{r}^{(1)}, \ldots, \mathbf{r}^{(N)}). \tag{9-21}$$

Assuming a solution of the form

$$\Psi_\alpha = \psi_i^{(1)}(\mathbf{r}^{(1)}) \cdots \psi_j^{(N)}(\mathbf{r}^{(N)}) \tag{9-22}$$

the equation separates into N one-particle equations of the form

$$\left\{ -\frac{\hbar^2}{2m^{(q)}} \nabla^{(q)2} + V^{(q)} \right\} \psi_i^{(q)}(\mathbf{r}_i^{(q)}) = \varepsilon_i^{(q)} \psi_i^{(q)}(\mathbf{r}_i^{(q)}) \qquad (q = 1, \ldots, N) \tag{9-23}$$

with

$$E_\alpha = \sum_{q=1}^{N} \varepsilon_i^{(q)}.$$

In other words, when interactions are neglected, the Schroedinger equation (9-21) separates and each particle has its own set of single-particle states $\psi_i^{(q)}$ and energies $\varepsilon_i^{(q)}$ which satisfies a one-particle equation. These single-particle states are sometimes called "orbitals." We shall discuss orbitals in atoms and molecules in Chapter 10. Since, in general, each particle is governed by a different Schroedinger equation (that is, $m^{(q)} \neq m^{(t)}$, $V^{(q)} \neq V^{(t)}$), the nature of the spectra and orbitals varies from particle to particle. We are reminded of this by the superscripts in $\psi_i^{(q)}$ and $\varepsilon_i^{(q)}$. Since the orbital index i stands for three quantum numbers (four including spin), the total state index α is therefore an abbreviation for $3N$ (or $4N$) quantum numbers.

The total state is known when the orbitals occupied by each of the particles is given. *The total state function is a simple product of these orbital functions, while the total energy is just the sum of the corresponding orbital energies.*

As an example, we consider two *distinguishable* spinless particles of mass m_1 and m_2 confined to an infinite one-dimensional well of width a. The Schroedinger equation is

$$\left[\left\{-\frac{\hbar^2}{2m_1}\frac{d^2}{dx^{(1)2}} - \frac{\hbar^2}{2m_2}\frac{d^2}{dx^{(2)2}}\right\} + \{V^{(1)} + V^{(2)}\}\right]\Psi_\alpha(x^{(1)}, x^{(2)})$$

$$= E_\alpha \Psi_\alpha(x^{(1)}, x^{(2)}),$$ (9-24)

where

$$V^{(1)} = V^{(2)} = 0 \qquad \text{for} \quad 0 \leqslant x \leqslant a$$

$$V^{(1)} = V^{(2)} = \infty \qquad \text{otherwise.}$$

Using

$$\Psi_\alpha = \psi_n^{(1)}(x^{(1)})\psi_{n'}^{(2)}(x^{(2)})$$

the equation separates into

$$\left\{-\frac{\hbar^2}{2m_1}\frac{d^2}{dx^{(1)2}} + V^{(1)}\right\}\psi_n^{(1)}(x^{(1)}) = \varepsilon_n^{(1)}\psi_n^{(1)}(x^{(1)})$$

and

$$\left\{-\frac{\hbar^2}{2m_2}\frac{d^2}{dx^{(2)2}} + V^{(2)}\right\}\psi_{n'}^{(2)}(x^{(2)}) = \varepsilon_{n'}^{(2)}\psi_{n'}^{(2)}(x^{(2)}).$$

The already familiar solutions are

$$\psi_n(x^{(1)}) = \left(\frac{2}{a}\right)^{1/2} \sin\frac{n\pi}{a} x^{(1)}, \qquad \varepsilon_n^{(1)} = \frac{\hbar^2}{2m_1}\frac{\pi^2}{a^2} n^2$$

and $(n, n' = 1, 2, \ldots).$

$$\psi_{n'}(x^{(2)}) = \left(\frac{2}{a}\right)^{1/2} \sin\frac{n'\pi}{a} x^{(2)}, \qquad \varepsilon_{n'}^{(2)} = \frac{\hbar^2}{2m_2}\frac{\pi^2}{a^2} n'^2$$

The total energy eigenstates and eigenvalues take the form

$$\Psi_\alpha = \Psi_{n,n'}(x^{(1)}, x^{(2)}) = \frac{2}{a} \sin\frac{n\pi}{a} x^{(1)} \sin\frac{n'\pi}{a} x^{(2)}$$ (9-25a)

with

$$E_\alpha = E_{n,n'} = \frac{\hbar^2\pi^2}{2a^2}\left(\frac{n^2}{m_1} + \frac{n'^2}{m_2}\right).$$ (9-25b)

All that must be indicated are the quantum numbers of the states n and n' occupied by the particles m_1 and m_2. This prescription is called a *configuration*. For example, the ground state configuration is $n = 1$, $n' = 1$ so that

$$\Psi_\alpha = \Psi_{1,\,1} = \frac{2}{a} \sin \frac{\pi}{a} x^{(1)} \sin \frac{\pi}{a} x^{(2)}$$

and

$$E_\alpha = E_{1,\,1} = \frac{\hbar^2 \pi^2}{2a^2} \left(\frac{1}{m_1} + \frac{1}{m_2} \right).$$

This example can easily be generalized to N noninteracting particles. The configuration is labeled by a set of quantum numbers n, n', n'', ..., etc., which indicates the orbital states occupied by each particle. We emphasize that a configuration has meaning only when the original many-particle Schroedinger equation is separable. This occurs when the system is ideal. For *real* systems, configurations of independent orbitals cease to be well defined, but serve a useful purpose in variational and other approximation methods.

VI Indistinguishable Ideal Systems

When a system is composed of indistinguishable noninteracting particles (that is, $m_q = m$ and $V^{(q)} = V$), the one-particle Schroedinger equations are all identical and of the form

$$\left\{ -\frac{\hbar^2}{2m} \nabla^{(q)2} + V(\mathbf{r}^{(q)}) \right\} \psi_i(\mathbf{r}^{(q)}) = \varepsilon_i \psi_i(\mathbf{r}^{(q)}). \tag{9-26}$$

Since the orbitals and their eigenvalues are the same for all the particles, the superscripts in ψ_i and ε_i have been removed. The total state functions analogous to (9-22) become

$$\Psi_\alpha = \psi_i(\mathbf{r}^{(1)}) \cdots \psi_j(\mathbf{r}^{(N)}) \tag{9-27a}$$

with

$$E_\alpha = \sum_i \varepsilon_i; \tag{9-27b}$$

the product and sums are over occupied orbitals. Unfortunately, the simple product in (9-27) violates the indistinguishability requirement since Ψ_α is not of definite symmetry with respect to permutations. However, we can always construct a properly symmetrized function from (9-27). Note that the coordinate indices in (9-27) can be permuted $N!$ ways. Adding these permuted products and dividing by $(N!)^{1/2}$ for normalization, we obtain a $\Psi_\alpha{}^s$ with the same energy E_α which is symmetric under permutations.

We shall denote this operation using the *symmetrization operator* which has precisely the effect mentioned above, that is,

$$\Psi_\alpha{}^s = \frac{1}{(N!)^{1/2}} \, \mathscr{S} \psi_i(\mathbf{r}^{(1)}) \cdots \psi_j(\mathbf{r}^{(N)}). \tag{9-28}$$

When $\mathscr{S}$ operates on a simple product as in (9-27a), it generates a sum of $N!$ permuted products which is automatically a symmetric state function.

Similarly, we can construct an antisymmetrized function using the *antisymmetrization operator* $\mathscr{A}$, with the property

$$\Psi_\alpha{}^A = \frac{1}{(N!)^{1/2}} \, \mathscr{A} \psi_i(\mathbf{r}^{(1)}) \cdots \psi_j(\mathbf{r}^{(N)}). \tag{9-29}$$

This operation is almost identical to that in (9-28) except that the terms involving an odd number of permutations from the original product are *subtracted* instead of added. The net effect is to generate an eigenfunction that is antisymmetric with respect to permutations.

We shall illustrate the symmetrization procedure using the two particles in the infinite well discussed in Section V. This time, however, we shall assume that they are *indistinguishable* (that is, $m_1 = m_2$) and we ignore spin.

The total state function is no longer a simple product as in (9-25). Proper symmetrization gives‡

$$\Psi_\alpha = \Psi_{nn'} = \frac{1}{\sqrt{2}} \left\{ \frac{2}{a} \sin \frac{n\pi}{a} x^{(1)} \sin \frac{n'\pi}{a} x^{(2)} \pm \frac{2}{a} \sin \frac{n\pi}{a} x^{(2)} \sin \frac{n'\pi}{a} x^{(1)} \right\}$$

$$\tag{9-30a}$$

with

$$E_{nn'} = \frac{\hbar^2}{2m} \frac{\pi^2}{a^2} (n^2 + n'^2). \tag{9-30b}$$

The upper sign is for bosons while the lower is for fermions. Note that while the configuration is given by the two quantum numbers n and n', neither is associated with either one of the particles. Since they are now indistinguishable, we simply say that the particles occupy the orbital states n and n'. We can no longer specify which one is in which state.

---

‡ We suppress the spin indices for brevity and assume both particles to be in the same spin state.

Another representation for an antisymmetric state is obtained by constructing the *Slater determinant* of the occupied orbitals

$$
\Psi_\alpha{}^A = \frac{1}{(N!)^{1/2}}
\begin{vmatrix}
\psi_i(\mathbf{r}^{(1)}) & \cdots & \psi_j(\mathbf{r}^{(1)}) \\
\vdots & \vdots & \vdots \\
\psi_i(\mathbf{r}^{(N)}) & \cdots & \psi_j(\mathbf{r}^{(N)})
\end{vmatrix}.
\tag{9-31}
$$

Equation (9-31) will be seen to be identical to (9-29), as can be verified by expanding the determinant. The Slater determinant is a convenient way of representing a state of ideal indistinguishable fermions (for example, conduction electrons in a metal).

If two particles are at the same point in space, then two of the rows in (9-31) will be identical and the total eigenfunction will vanish. This verifies the general form of Pauli's principle. However, note also that if two fermions occupy the same orbital, then two columns are equal and the determinant again vanishes. This leads to the special form of Pauli's principle:

Pauli's Exclusion Principle (for ideal fermions) In any configuration, no two indistinguishable fermions can be in the same single-particle state. Equivalently, no two fermions can be assigned the same set of quantum numbers.

VII Statistical Correlations in Ideal Bose and Fermi Systems

It has been pointed out that even in ideal quantum systems, indistinguishable bosons and fermions can, respectively, attract and repel each other. We shall illustrate this for the case of two free, *noninteracting*, indistinguishable, particles in one dimension. Since the single-particle states, ignoring spin, are of the form e^{ikx}, the total properly symmetrized state of the system is‡ (except for normalization)

$$
\Psi_\alpha = \Psi_{kk'}(x_1, x_2) = e^{ikx_1}e^{ik'x_2} \pm e^{ikx_2}e^{ik'x_1} \quad \left(\begin{matrix} \text{bosons,} \\ \text{fermions} \end{matrix}\right)
\tag{9-32}
$$

Introducing the variables $x = x_2 - x_1$ and $X = \tfrac{1}{2}(x_2 + x_1)$, the state function in (9-32) becomes

$$
\Psi_{kk'}(x, X) = e^{i(k+k')X}[e^{i(k-k')x/2} \pm e^{-i(k-k')x/2}].
\tag{9-33}
$$

---

‡ We again are assuming that the two particles are in the same spin state and suppress the spin indices.

The variables x and X represent the relative separation and the center of mass of the two particles respectively. Squaring (9-33) (multiplying by its complex conjugate), it follows that the probability density of finding the system with an interparticle separation x is (Figure 9-2)

$$\mathscr{P}_{kk'}(x) \propto \cos^2 \left(\frac{k-k'}{2}\right)x \qquad \text{(bosons)}$$

$$\propto \sin^2 \left(\frac{k-k'}{2}\right)x \qquad \text{(fermions)}. \tag{9-34}$$

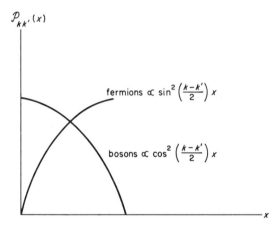

Figure 9-2 A plot of the functions representing the probability of finding two free indistinguishable particles with a separation of x in one dimension.

The probability density at zero separation is a maximum for bosons and zero for the fermions (Pauli principle). In the statistical sense, the bosons tend to be found near one another while the fermions move in such a way as to exclude other fermions.‡

In practice, no system of particles is truly ideal. An important question to be answered is whether or not the interactions in a real system are sufficiently small to permit the ideal system to serve as an accurate zeroth-order approximation. We shall examine a typical two-electron problem in the ideal limit by assuming the Coulomb interaction to be negligible.

---

‡ The statistical correlations are particularly significant when $(k-k')x \ll 1$ or $x \ll 1/(k-k')$. Noting that the de Broglie wavelength associated with the relative momentum of the particles is $\lambda = 2\pi/(k-k')$, we observe that the above condition can be written $x \ll \lambda$. Note also that for fermions, the probability function vanishes identically when the two particles are in the same momentum state (that is, $k = k'$).

VIII   The "Ideal" Helium Atom

The helium atom consists of two electrons bound to a nucleus of charge Ze ($Z = 2$) by a Coulomb force. The electrostatic Hamiltonian is

$$\mathscr{H} = -\frac{\hbar^2}{2m}\nabla^{(1)2} - \frac{\hbar^2}{2m}\nabla^{(2)2} - \frac{Ze^2}{r^{(1)}} - \frac{Ze^2}{r^{(2)}} + \frac{e^2}{|\mathbf{r}^{(2)} - \mathbf{r}^{(1)}|}. \qquad (9\text{-}35)$$

If we neglect the interaction term, the Schroedinger equation separates into two hydrogenlike equations, each of which is of the form

$$\left\{ -\frac{\hbar^2}{2m}\nabla^2 - \frac{Ze^2}{r} \right\}\psi_{nlm_lm_s} = \varepsilon_{nl}\,\psi_{nlm_lm_s} \qquad (9\text{-}36)$$

where

$$\psi_{nlm_lm_s} = R_{nl}(r)Y_{lm_l}(\theta, \phi)|m_s\rangle \qquad (9\text{-}37)$$

and‡

$$\varepsilon_{nl} = -\frac{Z^2(13.6)}{n^2} = -\frac{54}{n^2}\ \text{eV}.$$

The hydrogenlike wave functions are obtained from (5-74) by making the replacement $a \to a/Z$. Since the electrons are fermions, the total state of the ideal helium atom can be represented by the Slater determinant

$$\Psi_\alpha = \Psi_{\substack{nlm_lm_s \\ n'l'm_l'm_s'}} = \frac{1}{\sqrt{2}}\begin{vmatrix} \psi_{nlm_lm_s}(1) & \psi_{n'l'm_l'm_s'}(1) \\ \psi_{nlm_lm_s}(2) & \psi_{n'_l'm_l'm_s'}(2) \end{vmatrix}. \qquad (9\text{-}38a)$$

The corresponding energy‡ is

$$E_\alpha = E_{nln'l'} = \varepsilon_{nl} + \varepsilon_{n'l'} = -54\left(\frac{1}{n^2} + \frac{1}{n'^2}\right). \qquad (9\text{-}38b)$$

Note that while the configurational state of ideal helium requires eight quantum numbers, the corresponding energy depends on only two of these numbers. These configurations can therefore be highly degenerate.

The ground state energy has $n = n' = 1$. Since in this case the quantum numbers l, l', m_l, and m_l' must all be zero, it follows from Pauli's principle that the spin quantum numbers must be different, that is $m_s = \tfrac{1}{2}$, $m_s' = -\tfrac{1}{2}$ (Figure 9-3). The ground state of ideal helium is nondegenerate and unique

---

‡ Since Coulomb eigenfunctions reflect an accidental degeneracy, the energies are independent of l. However, to be more general we will treat the energy as though an l dependence existed. This is indeed the case for real atoms, as we shall see in Chapter 10.

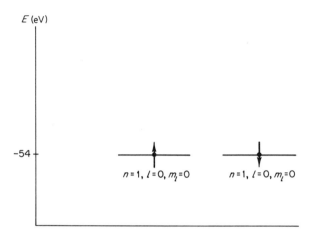

Figure 9-3 The configuration of helium's electrons in the ground state. The ground state energy neglecting interactions between the electrons is $E = -54 + (-54) = -108$ eV.

and can be expressed as

$$\Psi_{\substack{100\ \frac{1}{2} \\ 100-\frac{1}{2}}} = \frac{1}{\sqrt{2}} \psi_{100}(1)\psi_{100}(2) \begin{vmatrix} |\tfrac{1}{2}\rangle^{(1)} & |-\tfrac{1}{2}\rangle^{(1)} \\ |\tfrac{1}{2}\rangle^{(2)} & |-\tfrac{1}{2}\rangle^{(2)} \end{vmatrix}$$

$$= \frac{1}{\sqrt{2}}\psi_{100}(1)\psi_{100}(2)[|\tfrac{1}{2}\rangle^{(1)}|-\tfrac{1}{2}\rangle^{(2)} - |\tfrac{1}{2}\rangle^{(2)}|-\tfrac{1}{2}\rangle^{(1)}] \quad (9\text{-}39a)$$

with

$$E_{nn'} = E_{11} = -54\left[\frac{1}{1^2} + \frac{1}{1^2}\right] = -108 \quad \text{eV} \quad \text{(ground state)}. \quad (9\text{-}39b)$$

The *singly ionized* state of helium corresponds to one electron in the state $n = 1$ with the other electron in the free state $n' = \infty$. The energy of this state is

$$E_{mm'} = E_{1\infty} = -54\left[\frac{1}{1^2} + \frac{1}{\infty^2}\right] = -54 \quad \text{eV}.$$

The first ionization energy (that is, the energy required to singly ionize an atom) of ideal helium is therefore

$$(E_{1\infty} - E_{11}) = -54 - (-108) = 54 \quad \text{eV}.$$

Experimentally, it is found that only ~ 24 eV are required to singly ionize helium. The discrepancy is related to the fact that the electron–electron repulsion, which assists the ionization process, has been neglected. Clearly the order of magnitude of this difference makes it doubtful that the eigenstates of the ideal helium atom can in any way resemble those of the real atom.

Nevertheless, as we shall show in Chapter 10, these ideal states can be improved upon by using variational and other approximation methods.

IX   Excited States in Helium

The Slater determinant in (9-39) represents a state in which the electrons occupy the orbitals $nlm_l m_s$ and $n'l'm_l'm_s'$. We shall use the ket

$$|nn'll'm_l m_l'm_s m_s'\rangle$$

to denote this determinant. Since the magnetic quantum numbers m_l, m_l', m_s, and m_s' do not appear in the expression for the energy, the energy of the configuration is at most determined by the n–l values. We shall label an n–l configuration using the scheme:

$(1s^2)$ both electrons in $n = 1$, $l = 0$ level,

$(1s^1, 2s^1)$ one electron in $n = 1$, $l = 0$ level,
              the other in $n' = 2$, $l' = 0$ level,

$(1s^1, 2p^1)$ one electron in $n = 1$, $l = 0$ level,
              the other in $n' = 2$, $l' = 1$, level, etc.

The ground-state configuration of helium is a $(1s^2)$ configuration and, as we have seen, it is nondegenerate with $E_{11} = -108$ eV. The first excited configuration $(1s^1, 2s^1)$ has energy

$$E_{12} = -54\left(\frac{1}{1^2} + \frac{1}{2^2}\right) = -68 \quad \text{eV}$$

and is fourfold degenerate. The magnetic subconfigurations

$$|1,2,0,0,m_l m_l' m_s m_{s'}\rangle$$

are (Figure 9-4)

$$\Psi_a = \frac{1}{\sqrt{2}}\begin{vmatrix} \psi_{100\frac{1}{2}}(1) & \psi_{200\frac{1}{2}}(1) \\ \psi_{100\frac{1}{2}}(2) & \psi_{200\frac{1}{2}}(2) \end{vmatrix}$$

$$\Psi_b = \frac{1}{\sqrt{2}}\begin{vmatrix} \psi_{100-\frac{1}{2}}(1) & \psi_{200\frac{1}{2}}(1) \\ \psi_{100-\frac{1}{2}}(2) & \psi_{200\frac{1}{2}}(2) \end{vmatrix}$$

$$\Psi_c = \frac{1}{\sqrt{2}}\begin{vmatrix} \psi_{100\frac{1}{2}}(1) & \psi_{200-\frac{1}{2}}(1) \\ \psi_{100\frac{1}{2}}(2) & \psi_{200-\frac{1}{2}}(2) \end{vmatrix} \qquad (9\text{-}40)$$

$$\Psi_d = \frac{1}{\sqrt{2}}\begin{vmatrix} \psi_{100-\frac{1}{2}}(1) & \psi_{200-\frac{1}{2}}(1) \\ \psi_{100-\frac{1}{2}}(2) & \psi_{200-\frac{1}{2}}(2) \end{vmatrix}.$$

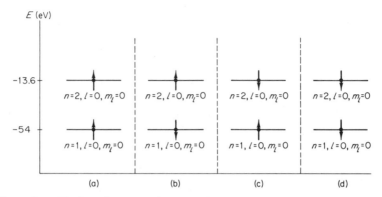

Figure 9-4   The four degenerate first excited configurations of helium's electrons. Each of these states has energy $E = \simeq -68$ eV.

In effect, the degeneracy of a level is equal to the number of ways in which the magnetic numbers can be arranged for a fixed *n–l* configuration, subject to the restrictions imposed by the exclusion principle. The reader should convince himself that the $(1s^1, 2p^1)$ configuration in helium is twelvefold degenerate.

A consequence of the degeneracy associated with a *n–l* configuration is the lack of uniqueness in the energy eigenstates. We would expect any linear combination of the form

$$|nn'll'\rangle = \sum_{\substack{m_l\ m_s \\ m_l'\ m_s'}} \sum_{n'l'm_l'm_s'} C_{nlm_lm_s} |nn'll'm_lm_l'm_sm_s'\rangle \tag{9-41}$$

to represent a new energy eigenstate of the ideal helium atom with

$$E_\alpha = E_{\substack{nl \\ n'l'}} = \varepsilon_{nl} + \varepsilon_{n'l'}. \tag{9-42}$$

Since we are interested in generating ideal atomic states which best approximate those for the *real* helium atom, we must construct states in (9-41) which diagonalize the perturbation matrix, in this case the electrostatic electron interaction,

$$\lambda \hat{V} = \frac{e^2}{|\mathbf{r}^{(2)} - \mathbf{r}^{(1)}|}. \tag{9-43}$$

The Slater determinants in (9-40) themselves do not satisfy this criterion. Since a straightforward diagonalization procedure is quite involved, we shall simplify matters using physical arguments.

It is convenient first to introduce the following many-body operators

$$\hat{\mathbf{L}}^{(T)} = \hat{\mathbf{L}}^{(1)} + \hat{\mathbf{L}}^{(2)} \quad \text{(total orbital momentum operator)}$$

$$\hat{\mathbf{S}}^{(T)} = \hat{\mathbf{S}}^{(1)} + \hat{\mathbf{S}}^{(2)} \quad \text{(total spin operator)}.$$

The components of these operators naturally satisfy

$$[\hat{L}_i^{(T)}, \hat{L}_j^{(T)}] = i\hbar\hat{L}_k^{(T)}$$

and $(ijk = xyz$ in cyclic order$).$ (9-44)

$$[\hat{S}_i^{(T)}, \hat{S}_j^{(T)}] = i\hbar\hat{S}_k^{(T)}$$

These operators, as one can easily verify, commute with the ideal helium Hamiltonian, that is, with the interaction term set equal to zero. Consequently, we may assume the existence of a set of eigenstates with the properties:

$$\mathcal{H}_0 | E_\alpha LSM_L M_S \rangle = E_\alpha | E_\alpha LSM_L M_S \rangle \qquad (E_\alpha = \varepsilon_{nl} + \varepsilon_{n'l'})$$

$$\hat{L}^{(T)2} | E_\alpha LSM_L M_S \rangle = L(L+1)\hbar^2 | E_\alpha LSM_L M_S \rangle$$

$$\hat{S}^{(T)2} | E_\alpha LSM_L M_S \rangle = S(S+1)\hbar^2 | E_\alpha LSM_L M_S \rangle \qquad (9\text{-}45)$$

$$\hat{L}_Z^{(T)} | E_\alpha LSM_L M_S \rangle = M_L \hbar | E_\alpha L SM_L M_S \rangle \qquad (|M_L| \leqslant L)$$

$$\hat{S}_Z^{(T)} | E_\alpha LSM_L M_S \rangle = M_S \hbar | E_\alpha LSM_L M_S \rangle \qquad (|M_S| \leqslant S).$$

The values of L and S remain to be determined. All we can say, based on (9-44), is that they are either integral or half-integral.

What is unusual about the $\hat{\mathbf{L}}^{(T)}$ and $\hat{\mathbf{S}}^{(T)}$ operators is that they also commute with the *total* Hamiltonian in (9-35). Equivalently, these observables are conserved even in the presence of the electrostatic repulsion between the electrons. Thus the atomic states of the real helium atom are simultaneously eigenstates of $\hat{\mathbf{L}}^{(T)}$ and $\hat{\mathbf{S}}^{(T)}$ and it seems plausible that the states listed above, (9-45), are suitable zeroth-order approximations. A direct, but not so simple, calculation would verify that the states in (9-45) indeed diagonalize the electrostatic perturbation term. What remains is to find the coefficients that couple Slater determinants into atomic states, namely,

$$|E_\alpha LSM_L M_S \rangle = \sum_{\substack{m_l \ m_s \\ m_l' \ m_s'}} \sum_{n'l'm_l'm_s'} C_{nlm_lm_s} \ |nn'll'm_l m_l' \ m_s m_s' \rangle. \qquad (9\text{-}46)$$

These Clebsch–Gordan coefficients are too complicated to evaluate here, but the calculations can be simplified somewhat using the methods of group theory. We shall content ourselves merely with the knowledge that these coefficients exist in tables.‡

Constructing atomic eigenstates of $\hat{\mathbf{L}}^{(T)}$ and $\hat{\mathbf{S}}^{(T)}$ as in (9-46) is known as *L–S* or *Russell–Saunders* coupling. It is possible to deduce the allowed values of L and S for a given n–l configuration using the vector model introduced

‡ See, for example, A. Messiah, "Quantum Mechanics," Volume II, p. 1054. Wiley, New York, 1962.

in Chapter 6. This model predicts that the allowed values of L and S are integrally spaced and restricted by the relations

$$|l - l'| \leqslant L \leqslant l + l' \qquad \text{and} \qquad |s - s'| \leqslant S \leqslant s + s'. \qquad (9\text{-}47)$$

Since for electrons we have $s = s' = \frac{1}{2}$, the total spin quantum number for atomic states of helium is either $S = 0$ or $S = 1$. In a $(1s^1, 2p^1)$ configuration, that is, $l = 0$ and $l' = 1$, the allowed values of L are $L = 0$ and $L = 1$. A word of caution is in order here. Recall that the sum in (9-46) is over only those configurations consistent with the Pauli exclusion principle. The vector model, however, generates some states that can only be constructed from configurations violating this principle. Hence the model actually gives an upper bound to the allowed values of L and S. Some of these values must be ruled out by the Pauli principle. The exclusion principle is especially effective when we couple configurations in which the electrons have the same principal (n) and orbital (l) quantum numbers. In this case, the electrons are said to be *equivalent*. Care must therefore be taken to rule out certain states predicted by the vector model when coupling equivalent electrons.

The ground state configuration of helium is $(1s^2)$. In this state, we are dealing with equivalent electrons since $n = n' = 1$ and $l = l' = 0$. Since the Pauli principle allows only one magnetic subconfiguration (namely, $m_l = m_l' = 0$, $m_s = \frac{1}{2}$, $m_s' = -\frac{1}{2}$), this level is *nondegenerate* and *unique*. According to the vector model (9-47), atomic states with $L = 0$, $S = 0$ and $L = 0$, $S = 1$ are possible. However, the latter has three magnetic substates ($M_s = 0$, $M_s = \pm 1$) that are *degenerate* and therefore cannot represent the ground state. We can therefore make the identification for the ground state of helium [see (9-39)],

$$|(1s^2)\rangle = |ELSM_L M_S\rangle = |-108, 0, 0, 0, 0\rangle$$
$$= [1/\sqrt{2}]\psi_{100}(1)\psi_{100}(2)$$
$$\times \{|\tfrac{1}{2}\rangle^{(1)}| -\tfrac{1}{2}\rangle^{(2)} - |\tfrac{1}{2}\rangle^{(2)}| -\tfrac{1}{2}\rangle^{(1)}\}. \qquad (9\text{-}48)$$

In other words, since the ground state of helium is nondegenerate, the Slater determinant on the right side of (9-48) is also an eigenstate of $\hat{L}^{(T)2}$, $\hat{L}_z^{(T)}$, $\hat{S}_z^{(T)2}$, and $\hat{S}^{(T)}$ with eigenvalues $L = S = M_L = M_S = 0$.

The four excited Slater determinants in (9-40) corresponding to the $(1s^1, 2s^1)$ configuration may be coupled, according to the vector model, into $L = 0$, $S = 0$ or $L = 0$, $S = 1$. Since the first is nondegenerate and the second is triply degenerate, the total degeneracy of the level remains fourfold as expected. In this case, both states generated by the vector model are consistent with Pauli's principle. This is expected since the electrons in a $(1s^1, 2p^1)$ configuration are nonequivalent (that is, $m \neq m'$).

We state without proof that the Clebsch–Gordan coefficients in (9-41) for the $(1s^1, 2s^1)$ configuration give the following atomic states,

$$| E, L, S, M_L, M_S \rangle$$

$$S = 0 \begin{cases} | -68,0,0,0,0 \rangle \to \dfrac{\{\psi_{100}(1)\psi_{200}(2) + \psi_{100}(2)\psi_{200}(1)\}}{\sqrt{2}} \\[2ex] \qquad\qquad \times \dfrac{\{|\frac{1}{2}\rangle^{(1)}| -\frac{1}{2}\rangle^{(2)} - |\frac{1}{2}\rangle^{(2)}| -\frac{1}{2}\rangle^{(1)}\}}{\sqrt{2}} \end{cases}$$

$$S = 1 \begin{cases} | -68,0,1,0,1 \rangle \to \dfrac{\{\psi_{100}(1)\psi_{200}(2) - \psi_{100}(2)\psi_{200}(1)\}}{\sqrt{2}} |\frac{1}{2}\rangle^{(1)}|\frac{1}{2}\rangle^{(2)} \\[2ex] | -68,0,1,0,0 \rangle \to \dfrac{\{\psi_{100}(1)\psi_{200}(2) - \psi_{100}(2)\psi_{200}(1)\}}{\sqrt{2}} \\[2ex] \qquad\qquad \times \dfrac{\{|\frac{1}{2}\rangle^{(1)}| -\frac{1}{2}\rangle^{(2)} + |\frac{1}{2}\rangle^{(2)}| -\frac{1}{2}\rangle^{(1)}\}}{\sqrt{2}} \\[2ex] | -68,0,1,0,-1 \rangle \to \dfrac{\{\psi_{100}(1)\psi_{200}(2) - \psi_{100}(2)\psi_{200}(1)\}}{\sqrt{2}} | -\frac{1}{2}\rangle^{(1)}| -\frac{1}{2}\rangle^{(2)}. \end{cases}$$

$$(9\text{-}49)$$

These four excited states have the following properties:

(a) While they themselves are not Slater determinants, they are linear combinations of the four determinants given in (9-40).

(b) They are eigenfunctions of $\hat{L}^{(T)2}$, $\hat{S}^{(T)2}$, $\hat{L}_z^{(T)}$, and $\hat{S}_z^{(T)}$. The first state has $L = S = M_L = M_S = 0$; since the magnetic degeneracy $(2S + 1)$ is unity, it is a *singlet* state. The last three states belong to the value $S = 1$ and have, respectively, $M_S = +1, 0, -1$. The $S = 1$ state is therefore a *triplet* state. The spin degeneracy, $2S + 1$, is called the *multiplicity* of the state.

(c) They all have energy

$$E_{12} = -54 \left(\frac{1}{1^2} + \frac{1}{2^2} \right) = -68 \quad \text{eV}.$$

(d) They are composed of either a symmetric space part and an anti-symmetric spin part or vice versa; consequently, each is overall antisymmetric as required for fermions.

(e) Most important, they diagonalize the electrostatic interaction term in (9-35) and thus constitute a suitable zeroth-order approximation to the energy eigenstates of the real helium atom.

The L and S values of an atomic state are denoted spectroscopically using a capital letter to represent the L value and a superscript to represent the multiplicity $2S + 1$. For example, the state ^{1}S has $L = S = 0$ while ^{3}P has $L = 1$ and $S = 1$.

The configuration $(1s^1, 2p^1)$ in helium is twelvefold degenerate. According to the vector model, we can couple the magnetic configuration into atomic states with $L = 1$, $S = 0$ (^{1}P) and $L = 1$, $S = 1$ (^{3}P). Since the latter is ninefold degenerate and the former is threefold degenerate, the total degeneracy is twelvefold. Again, since the electrons are nonequivalent, none of the atomic states above are ruled out. Because of the accidental degeneracy, the $(1s^1, 2s^1)$ and $(1s^1, 2p^1)$ configurations both have the same energies; consequently, the ^{1}S, ^{3}S, ^{1}P, and ^{3}P states all have energy $E = -68$ eV. When first-order perturbation theory is applied to the electrostatic interaction (Chapter 10), the energies of these states shift according to their L and S values (Figure 9-5.) We shall show in Chapter 10 that as a rule (Hund's rule) the Pauli exclusion principle causes the higher multiplicity states to be of lower energy for a given configuration. Hence in the $(1s^1, 2s^1)$ configuration, the ^{3}S is of lower energy than the ^{1}S state. Also in the $(1s^1, 2p^1)$ configuration, the ^{3}P resides below the ^{1}P state (see Figure 9-5).

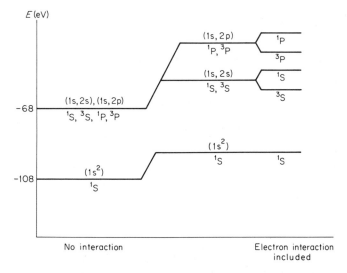

Figure 9-5   The shifts in energy of the ideal atomic states in helium when the electron interaction is taken into account.

While the discussion above is relevant to the helium atom, the method can be generalized to a Z-electron atom. If the dominant perturbation is the electrostatic interaction, then we must couple the Slater determinants into eigenstates of $\hat{\mathbf{L}}^{(T)} = \sum_{i=1}^{Z} \hat{\mathbf{L}}^{(i)}$ and $\hat{\mathbf{S}}^{(T)} = \sum_{i=1}^{Z} \hat{\mathbf{S}}^{(i)}$. The vector model for

L–S coupled states leads to the following restrictions on the values of L and S:

$$|\mathbf{l} + \mathbf{l'} + \mathbf{l''} \cdots|_{min} \leqslant L \leqslant (l + l' + l'' + \cdots)$$

$$|\mathbf{s} + \mathbf{s'} + \mathbf{s''} + \cdots|_{min} \leqslant S \leqslant (s + s' + s'' + \cdots) \quad (s = s' = s'' = \cdots = \tfrac{1}{2}). \tag{9-50}$$

The term $|\ \ |_{min}$ means that we add the vectors to give a minimum magnitude. It would appear, at first glance, that as the atomic number Z increases, the number of atomic states possible for a given configuration increases monotonically. Actually, due to the Pauli principle, many of the L–S coupled states predicted by the vector model are absent. In fact, atoms in the ground state usually have many equivalent electrons. It can be shown that the number of magnetic subconfigurations consistent with the exclusion principle for a given n–l configuration is given by the formula

$$\text{degeneracy} = \frac{g!}{k!(g-k)!}$$

where g is the number of states in the outermost unfilled subshell, that is, $2(2l+1)$, and k is the number of electrons in that subshell. Thus, for example, if an atom in the ground state has a completely filled outer shell, that is, $g = k$, then the configuration is nondegenerate and can only be coupled into a ^1S state as was the case in helium. The number of atomic states (spectral terms) for a given ground-state configuration is therefore determined by the outermost subshell. In Table 9-1 we have shown the possible spectral terms for

Table 9-1

Illustrating the Allowed L–S Coupled States for Equivalent s and p Electrons in an Outer Unfilled Subshell

Degeneracy of magnetic sub-configurations $\dfrac{g!}{k!(g-k)!}$	n–l Configuration	Atomic states (spectral terms) in the L–S coupling scheme	Total degeneracy of atomic states $\sum_{\text{spectral terms}} (2S+1)(2L+1)$
2	$(\cdots ns^1)$	^2S	2
1	$(\cdots ns^2)$	^1S	1
6	$(\cdots np^1)$	^2P	6
15	$(\cdots np^2)$	$^1S, {}^1D, {}^3P$	15
20	$(\cdots np^3)$	$^2P, {}^2D, {}^4S$	20
15	$(\cdots np^4)$	$^1S, {}^1D, {}^3P$	15
6	$(\cdots np^5)$	^2P	6
1	$(\cdots np^6)$	^1S	1

equivalent s and p electrons in an outer subshell. Note that certain terms generated by the vector model are absent due to the exclusion principle. Note also that the degeneracy associated with the magnetic subconfigurations for a given $n - l$ configuration is equal to the total degeneracy of the atomic states generated from the L–S coupling scheme. The degeneracy for an n–l configuration is largest in atoms whose outer subshell is half filled. In this case, the coupling process is the most complicated.

It is also possible to further couple the states $|ELSM_L M_S\rangle$ into eigenstates of $\hat{J}^{(T)2} = |\hat{\mathbf{L}}^{(T)} + \hat{\mathbf{S}}^{(T)}|^2$ and $\hat{J}_Z^{(T)} = \hat{L}_Z^{(T)} + S_Z^{(T)}$. We denote the latter by $|ELSJM_J\rangle$. The vector model restricts J according to $|L - S| \leqslant J \leqslant L + S$ with $|M_J| \leqslant J$. Both sets are Russell–Saunders states and both diagonalize the electrostatic perturbation. The Clebsch–Gordan coefficients for this coupling are identical in form to those used in the case of a single hydrogenic electron. If the secondary perturbation is due to spin–orbit effects (no strong external fields) then the select set is $|ELSJM_J\rangle$. To denote this set spectroscopically we add a subscript to represent the J value; for example, $|ELSJM_J\rangle = |E,2,1,3,M_J\rangle$ would be written $^3D_3$.

For atoms with very large atomic numbers it turns out that the *dominant* perturbations are due to spin–orbit effects. In that case, the Russell–Saunders $(L$–$S)$ atomic states do not diagonalize the perturbation. An entirely different coupling scheme, known as j–j coupling, is required for atoms with large Z. A discussion of j–j coupling may be found in any standard text on atomic theory.‡

X   The Quantum Ideal Gas

In the theory of solids, liquids, and gases the number of particles involved is usually of the order of $\sim 10^{23}$. Since particles in solids and liquids generally interact strongly, these systems cannot be regarded as ideal. In the case of dilute gases, the interparticle separation is usually sufficiently large so that the dilute gas can be considered ideal. Nevertheless, in certain situations, the indistinguishable nature of the gas particle plays an important role in the macroscopic behavior of the gas. For example, ^{4}He (a boson) behaves entirely differently from ^{3}He (a fermion) at low temperatures. Since the two atoms are isotopes, they have the same chemical properties. The difference in their nuclear spins accounts for their entirely different thermodynamic behavior.

---

‡ See, for example, E. U. Condon and G. H. Shortley, "The Theory of Atomic Spectra," Chapter X. Cambridge Univ. Press, London and New York, 1963.

The Hamiltonian for a collection of ideal indistinguishable particles in a *large* box of volume $\mathscr{V}$ is

$$\mathscr{H} = \sum_{q=1}^{N} \frac{\hat{p}^{(q)2}}{2m} + \hat{V}^{(q)} \qquad \text{where} \quad \begin{array}{ll} \hat{V}^{(q)} = 0 & \text{inside} \\ \hat{V}^{(q)} = \infty & \text{outside.} \end{array} \qquad (9\text{-}51)$$

The (box-normalized) single-particle states and their energies are of the form

$$\psi_{\mathbf{k},m_s} = \frac{1}{\sqrt{\mathscr{V}}} e^{i\mathbf{k}\cdot\mathbf{r}}|m_s\rangle \qquad \text{and} \qquad \varepsilon_{\mathbf{k}} = \frac{\hbar^2 k^2}{2m} \qquad (9\text{-}52)$$

with the allowed values of the quantum vector $\mathbf{k}$ given by

$$\mathbf{k} = \frac{2\pi}{\mathscr{V}^{1/3}}(n_x\mathbf{i} + n_y\mathbf{j} + n_z\mathbf{k}), \qquad n_x, n_y, n_z = \text{integers.}$$

The total states for bosons are symmetrized products of the form

$$|\alpha\rangle = |\text{total}\rangle \rightarrow \frac{1}{(N!)^{1/2}} \mathscr{P}\psi_{\mathbf{k},m_s}(1)\psi_{\mathbf{k}',m_{s'}}(2)\cdots\psi_{\mathbf{k}^{(N)},m_s^{(N)}}(N).$$

$$(9\text{-}53)$$

States for fermions can be written in terms of $N \times N$ Slater determinants as

$$|\alpha\rangle = |\text{total}\rangle \rightarrow \frac{1}{(N!)^{1/2}} \begin{vmatrix} \psi_{\mathbf{k},m_s}(1) & \cdots & \psi_{\mathbf{k}^{(N)},m_s^{(N)}}(1) \\ \vdots & \vdots & \vdots \\ \psi_{\mathbf{k},m_s}(N) & \cdots & \psi_{\mathbf{k}^{(N)},m_s^{(N)}}(N) \end{vmatrix}. \qquad (9\text{-}54)$$

To abbreviate the notation for the many-body problem, we note that once the *occupation numbers* of the single-particle states, $n_{\mathbf{k},m_s}$, are given we can construct the state function of the proper symmetry. We set

$$|\alpha\rangle = |\{n_{\mathbf{k},m_s}\}\rangle$$

where the quantum number $\alpha = \{n_{\mathbf{k},m_s}\}$ represents a sequence of integers indicating the number of particles in the single-particle states $\psi_{\mathbf{k},m_s}$. For fermions, we have the additional restriction that $n_{\mathbf{k},m_s}$ must be either one or zero (Pauli principle). The corresponding total energy of the gas is

$$E_\alpha = \sum_{\mathbf{k},m_s} n_{\mathbf{k},m_s}\varepsilon_{\mathbf{k}} = \sum_{\mathbf{k},m_s} n_{\mathbf{k},m_s}\frac{\hbar^2 k^2}{2m}. \qquad (9\text{-}55)$$

Of course, the occupation numbers must satisfy

$$N = \sum_{\mathbf{k},m_s} n_{\mathbf{k},m_s}. \qquad (9\text{-}56)$$

In the ground state of an ideal Bose gas, each particle is in the state of lowest energy. On the other hand, an ideal Fermi gas, such as the electrons in a metal, has one particle in each of the N states of lowest energy. The most energetic level occupied in the ground state is called the *Fermi level*.

XI   The N-Representation, the Density Operator, and Quantum Statistics

A given configuration $\{n_{k,m_s}\}$ of a many-particle system can be labeled by its energy $E_\alpha = \sum n_{k,m_s} \varepsilon_k$ and by its total particle number $N = \sum n_{k,m_s}$. We shall make the identification (for ideal systems)

$$|E_\alpha, N\rangle = |\alpha\rangle = |\{n_{k,\,m_s}\}\rangle. \tag{9-57}$$

The ket $|E_\alpha, N\rangle$ represents a state of energy E_α and N particles. This level is highly degenerate since it is generally possible to rearrange the occupation numbers keeping the energy and total number of particles fixed. Note that when the particles interact, the configuration on the right ceases to have meaning. The energy and total particle number on the left of (9-57) are still well defined but no longer given by (9-55) and (9-56).

For many applications it is sufficient to treat N as a fixed parameter. However in cases where it is not conserved, we must regard N as a quantum number and consider transitions between states of different N. For example, in high-energy processes where particles are created or destroyed, N is not fixed. A more immediate reason for regarding N as a variable (N-representation) is that it lends itself to the theory of quantum statistical mechanics. The states $|E_\alpha, N_\alpha\rangle$ will now be regarded as eigenstates of $\mathscr{H}$ and $\hat{N}$, that is,

$$\begin{aligned} \mathscr{H}|E_\alpha, N_\alpha\rangle &= E_\alpha|E_\alpha, N_\alpha\rangle \\ \hat{N}|E_\alpha, N_\alpha\rangle &= N_\alpha|E_\alpha, N_\alpha\rangle \end{aligned} \qquad (N_\alpha = 0,1,2,\ldots). \tag{9-58}$$

If the many-body system is in one of its eigenstates, $|E_\alpha, N_\alpha\rangle$, then the expectation value of a many-body operator (for example, the kinetic energy operator)

$$\hat{A} = \hat{A}(\hat{r}^{(1)}\hat{p}^{(1)}, \ldots \hat{r}^{(N_\alpha)}, \hat{p}^{(N_\alpha)})$$

can be evaluated as

$$\langle \hat{A} \rangle_\alpha = \langle E_\alpha, N_\alpha| \hat{A} |E_\alpha, N_\alpha\rangle. \tag{9-59}$$

When the system is known to be in one of its eigenstates, then we say that the system is in a "pure" state.

In practice, we often deal with a "mixed" state and have only a statistical knowledge of the system. The statistical information is thermodynamic in nature and is represented by a weighting function W_α that gives the relative thermodynamic weight of finding the system in the state $|E_\alpha, N_\alpha\rangle$. The thermodynamic average is obtained using

$$\bar{A}_{\text{therm}} = \sum_\alpha W_\alpha\langle E_\alpha, N_\alpha| \hat{A} |E_\alpha, N_\alpha\rangle. \tag{9-60}$$

In a sense, we are calculating a weighted thermal average of quantum expectation values. A scheme of this type is called an *ensemble average*, and from probabilistic considerations, we expect the ensemble weight functions to be normalized, that is,

$$\sum_{\alpha} W_{\alpha} = 1.$$

Possibly the most important ensemble weight function (density) is that of the *grand canonical ensemble*, defined by

$$W_{\alpha} = \frac{e^{(\mu N_{\alpha} - E_{\alpha})/k_{B}T}}{\sum_{\alpha} e^{(\mu N_{\alpha} - E_{\alpha})/k_{B}T}} \qquad (k_{B} = \text{Boltzmann's constant}). \qquad (9\text{-}61)$$

The denominator is chosen to normalize the ensemble density. The parameters μ (chemical potential) and T (Kelvin temperature) characterize the thermodynamic state of the system. Once the chemical potential and the temperature of the system are known, the thermodynamic average of any many-body operator can be calculated using

$$\bar{A}(\mu, T) = \frac{\sum_{\alpha} e^{(\mu N_{\alpha} - E_{\alpha})/k_{B}T} \langle E_{\alpha}, N_{\alpha} | \hat{A} | E_{\alpha}, N_{\alpha} \rangle}{\sum_{\alpha} e^{(\mu N_{\alpha} - E_{\alpha})/k_{B}T}}. \qquad (9\text{-}62)$$

The above average can be expressed in a more compact form as

$$\bar{A}(\mu, T) = \frac{\text{Tr}\hat{\rho}\hat{A}}{\text{Tr}\hat{\rho}} \qquad (9\text{-}63)$$

where the grand canonical ensemble *density operator* is

$$\hat{\rho} = e^{(\mu \hat{N} - \mathcal{H})/k_{B}T}.$$

Since the trace is the sum of the diagonal elements (in any basis), we find

$$\text{Tr } \hat{\rho}\hat{A} = \sum_{\alpha} \langle E_{\alpha}, N_{\alpha} | e^{(\mu \hat{N} - \mathcal{H})/k_{B}T} \hat{A} | E_{\alpha}, N_{\alpha} \rangle$$

$$= \sum_{\alpha} e^{(\mu N_{\alpha} - E_{\alpha})/k_{B}T} \langle E_{\alpha}, N_{\alpha} | \hat{A} | E_{\alpha}, N_{\alpha} \rangle$$

and

$$\text{Tr } \hat{\rho} = \sum_{\alpha} \langle E_{\alpha}, N_{\alpha} | e^{(\mu \hat{N} - \mathcal{H})/k_{B}T} | E_{\alpha}, N_{\alpha} \rangle = \sum_{\alpha} e^{(\mu N_{\alpha} - E_{\alpha})/k_{B}T}.$$

Equations (9-63) and (9-62) are in fact identical. Of course, in order to evaluate the thermal average in (9-62) we must first solve the many-body problem and find each of the expectation values $\langle E_{\alpha}, N_{\alpha} | \hat{A} | E_{\alpha}, N_{\alpha} \rangle$.

We state without proof that for *ideal non-interacting systems* the average over many-body states, $|E_\alpha, N_\alpha\rangle$, reduces to an average over single-particle states of the form

$$\bar{A}(\mu, T) = \sum_{\mathbf{k}, m_s} \bar{n}_{\mathbf{k}, m_s}(\mu, T)\langle \mathbf{k}, m_s | \hat{A} | \mathbf{k}, m_s\rangle. \tag{9-64}$$

Here, $\langle A\rangle_{\mathbf{k}, m_s}$ is the expectation value of a *one-body* operator in a single-particle state and $\bar{n}_{\mathbf{k}, m_s}(\mu, T)$ is essentially a *thermal average* of the occupation numbers. These average occupation numbers are given by‡ (see problem 9-8)

$$\bar{n}_{\mathbf{k}, m_s} = \frac{1}{e^{(\varepsilon_{\mathbf{k}} - \mu)/k_B T} - 1} \qquad \begin{array}{l}\text{(Bose–Einstein distribution for}\\ \text{indistinguishable bosons)}\end{array} \tag{9-65a}$$

$$\bar{n}_{\mathbf{k}, m_s} = \frac{1}{e^{(\varepsilon_{\mathbf{k}} - \mu)/k_B T} + 1} \qquad \begin{array}{l}\text{(Fermi–Dirac distribution for}\\ \text{indistinguishable fermions)}.\end{array} \tag{9-65b}$$

The thermodynamic energy and particle number are respectively

$$\bar{E}(\mu, T) = \sum_{\mathbf{k}, m_s} \bar{n}_{\mathbf{k}, m_s}\langle \mathbf{k}, m_s | \mathscr{H} | \mathbf{k}, m_s\rangle = \sum_{\mathbf{k}, m_s} \bar{n}_{\mathbf{k}, m_s} \frac{\hbar^2 k^2}{2m} \tag{9-66}$$

and

$$\bar{N}(\mu, T) = \sum_{\mathbf{k}, m_s} \bar{n}_{\mathbf{k}, m_s}. \tag{9-67}$$

In practice, the particles are contained in a large box and the quantum number $\mathbf{k}$ becomes quasi-continuous. The sum can be replaced by an integral using (5-16), which gives

$$\sum_{\mathbf{k}, m_s} \to \Gamma_s \int \frac{d\mathbf{k}\,\mathscr{V}}{(2\pi)^3} = \Gamma_s \int \frac{d\mathbf{p}\,\mathscr{V}}{h^3} \qquad \left(\mathbf{k} = \frac{\mathbf{p}}{\hbar}\right). \tag{9-68}$$

The spin degeneracy factor $\Gamma_s = 2s + 1$ is two for the electron ($s = \frac{1}{2}$). Thermal averages of spin-independent observables of the form $\hat{A}(\hat{\mathbf{p}})$ take the form

$$\bar{A}(\mu, T) = \Gamma_s \frac{\mathscr{V}}{h^3} \int \bar{n}(\mathbf{p})A(\mathbf{p})\, d\mathbf{p}$$

where

$$\bar{n}(\mathbf{p}) = \frac{1}{e^{(p^2/2m - \mu)/k_B T} + 1}.$$

---

‡ For distinguishable particles, $\bar{n}$ is given by the quantum Boltzmann distribution

$$\bar{n}_{\mathbf{k}, m_s} = e^{\mu/k_B T} e^{-\varepsilon_{\mathbf{k}}/k_B T}.$$

For a derivation see, for example, J. Kestin and J. R. Dorfman, "A Course in Statistical Thermodynamics," pp. 343–349, Academic Press, New York, 1971; or C. Kittel, "Elementary Statistical Physics," p. 86, Wiley, New York, 1958.

For example, the thermodynamic energy and particle number in (9-66) and (9-67) become

$$\bar{E}(\mu, T) = \Gamma_s \frac{\mathscr{V}}{h^3} \int \bar{n}(\mathbf{p}) \frac{p^2}{2m} \, d\mathbf{p} \qquad \text{(all } \mathbf{p} \text{ space)} \qquad (9\text{-}69a)$$

and

$$\bar{N}(\mu, T) = \Gamma_s \frac{\mathscr{V}}{h^3} \int \bar{n}(\mathbf{p}) \, d\mathbf{p} \qquad \text{(all } \mathbf{p} \text{ space)}. \qquad (9\text{-}69b)$$

We must be particularly careful when replacing the sum by an integral for low-temperature bosons. Near absolute zero, most of the Bose particles tend to occupy the single-particle states in the neighborhood of $\mathbf{k} = 0$. These states do not become quasi-continuous even for large volumes and so an integral cannot be used in place of the sum. In fact, below a certain critical temperature T_0 the particles appear to rapidly populate the ground state $\mathbf{k} = 0$. This phenomenon is known as the Einstein condensation (Figure 9-6). Since for fermions the Pauli principle prevents this "crowding" about $\mathbf{k} = 0$, this condensation is not encountered with electrons.

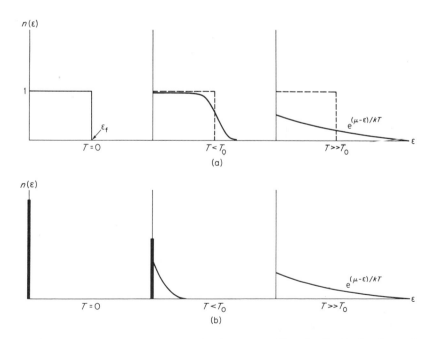

Figure 9-6 (a) The Fermi–Dirac and (b) the Bose–Einstein distribution for various temperatures. T_0 is the degeneracy temperature given in Eq. (9-73).

It is evident from (9-69) that thermal averages depend on the volume of the gas and should be written

$$\bar{E} = \bar{E}(\mu, T, \mathscr{V}) \qquad \text{and} \qquad \bar{N} = \bar{N}(\mu, T, \mathscr{V}).$$

It is also clear from Figure 9-6 that at $T = 0$ the Fermi–Dirac distribution has one particle in each state below the Fermi level‡ and none in those states with energy above ε_F. The Fermi level is therefore determined by the number of fermions in the system, or equivalently by the particle density. To relate ε_F to the density, we write (9-69b) as

$$\frac{\bar{N}}{\mathscr{V}} = \rho = \frac{\Gamma_s}{h^3} \int \bar{n}(\mathbf{p}) \, d\mathbf{p} \qquad (\Gamma_s = 2) \tag{9-70}$$

where (at $T = 0$)

$$\bar{n}(\mathbf{p}) = 1, \qquad |\mathbf{p}| \leqslant p_F$$
$$\bar{n}(\mathbf{p}) = 0, \qquad |\mathbf{p}| > p_F \qquad \left(\frac{p_F{}^2}{2m} = \varepsilon_F \right).$$

Transforming the integral to spherical coordinates, (9-70) becomes

$$\rho = \frac{2}{h^3} \int_0^{p_F} 4\pi p^2 \, dp = \frac{8\pi}{3h^3} p_F{}^3;$$

the Fermi momentum is simply

$$p_F = \left(\frac{3h^3}{8\pi} \rho \right)^{1/3}. \tag{9-71a}$$

Using $p_F = (2m\varepsilon_F)^{1/2}$, the Fermi energy becomes

$$\varepsilon_F = 3^{2/3}\pi^{4/3} \frac{\hbar^2}{2m} \rho^{2/3}. \tag{9-71b}$$

Summarizing, we note that the essential difference between the classical and quantum ideal many-body problems rests with statistical correlation effects. These effects are reflected in the one-particle statistical distribution function

$$\bar{n}_k = \frac{1}{e^{(\varepsilon_k - \mu)/k_B T} \pm 1}. \tag{9-72}$$

The ± 1 term in (9-72), resulting from these correlations and absent in the statistics of distinguishable particles, influences the thermodynamic behavior of fermions and bosons. When the average interparticle spacing between indistinguishable particles becomes large compared to their mean deBroglie

---

‡ The Fermi energy is equal to the chemical potential at $T = 0$, that is, $\varepsilon_F = \mu|_{T=0}$.

wavelengths, we expect these particles to acquire some degree of spatial distinguishability. If we assume that the average energy of a particle in a gas is $\bar{\varepsilon} \sim \pi k_B T$, then

$$\bar{\lambda} = \frac{h}{\bar{p}} = \frac{h}{(2m\bar{\varepsilon})^{1/2}} \simeq \frac{h}{(2\pi m k_B T)^{1/2}} .$$

The interparticle spacing is roughly given by $d = 1/\rho^{1/3}$. Consequently, we expect the particles to become quasi-distinguishable when

$$\bar{\lambda} \ll d, \qquad \frac{h}{(2\pi m k_B T)^{1/2}} \ll \frac{1}{\rho^{1/3}}$$

or

$$T_0 \ll T \qquad \text{where} \qquad T_0 = \frac{\rho^{2/3} h^2}{2\pi m k_B} . \qquad (9\text{-}73)$$

Equivalently, the Fermi–Dirac and Bose–Einstein distributions should reduce to a Boltzmann distribution, $e^{(\mu - \varepsilon_k)kT_B}$, when the temperature is much greater than the *degeneracy temperature*, T_0. This is indeed the case, as can be seen from Figure 9-6. On the other hand, below this temperature, quantum effects dominate. For fermions, the system is almost in the ground state. For bosons, we observe the onset of Einstein condensation.

For dilute systems, the density is low and T_0 in (9-73) is very close to absolute zero. On the other hand, the free electron gas in a metal has $\sim 10^{23}$ particles/cm^3. The corresponding degeneracy temperature is $T_0 \sim 10,000°$ K so that even at room temperature most of the electrons in a metal essentially occupy the lowest states permissible by Pauli's principle.

Our discussion thus far has been focused primarily on ideal systems. When interactions are considered, an exact solution is impossible except in rare instances. The relative strengths and ranges of interactions of physical interest often rule out the possibility of a direct perturbation treatment. In Chapter 10, we shall develop some techniques that have been found useful in treating real systems of particles.

Suggested Reading

Borowitz, S., "Fundamentals of Quantum Mechanics." Benjamin, New York, 1967.
Condon, E. U., and Shortley, G. H., "The Theory of Atomic Spectra," Chapter VI. Cambridge Univ. Press, London and New York, 1963.
Heitler, W., "Elementary Wave Mechanics," Chapter V. Oxford Univ. Press, London and New York, 1956.
Kestin, J., and Dorfman, J. R., "A Course in Statistical Thermodynamics," Chapter 8. Academic Press, New York, 1971.

Kittel, C., "Elementary Statistical Physics," pp. 86–102. Wiley, New York, 1958.
Merzbacher, E., "Quantum Mechanics," 2nd ed., Chapter 18. Wiley, New York, 1970.
Pauling, L., and Wilson, E. B., "Introduction to Quantum Mechanics," Chapter VIII. McGraw-Hill, New York, 1935.
Sommerfeld, A., "Thermodynamics and Statistical Mechanics," Section 37. Academic Press, New York, 1964.

Problems

9-1. (a)   Consider a collection of N particles, each of mass m_i and each experiencing an external potential $V_i(\mathbf{r})$. The interactions between the particles can be written $V_{ij} = V(|\mathbf{r}_i - \mathbf{r}_j|)$. Write the Hamiltonian for the system and deduce Hamilton's equations of motion for the ith particle.
(b)   Write the quantum-mechanical Hamiltonian in wave mechanics and show that when the interactions are neglected, the Schroedinger equation is separable into one-particle equations.

9-2. Consider two noninteracting particles (m_1 and m_2) in a one-dimensional oscillator potential $V = \frac{1}{2}Kx^2$ and in the same spin state.
(a)   Find the general expression for the quantized energies of the system.
(b)   If the particles are distinguishable, that is, $m_1 \neq m_2$, find the corresponding eigenfunctions $\psi_{n,n'}(x_1, x_2)$.
(c)   Repeat (b) for the case in which the particles are indistinguishable fermions ($m_1 = m_2$). Repeat for the case of bosons.
(d)   From part (c), calculate the probability of finding the particles with a given separation $x = x_2 - x_1$ when the system is in the state $n = 0$, $n' = 1$. Do this both for fermions and bosons.

9-3. (a)   Using (9-17) show that the effect of the exchange operator $\hat{P}^{(qt)}$ on a position eigenket is

$$\hat{P}^{(qt)}|\mathbf{r}^{(1)} \cdots \mathbf{r}^{(q)}, \mathbf{r}^{(t)} \cdots \mathbf{r}^{(N)}\rangle = |\mathbf{r}^{(1)} \cdots \mathbf{r}^{(t)}, \mathbf{r}^{(q)} \cdots \mathbf{r}^{(N)}\rangle.$$

(b)   Show using (a) that if $|\alpha^S\rangle$ and $|\alpha^A\rangle$ are eigenstates of $\hat{P}^{(qt)}$ with the property

$$\hat{P}^{(qt)}|\alpha^S\rangle = |\alpha^S\rangle, \qquad \hat{P}^{(qt)}|\alpha^A\rangle = -|\alpha^A\rangle,$$

then the corresponding state functions have the property

$$\Psi_\alpha^S(1, 2, \ldots, q, t, \ldots, N) = \Psi_\alpha^S(1, 2, \ldots, t, q, \ldots, N)$$

$$\Psi_\alpha^A(1, 2, \ldots, q, t, \ldots, N) = -\Psi_\alpha^A(1, 2, \ldots, t, q, \ldots, N).$$

(Hint: Use the procedures developed for parity in Chapter 4.)

9-4. (a) Find the degeneracy associated with a $(1s, 3d)$ configuration in helium and list the L–S coupled states. Are the electrons equivalent in this configuration?
(b) Verify that the degeneracy of the spectral terms equals that of the configuration.

9-5. (a) Find the degeneracy associated with a $(1s^2, 2s^2, 2p^3)$ configuration (that is, the ground state of the nitrogen atom).
(b) Find the possible L–S atomic states for this configuration. (Hint: Use Table 9-1.)
(c) The spectral term corresponding to the ground state can be found using the following (Hund's) rules:
  (1) For a given configuration the term with the greatest multiplicity (highest S value) is lowest in energy.
  (2) For this multiplicity, the term with the largest L value is lowest in energy.
From the terms obtained in (b) select the ground-state spectral term for the nitrogen atom.

9-6. Consider a single electron confined to an infinite spherical well of radius R, that is, $V = 0$ for $r < R$ and $V = \infty$ otherwise.
(a) Find expressions for the eigenfunctions and energies for the ground and first excited states using spherical coordinates.
(b) Assume next that *two* electrons are confined to this well. Neglecting the Coulomb interactions, find the Slater determinants (including spin) and energy eigenvalues corresponding to the ground and first excited states of the system.
(c) Taking the Coulomb interaction to be the dominant perturbation, find the select L–S coupled ground and first excited states and write the expressions for the first-order shift in energy for these states. (Hint: Follow the development of the helium atom.)

9-7. A configuration for a three-particle system has one particle in each of the following orbital states:

$$\psi_1 = N_1 \exp - m_1{}^2 r_1{}^2; \qquad \psi_2 = N_2 \sin m_2 r_2; \qquad \psi_3 = N_3 \exp - m_3 r_3;$$

m_i refers to the mass of each particle and N_i is a normalization constant.
(a) If the particles are distinguishable (that is, $m_1 \neq m_2 \neq m_3$), construct the total state function of the system.
(b) If the particles are indistinguishable (that is, $m_1 = m_2 = m_3$), construct the state functions of even and odd symmetry.

9-8. (a) Equation (9-62) can be applied to the thermodynamic particle number $\bar{N}$ by setting $\hat{A} = \hat{N}$. Show that $\bar{N}$ can be written in the form

$$\bar{N} = k_B T \frac{\partial}{\partial \mu} \ln \sum_\alpha \exp \left[\frac{(\mu N_\alpha - E_\alpha)}{k_B T} \right].$$

(Hint: $\hat{N} | E_\alpha, N_\alpha \rangle = N_\alpha | E_\alpha, N_\alpha \rangle$.)

(b) For *ideal* systems we have $N_\alpha = \sum_i n_i$ and $E_\alpha = \sum_i n_i \varepsilon_i$ where n_i and ε_i are respectively occupation numbers and energies of the single-particle states. Show that the above result takes the form

$$\bar{N} = k_B T \frac{\partial}{\partial \mu} \ln \sum_{\{n_i\}} \prod_i \exp \left[\frac{n_i (\mu - \varepsilon_i)}{k_B T} \right]$$

where the sum is over all possible configurations.

(c) Show by interchanging the sum and product that the above result takes the form

$$\bar{N} = \sum_i \bar{n}_i$$

where

$$\bar{n}_i = \frac{1}{\exp[(\varepsilon_i - \mu)/k_B T] \pm 1}$$

where the upper sign refers to fermions. (Hint: For fermions we have the restriction $n_i = 0, 1$.)

9-9. A neutron star is believed to contain neutrons at a density of 10^{14} gm/cm^3. Find the degeneracy temperature for such a star. The temperature of neutron stars is believed to be well below the degeneracy temperature. What can you say about the neutron distribution?

10 | Interacting Many-Particle Systems

When interactions are neglected in a many-particle system, the particles retain their kinematic identity and remain uncorrelated except for the exchange effects associated with indistinguishability. In real systems, where the interactions play an important role, the many-particle Schroedinger equation is no longer separable. Consequently, the many-body problem does not reduce to N one-body problems and the notion of single particles ceases to have meaning quantum-mechanically.

It is sometimes possible to perform a canonical transformation to a new set of coordinates and momenta in such a way that the new Hamiltonian no longer explicitly contains two-body interactions. When we "transform away" interactions in this way, we are led to the notion of *quasi-particles* each of which has one of the new canonical variables assigned to it. The quasi-particles constitute an ideal system and may be treated using the techniques developed in the previous chapter. Of course, the physical significance of these quasi-particles needs special interpretation, but by using the inverse transformation we can always relate their properties to those of the original particles. As an example, we shall demonstrate that the normal-coordinate transformation, when applied to a set of mass points coupled by springs, generates a new Hamiltonian representing an ideal gas of quasi-particles called *phonons*.‡ Each phonon (quantum of elastic vibration) can be shown to be a boson. Furthermore, these phonons can be created or destroyed.

This approach to the many-body problem is contingent upon finding the proper canonical transformation. Otherwise, approximation methods, for example, perturbation theory, variational techniques, and so forth, must be used.

As a starting point, we shall consider the problem of two particles and develop a transformation which will always render it soluble.

I   The Isolated Two–Body Problem

The hydrogen atom (proton plus electron) is an important example of an isolated (no external forces) two-body problem. In this case the particles interact via the Coulomb forces. In our previous discussion of hydrogen (Chapter 5), we had assumed that the nucleus remained stationary. When nuclear motion is taken into account, the Schroedinger equation takes the form

$$\left\{ -\frac{\hbar^2}{2M} \nabla^{(1)2} - \frac{\hbar^2}{2m} \nabla^{(2)2} - \frac{e^2}{|\mathbf{r}^{(2)} - \mathbf{r}^{(1)}|} \right\} \Psi_\alpha(\mathbf{r}^{(1)}, \mathbf{r}^{(2)}) = E_\alpha \Psi_\alpha(\mathbf{r}^{(1)}, \mathbf{r}^{(2)}) \quad (10\text{-}1)$$

where M is the proton mass, m the electron mass, $\mathbf{r}^{(1)}$ the proton coordinate, and $\mathbf{r}^{(2)}$ the electron coordinate.

Since the particles are distinguishable ($M \neq m$), we expect that $\Psi_\alpha(\mathbf{r}^{(1)}, \mathbf{r}^{(2)})$ will be neither symmetric nor antisymmetric under the exchange operation $1 \leftrightarrow 2$. As it stands, (10-1) is not separable into an electron and a proton part. To effect a separation, we introduce the center of mass (c.m.) transformation (Figure 10-1)

$$\mathbf{R} = \frac{M\mathbf{r}^{(1)} + m\mathbf{r}^{(2)}}{M + m} \qquad \text{(c.m. vector)}$$

$$\mathbf{r} = \mathbf{r}^{(2)} - \mathbf{r}^{(1)} \qquad \text{(relative vector)}.$$

‡ Strictly speaking, the term quasi-particle is usually reserved for the case in which there is some correspondence to the original particle. The phonon actually corresponds to a situation where a large group of particles are excited and is therefore termed a "collective excitation." We shall, however, use the term quasi-particle in the looser sense and apply it to the phonon.

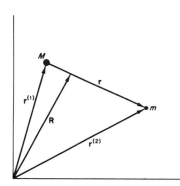

Figure 10-1 The center of mass vector **R** and relative vector **r** for a two-body system.

Using this transformation, (10-1) becomes, after some simplification,

$$\left\{-\frac{\hbar^2}{2(M+m)}\nabla^{(R)2}-\frac{\hbar^2}{2\mu}\nabla^{(r)2}-\frac{e^2}{r}\right\}\Psi_\alpha(\mathbf{R},\mathbf{r})=E_\alpha\Psi_\alpha(\mathbf{R},\mathbf{r}) \qquad (10\text{-}2)$$

where

$$\mu=\frac{mM}{M+m}=\frac{m}{1+m/M} \qquad \text{(reduced mass of system).}$$

The quantity $|\Psi_\alpha(\mathbf{R},\mathbf{r})|^2$ represents the probability density of finding the hydrogen atom with its center of mass at the point **R** in space and with an electron–proton separation **r**. Since this two-particle problem has six degrees of freedom (neglecting spin), we expect the state index α to represent a set of six quantum numbers.

Using a trial solution of the form $\Psi_\alpha=\xi_\mathbf{k}(\mathbf{R})\psi_i(\mathbf{r})$, we find that (10-2) separates into

$$-\frac{\hbar^2}{2(M+m)}\nabla^{(R)2}\xi_\mathbf{k}(\mathbf{R})=\varepsilon_\mathbf{k}^{(\mathrm{c.m.})}\xi_\mathbf{k}(\mathbf{R}) \qquad (10\text{-}3)$$

and

$$\left\{-\frac{\hbar^2}{2\mu}\nabla^{(r)2}-\frac{e^2}{r}\right\}\psi_i(\mathbf{r})=\varepsilon_i^{(\mathrm{int})}\psi_i(\mathbf{r}) \qquad (10\text{-}4)$$

with

$$E_\alpha=\varepsilon_\mathbf{k}^{(\mathrm{c.m.})}+\varepsilon_i^{(\mathrm{int})}$$

where $\varepsilon_i^{(\mathrm{int})}$ refers to the internal energy of the atom. This separation results only if the system is isolated, that is, if external forces are absent. The c.m. equation (10-3) is equivalent to that of a free particle and represents a pure translation. The spectrum is continuous with

$$\xi_\mathbf{k}=e^{i\mathbf{k}\cdot\mathbf{R}}, \qquad \varepsilon_\mathbf{k}^{(\mathrm{c.m.})}=\frac{\hbar^2k^2}{2(M+m)}.$$

Since it does not affect the internal behavior, the translation part can be neglected when calculating the quantized hydrogenic levels. The second equation (10-4) governs the quantum behavior of the relative motion of the particles and determines the nature of the internal spectrum of the atom. Except for the replacement

$$m \to \mu = \frac{m}{1 + m/M}$$

(10-4) is identical to (5-1) in Chapter 5. Consequently, we expect the eigensolutions to be of the form

$$\psi_i = \psi_{nlm_l} = R_{nl}(r)Y_{lm_l}(\theta, \phi)$$

and

$$\varepsilon_i^{(\text{int})} = \varepsilon_n = -\frac{\mu e^4}{2\hbar^2 n^2}.$$

Transitions between these levels lead to an emission spectrum essentially similar to the one found when the proton is at rest; the only difference is that the Rydberg constant contains μ instead of m. The reduced mass corrections in hydrogen are observed with high-resolution spectroscopes even though they are only of the order $m/M \sim \frac{1}{2} \times 10^{-3}$. In fact, it is possible to distinguish between the emission spectra of hydrogen and deuterium (proton + neutron in nucleus); the spectra differ as a result of the different m/M ratios.

Unfortunately, no one has yet found a transformation that can be applied to the general three-body problem. Thus, for example, an exact solution to the helium atom (nucleus plus two electrons) does not exist.

II Scattering from a Mobile Target

The analysis of the scattering process developed in Chapter 8 was restricted to problems in which the target remained stationary throughout the collision. The target's recoil is in fact small when its mass is much larger than that of the projectile. If, on the other hand, the target and projectile masses are comparable, the target's motion is important and we have a two-body scattering problem.

The two-body scattering problem can be solved using the center of mass transformation. If the scattering process is observed from a frame moving with the center of mass (c.m. frame), then the scattering angle observed is the same as if the target mass M were fixed and the projectile mass m were replaced by the reduced mass $\mu = [m/(1 + m/M)]$. The scattering angle $\theta_{\text{c.m.}}$ and the cross section $\sigma_{\text{c.m.}}$, as observed in the c.m. frame, can be calculated using the methods developed in Chapter 8, replacing m by μ.

Unfortunately for theorists, experimental scattering data is taken with respect to a frame fixed in the laboratory. The laboratory cross section σ_{lab} differs from $\sigma_{\text{c.m.}}$ in that the scattering angles appear different when viewed from the two frames‡ (Figure 10-2). The number of particles scattered into a

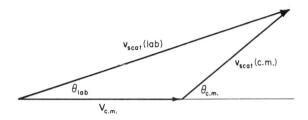

Figure 10-2   Using the relation $v_{\text{scat}}(\text{lab}) = v_{\text{scat}}(\text{c.m.}) + V_{\text{c.m.}}$ to illustrate the relation between Θ_{lab} and $\Theta_{\text{c.m.}}$.

given solid angle must be the same regardless of the frame used. For central force scattering problems, this number is

$$dN = I_0\,\sigma_{\text{c.m.}}\;d\Omega_{\text{c.m.}} = I_0\,\sigma_{\text{lab}}\;d\Omega_{\text{lab}}$$

from which it follows that

$$\sigma_{\text{lab}} = \sigma_{\text{c.m.}}\,\frac{d\Omega_{\text{c.m.}}}{d\Omega_{\text{lab}}} = \frac{\sin\theta_{\text{c.m.}}}{\sin\theta_{\text{lab}}}\,\frac{d\theta_{\text{c.m.}}}{d\theta_{\text{lab}}}\,\sigma_{\text{c.m.}} \tag{10-5}$$

where $\sigma_{\text{c.m.}} = |f(\theta_{\text{c.m.}})|^2$.

Once the relation between $\theta_{\text{c.m.}}$ and θ_{lab} is known, the c.m. cross section (obtained theoretically) can be transformed to the lab cross section (observed experimentally). The geometrical relation between these angles is illustrated in Figure 10-2, from which we obtain the relation

$$\tan\theta_{\text{lab}} = \frac{v_{\text{scat}}(\text{c.m.})\,\sin\theta_{\text{c.m.}}}{v_{\text{scat}}(\text{c.m.})\,\cos\theta_{\text{c.m.}} + V_{\text{c.m.}}}. \tag{10-6}$$

If the target is initially stationary, then the initial projectile speeds as observed from the c.m. and lab frames are related by

$$v_0(\text{c.m.}) = \frac{\mu}{m}\,v_0(\text{lab}).$$

‡ Within the last decade "colliding beam" experiments have been developed in which identical masses with equal and opposite velocities collide. Here the c.m. and lab frames coincide.

Furthermore, since energy is conserved in the c.m. frame, we have

$$v_0(\text{c.m.}) = \frac{\mu}{m} v_0(\text{lab}) = v_{\text{scat}}(\text{c.m.}). \qquad (10\text{-}7)$$

The velocity of the center of mass can be expressed as

$$(M + m)\mathbf{V}_{\text{c.m.}} = mv_0(\text{lab}) \qquad \text{or} \qquad \mathbf{V}_{\text{c.m.}} = \frac{\mu}{M} v_0(\text{lab}). \qquad (10\text{-}8)$$

Using (10-7) and (10-8) in (10-6), we have finally

$$\tan \theta_{\text{lab}} = \frac{\sin \theta_{\text{c.m.}}}{\cos \theta_{\text{c.m.}} + (m/M)}$$

which is the required relation between the angles (Problem 10-3b). Note that as $m/M \to 0$, the two angles, and consequently the two cross sections, become identical as expected.

We have assumed above that the particles were distinguishable (that is, $M \neq m$). However, if the two particles are indistinguishable, then $\sigma_{\text{c.m.}}$ must reflect the proper exchange symmetry. If the masses are identical, then, when observed from the c.m. frame, the particles appear to approach each other with equal and opposite velocities. In order to conserve momentum, they must scatter with equal and opposite velocities (Figure 10-3). Consequently, the targets recoil at an angle of 180° with respect to the projectiles.‡ The scattering amplitudes for the target and projectile are related by

$$f_{\text{target}}(\pi - \theta_{\text{c.m.}}) = f_{\text{projectile}}(\theta_{\text{c.m.}}).$$

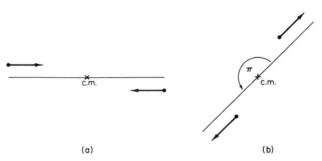

(a)   (b)

Figure 10-3 The kinematics (a) before and (b) after a collision of two masses as observed from the center of mass.

---

‡ For two-body scattering, momentum conservation will always imply that scattered particles move 180° with respect to each other in the c.m. frame even when $m \neq M$.

However, if the target and projectile are indistinguishable spinless bosons, the total cross section must be symmetrized and taken to be‡

$$\sigma_{c.m.} = |f_{tot}(\theta_{c.m.})|^2 = |f(\theta_{c.m.}) + f(\pi - \theta_{c.m.})|^2$$
$$= |f(\theta_{c.m.})|^2 + |f(\pi - \theta_{c.m.})|^2 + 2\,\text{Re}\,f(\theta_{c.m.})f(\pi - \theta_{c.m.})$$
$$\text{(spinless bosons).} \quad (10\text{-}9)$$

The first two terms on the right are expected on classical grounds. Since the particles are identical, the total classical cross section is *redefined* to include the recoiling targets. The third term however is purely quantum-mechanical and is due to exchange effects.

If we assume α particles to be spinless bosons, the c.m. cross section for α–α scattering can be calculated directly. Using the Rutherford result, (8-14), which is

$$f(\theta) = -\frac{\Gamma(1 + i\gamma)}{\Gamma(1 - i\gamma)}\frac{\gamma}{k}\frac{\exp(-i\gamma \ln \sin^2 \theta/2)}{2\sin^2 \theta/2}$$

(10-9) becomes after some lengthy algebra

$$\sigma_{c.m.} = \left(\frac{Z^2 e^2}{4\varepsilon}\right)^2 \left(\frac{1}{\sin^4 \theta/2} + \frac{1}{\cos^4 \theta/2} + \frac{2\cos\gamma(\ln\tan^2 \theta/2)}{\sin^2 \theta/2 \cos^2 \theta/2}\right)$$

$$(Z = 2; \quad \alpha\text{–}\alpha \text{ scattering}) \quad (10\text{-}10)$$

where $\gamma = Z^2 e^2 \mu/\hbar(2\mu\varepsilon)^{1/2}$. The first two terms on the right do not contain h and are also obtained classically. The third term is the *exchange* term§ (Mott scattering) and is completely quantum-mechanical in nature.

For spin one-half fermions (for example, electron–electron scattering), the scattering can occur in either a singlet ($S = 0$) or a triplet ($S = 1$) state. These states have symmetric and antisymmetric space parts respectively [see (9-49)]. If the beam is unpolarized, we generate an average cross section by weighting each state ($S = 0$ or $S = 1$) by its multiplicity, namely,

$$\bar{\sigma}_{c.m.} = \tfrac{1}{4}|f(\theta) + f(\pi - \theta)|^2 + \tfrac{3}{4}|f(\theta) - f(\pi - \theta)|^2$$
$$= |f(\theta)|^2 + |f(\pi - \theta)|^2 - \text{Re}\,f(\theta)f(\pi - \theta) \quad \text{(spin $\tfrac{1}{2}$ fermions).}$$
$$(10\text{-}11)$$

Both (10-9) and (10-11) can eventually be transformed to the lab frame using (10-5).

---

‡ The reduced mass $\mu = \tfrac{1}{2}m$ is to be used in this relation.
§ As $h \to 0$ ($\gamma \to \infty$), the numerator oscillates very rapidly with variations in θ and becomes immeasurable. The exchange term then becomes negligible and we recover the classical result.

III The Helium Atom—A Perturbation Treatment

In Chapter 9, we found the proper zeroth-order states of the ideal helium atom to be of the form $|E_\alpha LSM_L M_S\rangle$. Using the perturbation methods of Chapter 7, the first-order corrections to the energy are given by

$$E_{\alpha LS} = E_\alpha + \langle E_\alpha LSM_L M_S | \frac{e^2}{|\mathbf{r}^{(2)} - \mathbf{r}^{(1)}|} | E_\alpha LSM_L M_S\rangle$$

$$= -54\left(\frac{1}{n^2} + \frac{1}{n'^2}\right) + \int d\mathbf{r}^{(1)} \, d\mathbf{r}^{(2)} \Psi^*_{E_\alpha LSM_{LM_S}}(\mathbf{r}^{(1)}, \mathbf{r}^{(2)}) \left[\frac{e^2}{|\mathbf{r}^{(2)} - \mathbf{r}^{(1)}|}\right]$$

$$\times \, \Psi_{E_\alpha LSM_{LM_S}}(\mathbf{r}^{(1)}, \mathbf{r}^{(2)}). \tag{10-12}$$

Since the perturbation only involves spatial variables, the normalized spin part of $\Psi_{E_\alpha L_L SM_{L_S}}$ contributes unity to the integral and can be neglected in the calculations that follow.

The energy correction to the ^1S ground state ($n = n' = 1$) of helium gives, using (9-48),

$$E_{\alpha 00} = -108 + \int d\mathbf{r}^{(1)} \, d\mathbf{r}^{(2)} \psi^*_{100}(\mathbf{r}^{(1)}) \psi^*_{100}(\mathbf{r}^{(2)}) \frac{e^2}{|\mathbf{r}^{(2)} - \mathbf{r}^{(1)}|}$$

$$\times \, \psi_{100}(\mathbf{r}^{(1)}) \psi_{100}(\mathbf{r}^{(2)}).$$

The orbital functions are hydrogenlike, that is,

$$\psi_{100}(r) = \frac{1}{\sqrt{\pi}} \left(\frac{Z}{a}\right)^{3/2} e^{-Zr/a} \qquad (Z = 2)$$

and the integral above can be shown to be $\Delta E \sim 34$ eV. The ground-state energy corrected to first order is (Figure 10-4)

$$E_{\alpha 00} = -108 + 34 = -74 \quad \text{eV} \qquad \text{(ground state).}$$

The first ionization energy of helium becomes

$$E_{\text{ion}} = E_{100} - E_{\alpha 00} = -54 - (-74) = 20 \quad \text{eV.}$$

This compares favorably with the observed value of ~ 24 eV. However the relative size of the energy shift gives

$$\left|\frac{\Delta E}{E_\alpha}\right| \simeq \frac{34}{108} \simeq 0.3$$

which is considerable; consequently the validity of the perturbation theory as presented remains questionable.

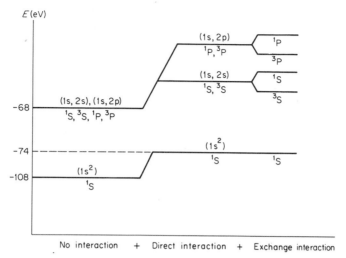

Figure 10-4 The shifts in energy of the ideal atomic states in helium when the electron interaction is taken into account.

The shifts in the first excited ^1S and ^3S states are found in a similar manner. Using (9-49), these shifts are found to be

$$\Delta E = \tfrac{1}{2} \int d\mathbf{r}^{(1)}\, d\mathbf{r}^{(2)} \{\psi_{100}^*(\mathbf{r}^{(1)})\psi_{200}^*(\mathbf{r}^{(2)}) \pm \psi_{100}^*(\mathbf{r}^{(2)})\psi_{200}^*(\mathbf{r}^{(1)})\}$$

$$\times \frac{e^2}{|\mathbf{r}^{(2)} - \mathbf{r}^{(1)}|} \{\psi_{100}(\mathbf{r}^{(1)})\psi_{200}(\mathbf{r}^{(2)}) \pm \psi_{100}(\mathbf{r}^{(2)})\psi_{200}(\mathbf{r}^{(1)})\}$$

with the upper and lower signs taken for the ^1S and ^3S states, respectively. The energy shifts can be decomposed as follows:

$$\Delta E = \tfrac{1}{2} \int d\mathbf{r}^{(1)}\, d\mathbf{r}^{(2)}\, \frac{e^2}{|\mathbf{r}^{(2)} - \mathbf{r}^{(1)}|}\, \{\,|\psi_{100}(\mathbf{r}^{(1)})|^2\, |\psi_{200}(\mathbf{r}^{(2)})|^2$$

$$+\, |\psi_{100}(\mathbf{r}^{(2)})|^2\, |\psi_{200}(\mathbf{r}^{(1)})|^2\}$$

(direct integral)

$$\pm \tfrac{1}{2} \int d\mathbf{r}^{(1)}\, d\mathbf{r}^{(2)}\, \frac{e^2}{|\mathbf{r}^{(2)} - \mathbf{r}^{(1)}|}\, \{2\, \mathrm{Re}\, \psi_{100}^*(\mathbf{r}^{(1)})\psi_{200}^*(\mathbf{r}^{(2)})$$

$$\times \psi_{100}(\mathbf{r}^{(2)})\psi_{200}(\mathbf{r}^{(1)})\}$$

(exchange integral).

The direct integral indicates that both first excited states (^1S and ^3S) are shifted as a result of the Coulomb repulsion of the electrons. The positive exchange integral for the ^1S state demonstrates a secondary upward shift for this state. The negative exchange integral for the ^{3}S state shows that the triplet is somewhat downshifted. Generalizing, we observe that the direct integral produces a shift according to the L value. Furthermore, a second splitting occurs according to the singlet or triplet nature of the state. The triplet state is always below the singlet state for a given configuration‡ (Figure 10-4).

The splitting of the singlet and triplet states can be explained in terms of the Pauli exclusion principle. Recall that the Slater determinant making up an $S = 0$ singlet state has the particles with opposite spin orientations. The spatial part is symmetric and repulsive correlations are absent. The particles can be near one another and the positive interaction energy can be large. On the other hand, the $S = 1$ state includes Slater determinants with like (parallel) spin orientations. Statistical repulsions associated with the antisymmetric space part keep the electrons apart and thus reduce the positive interaction energy. The ^{3}S ($S = 1$) state therefore resides below the ^{1}S ($S = 0$) state.

The electrostatic interaction does not remove the degeneracy completely since the perturbed energy in (10-12) is still independent of M_L and M_S. Consequently, any linear combinations of these magnetic substates are still suitable zeroth-order energy eigenstates of the electrostatic helium Hamiltonian. Since this Hamiltonian commutes with the operator $\hat{\mathbf{J}}^{(T)} = \hat{\mathbf{L}}^{(T)} + \hat{\mathbf{S}}^{(T)}$, we should be able to construct energy eigenstates of the form

$$|E_\alpha LSJM_J\rangle = \sum_{M_L, M_S} C_{M_L, M_S} |E_\alpha LSM_L M_S\rangle \qquad (10\text{-}13)$$

with the properties

$$\hat{J}^{(T)2} |E_\alpha LSJM_J\rangle = J(J+1)\hbar^2 |E_\alpha LSJM_J\rangle \qquad (|L-S| \leqslant J \leqslant L+S)$$

and

$$\hat{J}_z^{(T)} |E_\alpha LSJM_J\rangle = M_J \hbar |E_\alpha LSJM_J\rangle \qquad (|M_J| \leqslant J).$$

In other words, both the sets $|M_L M_S\rangle$ and $|JM_J\rangle$ diagonalize the electrostatic perturbation. If external fields are absent, then the dominant perturbation (after the electrostatic interaction) is due to fine-structure effects. The correct zeroth-order set is $|JM_J\rangle$, with the corresponding first-order energies reflecting a J dependence. If a weak **B** field is introduced, we observe a shift in the levels according to the value of M_J (anomalous Zeeman effect (7-57)).

If a strong **B** field is applied to the helium atom, fine-structure terms in the Hamiltonian can be neglected and the zeroth-order states are of the form $|M_L M_S\rangle$. The level shifts depend on M_L and M_S in accordance with a Paschen–Back pattern (6-38).

---

‡ A generalization of this principle to many-electron atoms is known as Hund's rule. It states that the higher the spin multiplicity (for a fixed configuration) the lower the energy (see Problem 9-5).

To distinguish between the states $|E_\alpha LSM_L M_S\rangle$ and $|E_\alpha LSJM_J\rangle$, a subscript (J value) is assigned to the latter's spectroscopic symbol. For example, ^3P would represent the state $|E_\alpha 1, 1, M_L M_S\rangle$ while $^3P_2$ would represent the state $|E_\alpha 1, 1, 2, M_J\rangle$.

The selection rules for radiative dipole transitions in L–S coupled atoms are analogous to those derived for hydrogen in Chapter 7, namely, $\Delta S = 0$ and $\Delta L = 0, \pm 1$, with $\Delta J = 0, \pm 1$, but $J = 0 \to J = 0$ forbidden. The rule $\Delta S = 0$ is particularly interesting since it implies that only singlet to singlet ($S = 0 \to S = 0$) or triplet to triplet ($S = 1 \to S = 1$) transitions are possible in the dipole approximation. If helium is in its first excited triplet state, ^3S, it cannot return to the ground state ^1S via a dipole transition. The first excited triplet state is thus a metastable (highly stable) state. A gas composed of helium atoms may be regarded as a two-component system. Those atoms in triplet states and singlet states constitute respectively *orthohelium* and *parahelium*. Each subsystem is spectroscopically distinct from the other.

We have already seen that the electrostatic interactions between the electrons in helium produce considerable corrections to the ideal (hydrogen-like) levels. Although the first-order corrections in energy are in agreement with experiment, there is no reason to believe that higher-order perturbation theory will improve matters. In fact, the perturbation series may not even converge. Fortunately, there are other methods available to treat real atoms, which yield finite results. These will be discussed below.

IV The Helium Atom—A Variational Approach

The variational method has been used with great success (Hylleraas, 1930) in estimating the energies of the helium atom. We shall begin by considering the single-parameter trial function of the form

$$\Psi_{\text{trial}}(\alpha, \mathbf{r}^{(1)}, \mathbf{r}^{(2)}) = \frac{\alpha^3}{\pi a^3} \exp\left(-\frac{\alpha(r^{(1)} + r^{(2)})}{a}\right) \qquad \left(a = \frac{\hbar^2}{me^2}\right). \quad (10\text{-}14)$$

This trial function is essentially a product of hydrogenlike functions, the variational parameter being α. The variational integral (using the helium Hamiltonian) is

$$I(\alpha) = \int d\mathbf{r}^{(1)} d\mathbf{r}^{(2)} \Psi_{\text{trial}}^*(\alpha, \mathbf{r}^{(1)}, \mathbf{r}^{(2)}) \left\{ -\frac{\hbar^2 \nabla^{(1)2}}{2m} - \frac{\hbar^2 \nabla^{(2)2}}{2m} - \frac{2e^2}{r^{(1)}} \right.$$

$$\left. -\frac{2e^2}{r^{(2)}} + \frac{e^2}{|\mathbf{r}^{(2)} - \mathbf{r}^{(1)}|} \right\} \Psi_{\text{trial}}(\alpha, \mathbf{r}^{(1)}, \mathbf{r}^{(2)})$$

$$= \langle T_1 \rangle + \langle T_2 \rangle + \langle V_1 \rangle + \langle V_2 \rangle + \langle V_{12} \rangle.$$

The integrals turn out to give

$$\langle T_1 \rangle = \langle T_2 \rangle = -\alpha^2 \varepsilon_1$$

$$\langle V_1 \rangle = \langle V_2 \rangle = 4\alpha\varepsilon_1 \qquad \left(\varepsilon_1 = \frac{-me^4}{2\hbar^2} = -13.6 \text{ eV} \right).$$

$$\langle V_{12} \rangle = -\tfrac{5}{4}\alpha\varepsilon_1$$

Differentiating the variational integral

$$I(\alpha) = -\varepsilon_1[2\alpha^2 - 8\alpha + \tfrac{5}{4}\alpha] = -\varepsilon_1[2\alpha^2 - \tfrac{27}{4}\alpha]$$

we obtain

$$\frac{dI}{d\alpha}\bigg|_{\alpha=\alpha_0} = -\varepsilon_1(4\alpha_0 - \tfrac{27}{4}) = 0 \qquad \text{or} \qquad \alpha_0 = \tfrac{27}{16}.$$

The variational energy and eigenfunction for the ground state of helium become

$$E_{\text{var}} = I(\alpha_0) = -\varepsilon_1[2(\tfrac{27}{16})^2 - \tfrac{27}{4}\tfrac{27}{16}] = 2(\tfrac{27}{16})^2\varepsilon_1 \sim -76.6 \quad eV$$

and

$$\Psi_{\text{var}} = \left[\frac{(\tfrac{27}{16})^3}{\pi a^3} \right] \exp\left[-(\tfrac{27}{16}) \frac{(r^{(1)} + r^{(2)})}{a} \right].$$

$$\tag{10-15}$$

This energy is lower than our first-order perturbation result ($E_{\text{pert}} \simeq -74$ eV) and is therefore more accurate. The observed value of the ground-state energy is

$$E_{\text{exp}} \simeq -78.6 \quad eV.$$

It is possible to improve on the results above by generalizing our trial function to contain more parameters. Hylleraas used a trial function of the form

$$\Psi_{\text{trial}} = e^{-\alpha s} P(s, t, u)$$

where P is a fourteen-term polynomial (with fourteen parameters) in the variables

$$s = \frac{r^{(1)} + r^{(2)}}{a}, \qquad t = \frac{r^{(1)} - r^{(2)}}{a}, \qquad \text{and} \qquad u = \frac{|\mathbf{r}^{(1)} - \mathbf{r}^{(2)}|}{a}.$$

The resulting variational energy for the ground state turns out to be 0.0016 eV *below* the observed value—an apparent contradiction to the variational theorem. Barring any numerical error, the discrepancy must be attributed to the omission of fine-structure terms in the electrostatic helium Hamiltonian. The order of the discrepancy is indeed that of fine structure.

It is also possible to apply the variational theorem to the excited states of helium. This requires placing subsidiary conditions (for example, orthogonality with respect to the ground-state function) on the excited state functions of helium and is beyond the scope of this text.‡

V The Statistical Model of Thomas and Fermi for Complex Atoms

The Z-electron atom, being a many-body system, cannot be solved exactly. We could attempt a perturbational approach using the atomic states of the ideal atom as a zeroth approximation. But as we have just seen, the corrections associated with the electron-interaction terms are often too large for this perturbation theory to be useful. We shall see that each atomic electron can be considered, approximately, as moving independently in an electrostatic potential resulting from a screening of the nuclear Coulomb field by the other electrons. It is therefore useful to have a model which estimates the potential and charge density within the electron cloud of an atom. Such a model was first developed independently by Thomas and by Fermi (1928).

The Thomas–Fermi (statistical) model assumes that the nucleus is surrounded by a spherically symmetric electron gas. The density is sufficiently great so that the electron cloud can be regarded as a degenerate Fermi gas. Furthermore, it is assumed that the density is a slowly varying function of the distance from the nucleus.§ Consequently, each volume element $d\mathbf{r}$ may be thought of as a box containing a homogeneous electron gas with local density $\rho(r)$. The total energy of an electron within this box is

$$\varepsilon(r) = \frac{p^2}{2m} - e\phi(r) \qquad (10\text{-}16)$$

where $\phi(r)$ is the local potential in the differential volume element. The maximum local kinetic energy is called the *local* Fermi level $\varepsilon_F(r) = p_F^2/2m$. Using the relation between the local density and ε_F, (9-71), we obtain

$$\varepsilon_{\max}(r) = \mu = \varepsilon_F - e\phi(r) = \frac{\hbar^2}{2m}[3\pi^2\rho(r)]^{2/3} - e\phi(r). \qquad (10\text{-}17)$$

---

‡ For a thorough treatment of the helium atom see H. A. Bethe and E. E. Salpeter, "Quantum Mechanics of One and Two Electron Atoms," pp. 118–165. Springer-Verlag, Berlin, 1957.

§ Specifically, the density must remain roughly constant over distances comparable to the de Broglie wavelengths of the electrons in that region. In atoms, this condition is generally violated both close to and far from the nucleus; the Thomas–Fermi model is therefore unreliable in these regions.

The maximum total local energy is just the chemical potential of the gas and is everywhere constant once equilibrium has been reached. For a neutral atom, $\phi(r)$ and $\rho(r)$ may be set equal to zero at infinity; it follows then that μ is zero at that point. Since it is a constant, μ must therefore vanish everywhere and (10-17) gives

$$\phi(r) = \frac{\hbar^2}{2me} [3\pi^2 \rho(r)]^{2/3}. \tag{10-18}$$

This establishes a relationship between the local density and the potential. Combining this with Poisson's equation, $\nabla^2 \phi = 4\pi e \rho$, we obtain

$$\nabla^2 \phi = \frac{1}{r^2} \frac{d}{dr}\left(r^2 \frac{d\phi}{dr}\right) = \frac{4\pi e}{3\pi^2} \frac{(2me)^{3/2}}{\hbar^3} \phi^{3/2}. \tag{10-19}$$

When $r \to 0$, Gauss's law implies that the electric field is due entirely to the nucleus, that is,

$$\lim_{r \to 0} E \to \frac{Ze}{r^2}. \tag{10-20a}$$

Equivalently, the potential as we approach the nucleus must have the form $\lim_{r \to 0} \phi(r) \sim Ze/r$. The boundary condition on ϕ is conveniently written $\lim_{r \to 0} r\phi = Ze$.

Far from the (neutral) atom, we expect the (screened) potential to fall off faster than $1/r$ since the net charge of the atom is zero. The other boundary condition for ϕ is therefore

$$\lim_{r \to \infty} r\phi \to 0. \tag{10-20b}$$

In order to make the analysis independent of Z and therefore valid for all atoms, we introduce the scaling transformation

$$r = \frac{1}{2}\left(\frac{3\pi}{4}\right)^{2/3} \frac{\hbar^2}{me^2} \frac{x}{Z^{1/3}} = 0.885\, a\, \frac{x}{Z^{1/3}} \tag{10-21}$$

where x is the (dimensionless) Thomas–Fermi variable. Furthermore, we define the Thomas–Fermi function χ by

$$\phi = \frac{Ze\chi}{r}. \tag{10-22}$$

Using the new variables, (10-19) becomes

$$x^{1/2} \frac{d^2\chi}{dx^2} = \chi^{3/2} \tag{10-23}$$

with the boundary conditions

$$\lim_{x \to 0} \chi \to 1 \quad \text{and} \quad \lim_{x \to \infty} \chi \to 0.$$

Equation (10-23) is the (nonlinear) Thomas–Fermi equation. Its solution has been developed numerically and has the form indicated in Figure 10-5. Once χ is determined, we obtain $\phi(r)$ and $\rho(r)$ from (10-22) and (10-18). Both the potential and density vary from atom to atom, according to the value of Z.

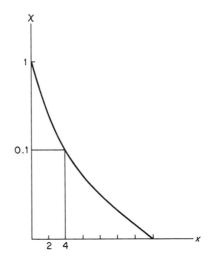

Figure 10-5 The Thomas–Fermi function $\chi(x)$. The vertical scale is not linear in order to magnify the curve for large x. Note that $\chi(4)$ is only one-tenth of $\chi(0) = 1$.

While the Thomas–Fermi model is semiclassical in nature and gives little insight into the details of atomic structure, it does lead to some results that remain valid even in a more sophisticated treatment. We shall define x_f as the radius of a sphere (in T–F units) about the nucleus which contains the fraction f of the Z electrons. Since x_f is independent of Z, it follows that the fraction within a sphere of radius r_f on the real distance scale is, from (10-21),

$$r_f = \frac{0.885a}{Z^{1/3}} x_f.$$

If we take the "size" of the atom to be the radial distance in which a given fraction (for example, 90%) of the electrons are situated, then this size decreases with increasing atomic number as $Z^{-1/3}$. This "shrinking" of the atom with increasing Z is a characteristic feature of atoms and appears even in more rigorous treatments.

The familiar concepts of valence and shell structure do not follow from the Thomas–Fermi statistical model. A more accurate quantum treatment of the Z-electron atom was originally developed by Hartree and by Fock. The

Hartree–Fock approximation generates results which have become fundamental to our understanding of the atom and which underlie the subject of
molecular chemistry.

VI   The Self-Consistent Field Method and the Hartree–Fock–Slater Approximation

The Z-electron atom—an interacting many-body system—cannot be
treated exactly, It would appear natural to ignore the interactions in a zeroth-
order approximation and apply perturbation theory. However, since the
electron interaction energy is considerable, it is doubtful that the ideal atomic
states formed from hydrogenlike orbital functions would even remotely
resemble eigenstates of real atoms.

One way in which to obtain more accurate orbital functions is to assume
that the nucleus is electrostatically screened by an average effective potential
of the electron cloud. Equivalently, the two-body interactions are replaced by
effective one-body potentials. The exact Hamiltonian for the atom can be re-
arranged by adding and subtracting the effective potential, $V_{\text{eff}}^{(q)}(\mathbf{r}^{(q)})$, that is,

$$\mathscr{H} = \left\{ \sum_{q=1}^{Z} -\frac{\hbar^2}{2m} \nabla^{(q)2} - \frac{Ze^2}{r^{(q)}} + V_{\text{eff}}^{(q)}(\mathbf{r}^{(q)}) \right\}$$

$$+ \left\{ \frac{1}{2} \sum_{q}^{Z} \sum_{t \neq q}^{Z} \frac{e^2}{|\mathbf{r}^{(q)} - \mathbf{r}^{(t)}|} - \sum_{q}^{Z} V_{\text{eff}}^{(q)}(\mathbf{r}^{(q)}) \right\}. \tag{10-24}$$

The first set of braces, which includes the kinetic energy, the electrostatic
potential of the nucleus, and the effective screening potential respectively, will
be regarded as the unperturbed Hamiltonian. The second set, *the correlation
energy*, represents the difference between the two-body interactions and the
effective one-body potentials and shall be treated as the perturbation; this
term is responsible for the *dynamical correlations* between the electrons. The
idea, of course, is to find effective potentials that best approximate the two-
body interactions because then the correlation term will be small enough so
that we can apply perturbation theory.

Since the unperturbed Hamiltonian is a sum of one-body operators, the
corresponding many-body Schroedinger equation is separable and leads to a
set of one-body eigenvalue equations of the form

$$\left\{ -\frac{\hbar^2}{2m} \nabla^2 - \frac{Ze^2}{r} + V_{\text{eff}}(\mathbf{r}) \right\} \psi_i(\mathbf{r}) = \varepsilon_i \psi_i(\mathbf{r}). \tag{10-25}$$

The ψ_i and ε_i are the eigenfunctions and energies of the orbitals. We have assumed, for simplicity, that each electron experiences the same effective potential and that this potential is central in character. An essential difference between these orbitals and the hydrogenlike states of the ideal atom is that the accidental (Coulomb) degeneracy is absent in the former. Consequently, the orbital energies depend on both n and l. *The total energy of the atom is the sum of the energies of the occupied states.*‡

The atomic states can be formed from Slater determinants of occupied orbitals. The charge density of the electron cloud in a given configuration is related to the individual particle probability densities by

$$ e\rho(\mathbf{r}) = e \sum_{i=1}^{Z} |\psi_i|^2 \tag{10-26} $$

where the sum is over occupied orbitals.

It seems plausible that the effective screening potential depends in some manner on this density and is therefore determined by the nature of the occupied orbital eigenfunctions. In a sense we are going in circles when we say ψ_i depends on $V_{\text{eff}}(\mathbf{r})$ in (10-25) and $V_{\text{eff}}(\mathbf{r})$ in turn depends on ψ_i via (10-26). This in fact is the idea behind the self-consistent field (SCF) method used to determine V_{eff} in a given configuration. We outline the procedure for ground-state configurations:

(a) Guess at an effective ground-state potential (for example, try the Thomas–Fermi potential).

(b) Solve (10-25) and find the orbital functions ψ_i and eigenvalues ε_i, recording the Z orbital functions of lowest energy.

(c) From (10-26), find the density of the electron cloud using the Z orbitals in step (b).

(d) Postulate a relationship between the cloud density and the effective potential (for example, Coulomb's law) and calculate $V_{\text{eff}}(\mathbf{r})$.

(e) If $V_{\text{eff}}(\mathbf{r})$ is not central in character, perform an average over a solid angle and obtain $V_{\text{eff}}(r)$.

(f) Return to step (b) and redo steps (b)–(f) until self-consistency is reached, that is, until V_{eff} no longer changes with each cycle.

The number of cycles required to reach a reasonable state of self-consistency depends on the original guess in step (a). In addition, the resulting effective potential depends critically on step (d).

---

‡ There are some exceptions to this rule for total energy, as in the case of the Hartree–Fock theory, but they will not be discussed here. See, for example, H. A. Bethe and R. W. Jackiw, "Intermediate Quantum Mechanics," 2nd ed., p. 63. Benjamin, New York, 1968.

Hartree originally suggested that the potential should be related to the charge density by Coulomb's law,‡ that is,

$$V_{\text{eff}}(\mathbf{r}) = e^2 \int \frac{d\mathbf{r}' \, \rho(\mathbf{r}')}{|\mathbf{r} - \mathbf{r}'|}. \tag{10-27}$$

In his analysis, Hartree completely neglected the effect of statistical correlations in (10-27). Fock, using a variational approach, generalized Hartree's theory to include exchange effects. Unfortunately, Fock's treatment leads to a set of equations which are rather difficult to handle numerically. Slater simplified the Hartree–Fock equations using an additional approximation based on the Thomas–Fermi theory. Slater's result is summarized below:

$$V_{\text{eff}}(\mathbf{r}) = e^2 \int \frac{d\mathbf{r}' \, \rho(\mathbf{r}')}{|\mathbf{r} - \mathbf{r}'|} - \frac{3}{2} e^2 \left(\frac{3}{\pi} \rho(\mathbf{r}) \right)^{1/3} \tag{10-28}$$

where

$$\rho = \sum_{\substack{\text{occupied} \\ \text{orbitals}}} |\psi_i|^2.$$

While the first term is essentially the Hartree potential,‡ the second represents Slater's approximation of Fock's exchange corrections.

Slater's approximation fails at large distances from the nucleus. Since the Coulomb integral in (10-28) corresponds to a source whose charge is $-Ze$, it follows from Gauss's law that for large r this potential cancels that of the nucleus. Equivalently, the net potential seen by an outer electron is essentially the Slater exchange potential, which falls off exponentially with distance. This is physically incorrect since we expect an outer electron to see a screened potential equivalent to that of the hydrogenic electron, namely, $-e^2/r$. The Slater correction must be modified at large r by replacing the $\rho^{1/3}$ term by a "Coulomb tail" (that is, $-e^2/r$) (Figure 10-6).

Since the HFS potential is usually not spherically symmetric, it is convenient to average over a solid angle, that is,

$$V_{\text{HFS}}(r) = \frac{1}{4\pi} \int_{4\pi} V_{\text{HFS}}(\mathbf{r}) \, d\Omega. \tag{10-29}$$

This averaging process is called the *central field approximation* and was mentioned in step (e) above.

---

‡ In the Hartree theory the Coulomb integral in (10-27) is evaluated omitting the effect of an electron on itself. In (10-28) the self-energy is included.

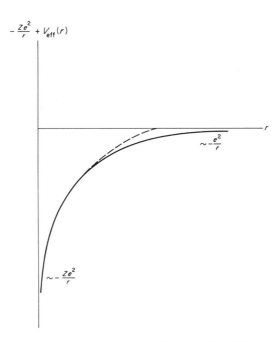

Figure 10-6 A typical potential energy function experienced by an atomic electron. The function plotted is obtained from a modified HFS potential. Note the attached Coulomb tail as $r \to \infty$. Near the nucleus, the potential is dominated by the nuclear charge. The dotted line indicates the unmodified HFS potential.

Computers have facilitated self-consistent HFS calculations. Ground-state potentials have been calculated and tabulated for all atoms of the periodic table.‡

The HFS theory as presented above must be modified in the case of the inner electrons of heavy atoms. Since these electrons have energies of the order of ~ 50 kV, relativistic effects are considerable.

It is usually assumed that the ground-state effective potential resembles the potential of low-lying excited states, that is, states in which only a single outer electron has been excited. The justification for this assumption is that the atomic density of an atom with large Z does not vary appreciably when only one electron has been moved to an adjacent orbital. The HFS orbital functions have the form

$$\psi_{nlm_l m_s} = R_{nl}(r) Y_{lm_l}(\theta, \phi) |m_s\rangle.$$

‡ See F. Herman and S. Skillman, "Atomic Structure Calculations." Prentice-Hall. Englewood Cliffs, New Jersey, 1963.

They are similar to the hydrogenlike functions of the ideal atom except that the radial part is no longer an associated Laguerre function. The energy of the orbital depends on l as well as on n and is therefore no longer Bohr-like.

The atomic states generated from the HFS orbitals represent zeroth-order approximations. Energetically these states coincide with those of the real atom. Equivalently, the *correlation energy* in (10-24) is quite small. This is not surprising since the Fock theory is based on a variational principle and accurate energies are not unexpected. On the other hand, the HFS state functions themselves may not resemble those of a real atom in other respects. Variational functions generally have little correlation with exact states. Where energies are concerned, for example, chemically or spectroscopically, the HFS approximation can be quite accurate. For other observables, the approximation can lead to totally meaningless results.

VII   Properties of Atoms in the HFS Approximation

The electrostatic Hamiltonian for a Z-electron atom in its ground state can be expressed as

$$\mathscr{H} = \mathscr{H}_0 + \lambda \hat{V}$$

where

$$\mathscr{H}_0 = \sum_{q=1}^{Z} \left\{ -\frac{\hbar^2}{2m} \nabla^{(q)2} - \frac{Ze}{r^{(q)}} + V_{\mathrm{HFS}}(r^{(q)}) \right\} \tag{10-30}$$

is the Hartree–Fock–Slater Hamiltonian and

$$\lambda V = \tfrac{1}{2} \sum_{q}^{Z} \sum_{t \neq q}^{Z} \frac{e^2}{|\mathbf{r}^{(q)} - \mathbf{r}^{(t)}|} - \sum_{q=1}^{Z} V_{\mathrm{HFS}}(r^{(q)}) \tag{10-31}$$

is the (dynamic) correlation term.

The unperturbed ideal states can be constructed from Slater determinants of HFS orbitals, $\psi_{nlm_lm_s}$. For a given n–l configuration of these orbitals, the vector model implies that ideal states can be constructed‡ using

$$|E_\alpha LS\rangle = \sum_{\substack{m_s \ m_l \\ m_s' \ m_l'}} \sum_{\substack{m_s m_s' \cdots}} C_{mm_l' \cdots} |nn' \cdots ll' \cdots m_l m_l' \cdots m_s m_s' \cdots\rangle \tag{10-32}$$
$$\vdots \quad \vdots$$

with the restrictions

$$|1 + 1' + \cdots|_{\min} \leqslant L \leqslant (l + l' + \cdots)$$

and $\tag{10-33}$

$$|s + s' + \cdots|_{\min} \leqslant S \leqslant (s + s' + \cdots).$$

---

‡ We are suppressing the remaining quantum numbers since they can be either M_L and M_S or J and M_J according to the coupling scheme.

The total energy of the n–l configuration is

$$E_{\text{HFS}} = \varepsilon_{nl} + \varepsilon_{n'l'} + \cdots. \tag{10-34}$$

The energy of the atom depends on the n and l values of the occupied HFS orbitals. Those electrons with the same principal quantum number n are said to be in the same *shell*. For fixed n, those with the same l are in the same *subshell*. The Pauli principle allows at most $2(2l + 1)$ electrons per subshell and a total of $2n^2$ electrons per shell. In the ground state of an atom, orbitals of lowest energy are occupied. For sufficiently large n and l, the l dependence may dominate over n in ε_{nl}. For example, in ^{36}Kr the 4s shell is of lower energy and is therefore filled before the 3d shell.

The radial density of an atom is determined by the radial eigenfunctions of the occupied orbitals according to

$$\mathscr{P}_r = \sum_{\substack{nl \\ \text{occupied}}}^{Z} R_{nl}^2(r) r^2. \tag{10-35}$$

The ground-state radial density for ^{18}Ar(1s^2, 2s^2, 2p^6, 3s^2, 3p^6) has been plotted in Figure 10-7. Note that this density appears to peak at characteristic distances from the nucleus, reflecting the spatial grouping of the electrons within shells.

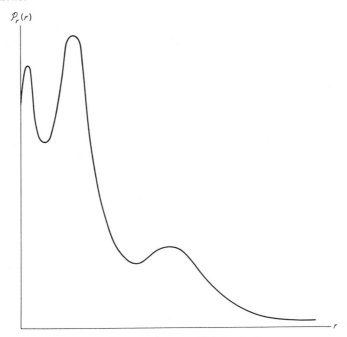

Figure 10-7 The ground-state radial density for argon.

The ground state of an atom with completely filled shells is nondegenerate since only one configuration of magnetic quantum numbers is possible without violating the exclusion principle. In fact the orbital magnetic substates (m_l) are populated symmetrically about the zero value. Also, half of the electrons have $m_s = \frac{1}{2}$ while the other half have $m_s = -\frac{1}{2}$. The ground state of such an atom is automatically a $^1S_0$ ($L = 0$, $S = 0$, $J = 0$) state. We can generalize this result somewhat by observing that a closed subshell does not contribute to the angular momentum of the atom. Consequently, the vector model need be applied only to electrons in unfilled subshells to determine the spectroscopic state of the atom.

Atoms with an outermost filled p^6 subshell usually require inordinately large amounts of energy to produce excitations. Thus, for example, the atoms $^{10}Ne(1s^2, 2s^2, 2p^6)$, $^{18}Ar(1s^2, 2s^2, 2p^6, 3s^2, 3p^6)$ and $^{36}Kr(1s^2, 2s^2, 2p^6, 3s^2, 3p^6, 4s^2, 3d^{10}, 4p^6)\ddagger$ are all inert noble gases.§ Furthermore, all have ^1S atomic ground states. These atoms are chemically inactive, relatively stable, and have high ionization energies.

Atoms with one electron less than the noble gases cited above find it energetically favorable to acquire an electron in a chemical reaction. These atoms, that is, $^9F(\cdots 2p^5)$, $^{17}Cl(\cdots 3p^5)$, $^{35}Br(\cdots 4p^5)$, and $^{53}I(\cdots 5p^5)$, are known as *halogens*. Chemically they behave as active nonmetals and produce high energies of reaction.

Those atoms having one electron more than the noble gases find it energetically favorable to contribute this *valence* electron in chemical reactions and exhibit a metal-like behavior. These *alkali* atoms include¶ $^{11}Na(\cdots 2p^6, 3s^1)$, $^{19}K(\cdots 3p^6, 4s^1)$, $^{37}Rb(\cdots 4p^6, 5s^1)$ and $^{55}Cs(\cdots 5p^6, 6s^1)$. The alkali atoms are of particular interest because their spectral lines are quite similar to those of hydrogen. Since the inner electrons in the alkalis are in closed subshells, the angular momentum characteristics are determined by the single valence s electron. The ground state of an alkali is a doublet $^2S_{\frac{1}{2}}$ ($L = 0$, $S = \frac{1}{2}$, $J = \frac{1}{2}$) state as in hydrogen. The essential difference between the valence electron in an alkali and the hydrogenic electron is that the former experiences a non-Coulombic, screened potential; consequently, the alkali energy spectrum is not Bohr-like.

A particularly useful empirical formula for the energy of the valence electron in an alkali is

$$\varepsilon_{nl} = \frac{\varepsilon_1}{[n - \mu(l)]^2} \qquad (\varepsilon_1 = -13.6 \quad eV). \qquad (10\text{-}36)$$

---

‡ Note the "inversion" between the 4s and 3d subshells.
§ ^{2}He with a $1s^2$ configuration is also inert.
¶ ^{3}Li($1s^2, 2s^1$) is also an alkali atom since it contains one more electron than ^{2}He($1s^2$).

The deviations from the Bohr formula are characterized by a set of parameters $\mu(l)$ known as "quantum defects." These parameters vary among the alkalis and account for the screening of the nucleus by the inner (core) electrons. For sodium ($n \geqslant 3$), the quantum defects are $\mu(0) = 1.35$, $\mu(1) = 0.86$, $\mu(2) = 0.01$, and $\mu(3) \simeq 0$. They decrease with increasing l. For large l (quasi-circular orbits), the electron does not appreciably penetrate the core and experiences a screened Coulomb potential equivalent to that of the hydrogenic electron. For low values, the orbit is highly eccentric; the valence electron penetrates the core and sees more of the nucleus. The orbital is more tightly bound to the nucleus and is of lower energy. Using (10-36) and the values of $\mu(l)$ for sodium, it is simple to verify that the 4s level is of lower energy than the 3d level.

The low-lying atomic states of an alkali are determined by the level occupied by the valence (optical) electron. The ground state ($3s^1$) of Na, for example, is a $^2S_{\frac{1}{2}}$ state. Equation (10-36) indicates that the energy of the valence electron is $\varepsilon_{30} \simeq -5.13$ eV. The $3p^1$ configuration can be coupled into $^3P_{\frac{1}{2}}$ and $^3P_{\frac{3}{2}}$ states, each of which has $\varepsilon_{31} \simeq -2.96$ eV. Actually, when fine-structure corrections are included, the $^2P_{\frac{3}{2}}$ resides slightly above the $^2P_{\frac{1}{2}}$. In fact, transitions to the ground $^2S_{\frac{1}{2}}$ state from the $^2P_{\frac{3}{2}}$ and $^2P_{\frac{1}{2}}$ states account for the characteristic yellow "P" doublet ($\lambda_1 = 5895.9$ Å, $\lambda_2 = 5889.9$ Å) in sodium's emission spectrum (Figure 10-8).

Atoms with partially filled subshells are considerably more difficult to handle theoretically because of the high degree of degeneracy present. Quite generally, the low-lying optical excitations in atoms are produced by the

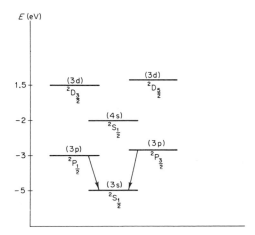

Figure 10-8   A level diagram for the low-lying excited states of sodium's valence electron illustrating the transitions responsible for the yellow doublet. The slight separation in the doublet P and doublet D states is fine structure.

transitions associated with the outer electrons. The inner electrons cannot make transitions to neighboring states since these states are normally occupied. Using high energy electrons or photons, it is possible to strip an atom of an inner electron. Transitions to fill the vacancy are usually accompanied by the emission of characteristic X rays.

VIII Diatomic Molecules—The Adiabatic Approximation

The quantum theory of molecules is considerably more complicated than that of atoms. In molecules, the electrons are bound to more than one center of force, in this case to the nuclei. The complexity is further increased by the fact that the nuclear system is in motion. There are, however, physical factors which make approximations feasible. It turns out that the frequency of the nuclear motion is much lower than the frequency associated with electrons.‡ This feature permits the application of the *adiabatic approximation*. We shall, discuss this approximation§ as it applies to diatomic molecules, that is, molecules composed of two atoms.

The Schroedinger equation for a diatomic molecule, neglecting spin, is

$$\left\{-\frac{\hbar^2}{2}\left[\frac{\nabla^{(A)2}}{M_A}+\frac{\nabla^{(B)2}}{M_B}\right]+\sum_{j=1}^{Z_A+Z_B}-\frac{\hbar^2}{2m}\nabla^{(j)2}+V_{tot}(\mathbf{r}^{(j)},\mathbf{R}^{(A)},\mathbf{R}^{(B)})\right\}$$

$$\times\Psi_\alpha(\mathbf{r}^{(j)},\mathbf{R}^{(A)},\mathbf{R}^{(B)})$$

$$=\varepsilon_\alpha\Psi_\alpha(\mathbf{r}^{(j)},\mathbf{R}^{(A)},\mathbf{R}^{(B)}) \tag{10-37}$$

where M_A, M_B is the mass of each nucleus, m the mass of each electron, $\mathbf{R}^{(A)}$, $\mathbf{R}^{(B)}$ the positions of the nuclei, $\mathbf{r}^{(j)}$ the position of the jth electron, and Z_A, Z_B the atomic number of each atom.

The total potential energy of the molecular system may be decomposed into

$$V_{tot}=\frac{Z_A Z_B e^2}{|\mathbf{R}^{(A)}-\mathbf{R}^{(B)}|}+\frac{1}{2}\sum_i\sum_{j\neq i}\frac{e^2}{|\mathbf{r}^{(i)}-\mathbf{r}^{(j)}|}$$

$$-e^2\sum_i\left\{\frac{Z_A}{|\mathbf{r}^{(i)}-\mathbf{R}^{(A)}|}+\frac{Z_B}{|\mathbf{r}^{(i)}-\mathbf{R}^{(B)}|}\right\}. \tag{10-38}$$

---

‡ This is due to the fact that the nuclei are considerably more massive than the electrons.
§ The adiabatic approximation was first applied to molecules by Born and Oppenheimer (1927).

These terms represent, respectively, the electrostatic repulsion energy between the nuclei, the repulsion energy between the electrons, and the energy of attraction of the electrons to the nuclei.

Since the kinetic energy is much smaller for the nuclei than for the electrons, we shall make the following assumptions:

(a) The molecular state function can be expressed as a product of a nuclear and an electronic part, namely,

$$\Psi_\alpha(\mathbf{r}^{(j)}, \mathbf{R}^{(A)}, \mathbf{R}^{(B)}) = \xi_{\text{nuc}}(\mathbf{R}^{(A)}, \mathbf{R}^{(B)})\psi_{\text{elec}}(\mathbf{r}^{(j)}). \qquad (10\text{-}39)$$

(b) The electronic part can be treated as though the nuclei remained stationary with $\mathbf{R}^{(A)}$ and $\mathbf{R}^{(B)}$ fixed.

We shall further assume that in the case of the diatomic molecule, the nuclear coordinates enter the molecular potential through the relative vector $\mathbf{R} = \mathbf{R}^{(A)} - \mathbf{R}^{(B)}$. This argument cannot be generalized easily to a triatomic molecule, the latter having a far more complicated structure.

Assumption (b) implies that the electronic part satisfies the equation

$$\left\{\sum_j -\frac{\hbar^2}{2m}\nabla^{(j)2} + V_{\text{tot}}(\mathbf{r}^{(j)}, \mathbf{R})\right\}\psi_\mathbf{n}(\mathbf{r}^{(j)}, \mathbf{R}) = E_\mathbf{n}(\mathbf{R})\psi_\mathbf{n}(\mathbf{r}^{(j)}, \mathbf{R}) \quad (10\text{-}40)$$

where $\mathbf{n}$ represents a set of electronic quantum numbers. It should be emphasized that $\mathbf{R}$ is not a variable in (10-40), but rather that it acts as a parameter. In other words, the electronic energy eigenvalues and eigenfunctions depend parametrically on the nuclear separation $\mathbf{R}$. To fully understand the implications of (10-40), let us assume that the nuclei remain fixed, with $\mathbf{R}$ equal to some constant $\mathbf{R}_0$. Furthermore assume that the electrons are in the ground state with $\mathbf{n} = 0$. Now let the nuclei move *adiabatically*, that is, let $\mathbf{R}$ vary slowly. It seems plausible that the electrons will remain in the ground state $(\mathbf{n} = 0)$, but that the ground-state function and energy will vary slowly. In other words, the ground-state energy and eigenfunction are determined by solving (10-40), using the instantaneous value of $\mathbf{R}$. Equivalently, as the molecule vibrates and rotates, the electronic cloud remains unexcited and adiabatically follows the nuclear motion. This intuitive justification for the adiabatic approximation is based on the adiabatic theorem[‡] for which a rigorous proof exists.

For our purposes, it is sufficient to consider the case in which the electrons are in the ground state. We must therefore solve (10-40), finding the ground-state function $\psi_0(\mathbf{r}^{(j)}, \mathbf{R})$ and energy $E_0(\mathbf{R})$ for all parametric values of the nuclear separation variable $\mathbf{R}$. This is in itself quite an involved numerical

---

‡ See A. Messiah, "Quantum Mechanics," Volume II, p. 747. Wiley, New York, 1962.

procedure. Substituting (10-39) into (10-37) and using (10-40) we are led to the nuclear equation

$$\left\{-\frac{\hbar^2}{2}\left[\frac{\nabla^{(A)^2}}{M_A} + \frac{\nabla^{(B)^2}}{M_B}\right] + E_0(\mathbf{R})\right\}\xi_v(\mathbf{R}^{(A)}, \mathbf{R}^{(B)}) = \varepsilon_v\,\xi_v(\mathbf{R}^{(A)}, \mathbf{R}^{(B)}). \quad (10\text{-}41)$$

Note that the electronic eigenvalue plays the role of an internuclear (molecular) potential. In general, the nuclear part should be written as $\xi_{n,v}$ and the energy as $\varepsilon_{n,v}$ since the molecular potential $E_n(\mathbf{R})$ depends on the electronic state.

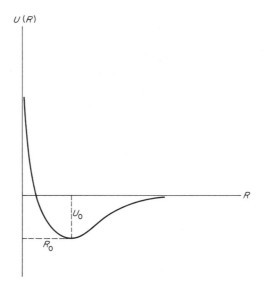

Figure 10-9   The Morse potential representing the molecular energy as a function of the internuclear separation R. The minimum energy U_0 occurs at R_0 and defines the equilibrium separation.

The general form of $E_0(\mathbf{R})$ for most diatomic molecules can be reasonably approximated by the Morse potential (Figure 10-9),

$$E_0(\mathbf{R}) = U(R) = -U_0[1 - e^{-(R-R_0)/a}]^2 + U_0$$

$$= U_0\{1 - [1 - e^{-(R-R_0)/a}]^2\} \qquad (U_0 < 0). \qquad (10\text{-}42)$$

The parameters U_0 and a represent strength and range parameters and vary among molecules. The parameter R_0 is the separation at which $U(R_0) = U_0$ is a minimum, that is, where $U'(R_0) = 0$. It therefore represents the equilibrium separation of the nuclei. Furthermore, U_0 can be regarded as the dissociation energy of the molecule.

Applying the c.m. transformation to (10-41), the equation of relative motion takes the form

$$\left\{-\frac{\hbar^2\nabla^2}{2\mu_M} + U(R)\right\}\xi_v(\mathbf{R}) = \varepsilon_v\,\xi_v(\mathbf{R}) \qquad (10\text{-}43)$$

where

$$\mu_M = \frac{M_A + M_B}{M_A\,M_B} \qquad \text{(reduced mass of nuclei)}.$$

Since the Morse potential (10-42) is central in character, (10-43) can be separated using

$$\xi_v(\mathbf{R}) = u_{v,K}(R)\,Y_{K,M_K}(\theta, \phi) \qquad (K = 0, 1, 2, \ldots\;;\quad |M_K| \leqslant K)$$

where the radial part satisfies

$$-\frac{\hbar^2}{2\mu_M}\frac{d^2u_{v,K}}{dR^2} + \left(U(R) + \frac{\hbar^2}{2\mu_M}\frac{K(K+1)}{R^2}\right)u_{v,K} = \varepsilon_{v,K}\,u_{v,K}. \qquad (10\text{-}44)$$

The effective molecular potential consists of the Morse potential and a "centrifugal potential." While the exact solution to (10-44) is rather complicated, it can be simplified by assuming that the displacement from equilibrium, $\rho = R - R_0$, is small compared with R_0. The Morse potential (or the exact molecular potential if it is known) can be expanded in a Taylor series about R_0 as

$$U(R) = U_0 + \tfrac{1}{2}U_0''\rho^2 + \cdots. \qquad (10\text{-}45)$$

The linear term is absent since $U_0' = 0$ by the definition of R_0. The second derivative of the Morse potential evaluated at $R = R_0$ plays the role of the elastic constant of a linear restoring force. If ρ is small, that is, the molecule is not highly excited, then the two terms in (10-45) approximate the Morse potential. Furthermore the centrifugal potential can be expanded (to order ρ^2) as

$$\frac{\hbar^2}{2\mu_M}\frac{K(K+1)}{(R_0+\rho)^2} = \sigma K(K+1)\left[1 - 2\frac{\rho}{R_0} + 3\frac{\rho^2}{R_0^2} + \cdots\right] \qquad (10\text{-}46)$$

where

$$\sigma = \frac{\hbar^2}{2\mu_M R_0^2}$$

roughly represents a "quantum" of rotation. The quantity $I_0 = \mu_M R_0^2$ will be recognized as the moment of inertia of the molecule about the c.m. when the molecule is in the equilibrium position.

We shall assume that the rotational energy $\sigma K(K + 1)$ is sufficiently small so that the terms involving ρ and ρ^2 in (10-46) can be treated as a perturbation. If they are neglected in a first approximation, the energy eigenvalues in (10-44) are found to be of the form

$$\varepsilon_{v,K} = U_0 + (v + \tfrac{1}{2})\hbar\omega + \sigma K(K + 1) \qquad (10\text{-}47)$$

where $\omega = (U_0''/\mu_M)^{1/2}$ and $v = 0, 1, 2, \ldots$. The first term is the ground-state electron energy at the equilibrium separation of the nuclei. The second and third terms represent the vibrational and rotational energies respectively.

When corrections (to order ρ^2) to the centrifugal potential in (10-46) are retained, quantum theory gives the vibrational and rotational energies approximately as

$$\varepsilon_{\text{vib}} \simeq (v + \tfrac{1}{2})\hbar\tilde{\omega}$$

$$\varepsilon_{\text{rot}} \simeq \sigma K(K + 1) - \frac{\{K(K + 1)\sigma\}^2}{3K(K + 1)\sigma + \tfrac{1}{2}U_0'' R_0{}^2}. \qquad (10\text{-}48)$$

The modified vibrational frequency is

$$\tilde{\omega} = \left(\frac{U_0'' R_0{}^2 + 6K(K + 1)\sigma}{\mu_M R_0{}^2}\right)^{1/2}. \qquad (10\text{-}49)$$

The modification of the natural frequency and the presence of the elastic parameter U_0'' in the rotational energy is due to "centrifugal stretching" of the molecule. If the molecule is relatively rigid and the rotational mode low ($\sigma K(K + 1) \ll U_0'' R_0{}^2$), then the molecule behaves independently as an oscillator and a rotor with (10-48) reducing to (10-47).

Of course the results in (10-48) are only valid for very low rotational and vibrational levels.‡ If the vibrational level is too high, then the displacement from equilibrium is sufficiently large to require retaining terms higher than ρ^2 in the expansion of the Morse potential in (10-45). In fact, (10-48) incorrectly implies that the vibrational levels are evenly spaced. For sufficiently high energies, the Morse potential supports unbound (dissociated) molecular states. This suggests that, at the very least, the molecular levels should converge toward a continuum as the energy is increased. More accurate treatments of molecules can be found elsewhere.§

For the special case of the diatomic *homonuclear* molecule (that is, one composed of identical atoms) the indistinguishability of the two nuclei must

---

‡ The restriction involved here is independent of the one required for the adiabatic approximation. For a perturbation treatment, see Bethe and Jackiw, "Intermediate Quantum Mechanics", 2nd. ed. p. 131. Benjamin, New York, 1968.

§ See L. Pauling and E. B. Wilson, Jr., "Introduction to Quantum Mechanics," Chapters 12 and 13. McGraw-Hill, New York, 1935.

be taken into account. We illustrate the situation for the hydrogen molecule where the nuclei (protons) are spin $\frac{1}{2}$ fermions. As in the case with helium's two electrons, if the protons have parallel spins we have a triplet state (with spin $S = 1$) for the total nuclear spin. Further, the total nuclear spin function is symmetric with respect to an interchange in nuclear spin coordinates. Since the total nuclear function for fermions must be overall antisymmetric it follows that the spatial part for the triplet state must change sign upon interchange of the nuclear spatial coordinates. This interchange is most easily accomplished by a parity reflection through the center of mass of the nuclei. To assure spatial antisymmetry, we require that the spatial part of the nuclear function be of odd parity. Since the rotational quantum number K determines the parity of the state, we require K to be odd. Thus in the triplet state of the two protons (*ortho*hydrogen) only *odd* rotational states are possible (that is, $K = 1, 3, 5 \ldots$). Similar reasoning shows that in the singlet state ($S = 0$) when the proton spins are antiparallel (*para*hydrogen) only the *even* rotational states ($K = 0, 2, 4 \ldots$) lead to overall antisymmetric state functions for the proton system. Hence the rotational quantization (odd or even K) depends on the total nuclear spin state. The likelihood of the two protons changing the relative orientations of their spins is small. Equivalently the total nuclear spin S is conserved during transitions and we have the selection rule $\Delta S = 0$. This means that transitions can occur either between para and para or between ortho and ortho states and that the two species of hydrogen gas remain spectroscopically distinct.

IX   The Hydrogen Molecule and the Covalent Bond (London–Heitler Theory)

The hydrogen molecule involves two electrons and is perhaps the simplest (neutral) diatomic molecule to treat. It is instructive to apply the techniques of the previous section to calculate the shape of the molecular potential $U(R)$ and to see how a stable covalent (homopolar) bond is formed.

According to (10-37), the molecular Hamiltonian for H_2 is (Figure 10-10)

$$\mathscr{H} = -\frac{\hbar^2}{2M} [\nabla^{(A)2} + \nabla^{(B)2}] - \frac{\hbar^2}{2m} [\nabla^{(1)2} + \nabla^{(2)2}] + \frac{e^2}{|\mathbf{R}^{(A)} - \mathbf{R}^{(B)}|}$$

$$+ \frac{e^2}{|\mathbf{r}^{(1)} - \mathbf{r}^{(2)}|} - \frac{e^2}{|\mathbf{r}^{(1)} - \mathbf{R}^{(A)}|} - \frac{e^2}{|\mathbf{r}^{(2)} - \mathbf{R}^{(A)}|}$$

$$- \frac{e^2}{|\mathbf{r}^{(1)} - \mathbf{R}^{(B)}|} - \frac{e^2}{|\mathbf{r}^{(2)} - \mathbf{R}^{(B)}|} . \tag{10-50}$$

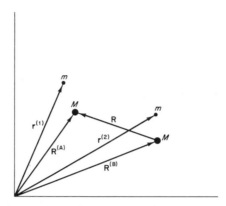

Figure 10-10 Illustrating the position vectors associated with the electrons (m) and protons (M) in the H_2 molecule.

Using the adiabatic approximation, we set up the eigenvalue equation for the electrons keeping $\mathbf{R}^{(A)}$ and $\mathbf{R}^{(B)}$ fixed and dropping $\nabla^{(A)}$ and $\nabla^{(B)}$ in (10-50). From (10-40) we have for the two electrons

$$\left\{-\frac{\hbar^2}{2m}[\nabla^{(1)2} + \nabla^{(2)2}] + V_{tot}(\mathbf{R}^{(A)}, \mathbf{R}^{(B)}, \mathbf{r}^{(1)}, \mathbf{r}^{(2)})\right\}\Psi_n(\mathbf{r}^{(1)}, \mathbf{r}^{(2)}, \mathbf{R}^{(A)}, \mathbf{R}^{(B)})$$

$$= E_n(R)\Psi_n(\mathbf{r}^{(1)}, \mathbf{r}^{(2)}, \mathbf{R}^{(A)}, \mathbf{R}^{(B)}) \tag{10-51}$$

where $\mathbf{n}$ represents the quantum numbers of the electronic state and $\mathbf{R} = \mathbf{R}^{(B)} - \mathbf{R}^{(A)}$. The term V_{tot} contains the potential energies given in (10-50).

We shall attempt to solve (10-51) using the perturbation methods of Chapter 7. We assume that the molecule is formed from two noninteracting hydrogen atoms in which electron 1 is on nucleus A and electron 2 is on nucleus B. The unperturbed Hamiltonian is then

$$\mathcal{H}_0 = -\frac{\hbar^2}{2m}\nabla^{(1)2} - \frac{e^2}{|\mathbf{r}^{(1)} - \mathbf{R}^{(A)}|} - \frac{\hbar^2}{2m}\nabla^{(2)2} - \frac{e^2}{|\mathbf{r}^{(2)} - \mathbf{R}^{(B)}|} \tag{10-52a}$$

whereas the perturbation represents the interaction between the two atoms, that is,

$$\lambda\hat{V} = \frac{e^2}{R} + \frac{e^2}{|\mathbf{r}^{(1)} - \mathbf{r}^{(2)}|} - \frac{e^2}{|\mathbf{r}^{(1)} - \mathbf{R}^{(B)}|} - \frac{e^2}{|\mathbf{r}^{(2)} - \mathbf{R}^{(A)}|}. \tag{10-52b}$$

The unperturbed Hamiltonian leads to a separable equation and the eigenfunctions represent a pair of isolated hydrogen atoms; the spatial part of the eigenfunction may be written as a product of the form

$$\Psi_{\mathbf{n}} = \psi^{(A)}_{nlm_l}(1)\psi^{(B)}_{n'l'm_{l'}}(2) \tag{10-53}$$

where the ψ functions are hydrogenic. The quantum numbers characterize the electron states of the two atoms. The superscripts remind us that the two functions are taken with respect to origins situated on *different nuclei*.

The total Hamiltonian in (10-51) is permutation symmetric with respect to the electrons. Consequently, the particles are indistinguishable when the molecule is formed and we are required to properly symmetrize (10-53) by writing

$$\Psi_{\mathbf{n}}^{\pm} = \psi^{(A)}_{nlm_l}(1)\psi^{(B)}_{n'l'm_{l'}}(2) \pm \psi^{(A)}_{nlm_l}(2)\psi^{(B)}_{n'l'm_{l'}}(1). \tag{10-54}$$

As in the case of the helium atom, the symmetric spatial part (plus sign) is to be multiplied by an antisymmetric spin part corresponding to an $S = 0$, singlet, state in which the two electron spins are antiparallel. Similarly, the antisymmetric state in (10-54) is multiplied by a symmetric spin part and represents an $S = 1$, triplet, state in which the electron spins couple in a parallel manner.

There is one minor difficulty arising from the fact that we have decomposed the permutation-symmetric Hamiltonian in (10-51) somewhat unsymmetrically as in (10-52). Note that neither (10-52a) nor (10-52b) is permutation symmetric with respect to the electron indices. Thus in principle (10-54) is not an eigenfunction of $\mathcal{H}_0$. However, this can be remedied by symmetrizing the perturbation equations.

It is convenient to adjust our reference level such that the energy of two noninteracting atoms (unperturbed system) is zero. The molecular potential is then simply given by the first-order correction due to the interaction energy, that is,

$$E_{\mathbf{n}}(R) = \frac{\lambda \int \hat{V} |\Psi_{\mathbf{n}}^{\pm}|^2 \, d\mathbf{r}_1 \, d\mathbf{r}_2}{\int |\Psi_{\mathbf{n}}^{\pm}|^2 \, d\mathbf{r}_1 \, d\mathbf{r}_2} . \tag{10-55}$$

The denominator normalizes the molecular functions. When integrating terms involving $|\psi^{(A)}_{nlm_l}(1)\psi^{(B)}_{n'l'm_{l'}}(2)|^2$ we shall use λV as given by (10-52b) in order to symmetrize (10-55). On the other hand, when we integrate terms of the form $|\psi^{(A)}_{nlm_l}(2)\psi^{(B)}_{n'l'm_{l'}}(1)|^2$ we also make the interchange $1 \leftrightarrow 2$ in $\lambda \hat{V}$. The cross (exchange) terms are unaffected by the interchange $1 \leftrightarrow 2$ in $\lambda \hat{V}$.

It is left as an exercise to show that (10-55) takes the form

$$E^{\pm}(R) = \frac{D \pm I}{1 \pm N} \tag{10-56}$$

where

$$D(R) = \lambda \int V(\mathbf{R}^{(A)}, \mathbf{R}^{(B)}, \mathbf{r}^{(1)}, \mathbf{r}^{(2)}) \, |\psi_{nlm_l}^{(A)}(1)\psi_{n'l'm_{l'}}^{(B)}(2)|^2 \, d\mathbf{r}_1 \, d\mathbf{r}_2$$

$$I(R) = \lambda \int V(\mathbf{R}^{(A)}, \mathbf{R}^{(B)}, \mathbf{r}^{(1)}, \mathbf{r}^{(2)}) \{ \mathrm{Re} \; \psi_{nlm_l}^{*(A)}(1)\psi_{n'l'm_{l'}}^{*(B)}(2)$$

$$\times \; \psi_{nlm_l}^{(A)}(2)\psi_{n'l'm_{l'}}^{(B)}(1) \} \, d\mathbf{r}_1 \, d\mathbf{r}_2$$

$$N(R) = \mathrm{Re} \int \psi_{nlm_l}^{*(A)}(1)\psi_{n'l'm_{l'}}^{*(B)}(2)\psi_{nlm_l}^{(A)}(2)\psi_{n'l'm_{l'}}^{(B)}(1) \, d\mathbf{r}_1 \, d\mathbf{r}_2 \,.$$

The integrals $D(R)$ and $I(R)$ represent respectively direct and exchange interactions between the two atoms. $N(R)$ is due to normalization; however, note that the integral $N(R)$ is *not* proportional to $\delta_{nn'}\delta_{ll'}\delta_{m_l m_{l'}}$ since the hydrogenic functions $\psi^{(A)}$ and $\psi^{(B)}$ are taken with respect to *different nuclear origins*.

The nature of $E^{\pm}(R)$ can be understood from a qualitative evaluation of D and I. For large R, the interatomic separation is large and both integrals, D and I, tend to zero. For small R the dominant effect in $\lambda \hat{V}$ is due to nuclear repulsion and both D and I are positive and large; this leads to a molecular potential of the form $E^{\pm}(R)_{R \to 0} \simeq e^2/R$.

For intermediate separations it turns out that both D and I are *negative*, but I is much larger in absolute value than D. Thus I, the exchange energy, plays the dominant role at intermediate interatomic distances. From (10-56) it follows that $E^+(R)$ is negative whereas $E^-(R)$ is positive for intermediate R (Figure 10-11). We are thus led to the following rules for the covalent bond:

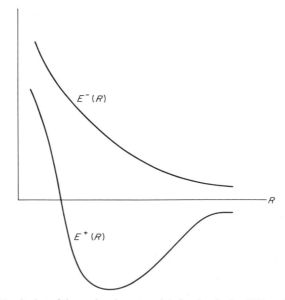

Figure 10-11   A plot of the molecular potentials for the singlet (E^+) and the triplet (E^-) states in the H_2 molecule.

(1)   Quantum exchange effects play a dominant role in the formation of a stable covalent or homopolar bond.

(2)   Two hydrogen atoms can form a stable molecule only if their electrons are antiparallel, that is, the molecule is formed in an $S = 0$, singlet, state.

Once $E^+(R) = U(R)$ (for example, in the ground electronic state) has been found, then the vibrational and rotational characteristics of the H_2 molecule can be found directly using the procedures of the previous section.

X   The Normal-Coordinate Transformation, the Linear Lattice, Phonons

The linear elastic chain or linear lattice is an example of a real many-body problem that can be solved exactly. This system is composed of a one-dimensional array of N identical mass points coupled by identical springs. We shall assume, for simplicity, that only longitudinal motions, that is, motions along the chain, are possible. The three-dimensional lattice, although somewhat more complicated than the linear lattice, can also be solved and plays an important role in the theory of lattice vibrations of solids. Although interatomic forces are not strictly elastic, they may be regarded so as long as the displacements of the atoms from their equilibrium sites are small.

The classical Hamiltonian for the chain (Figure 10-12) is

$$\mathcal{H} = \sum_{i=1}^{N} \frac{P_i{}^2}{2m} + \sum_{i=1}^{N} \frac{1}{2} K(\eta_{i+1} - \eta_i)^2 \tag{10-57}$$

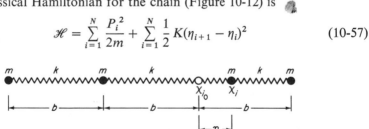

Figure 10-12   A linear elastic chain with the ith mass displaced from its equilibrium state.

where $n_i = x_i - x_{i_0}$ represents the displacement of the ith mass point from equilibrium (along the x axis) and K is the elastic constant of the springs. Do not confuse K with the rotational quantum number of Section VIII. It is convenient to define the equilibrium lattice spacing (lattice vector) by

$$b = x_{(i+1)_0} - x_{i_0}$$

and to assume that the chain is long,‡ that is, $L \gg b$ ($N \gg 1$).

---

‡ Note that the two end masses have only one nearest neighbor and are therefore dynamically different from the interior masses. However, when the chain is long the "end effects" are negligible.

Applying Hamilton's equations to (10-57), we obtain the following equations of motion for the ith mass:

$$\dot{\eta}_i = \frac{P_i}{m} \quad \text{and} \quad \dot{P}_i = K(\eta_{i+1} - \eta_i) - K(\eta_i - \eta_{i-1}). \tag{10-58}$$

When combined these give

$$m\ddot{\eta}_i = K(\eta_{i+1} - 2\eta_i + \eta_{i-1}) \quad (i = 1 \ldots N). \tag{10-59}$$

Equations (10-59) constitute a set of N coupled ordinary differential equations since the η_{i+1} and η_{i-1} appear in the ith differential equation. Equations of this form are familiar to electrical engineers who deal with linear circuit analysis. Solutions are usually obtained using a linear transformation (for example, Fourier analysis).

The coupling in (10-59) originates with the cross terms $\eta_{i+1}\eta_i$ in the Hamiltonian in (10-57). These terms in turn correspond to interactions between adjacent masses. If these terms were absent, the Hamiltonian would represent a system of uncoupled oscillators. It is possible to find a linear transformation to a new set of canonical variables, ξ_n and p_n, in terms of which the new Hamiltonian has no cross terms. In other words, it is possible to transform away the interactions and generate a new problem involving uncoupled oscillators. To accomplish this, we introduce the *normal-coordinate transformation*

$$\eta_i = \sum_{n=1}^{N} a_{in} \xi_n \tag{10-60a}$$

and

$$P_i = \sum_{n=1}^{N} m a_{in} p_n. \tag{10-60b}$$

Using the relation $P_i = m\dot{\eta}_i$, we find that $p_n = \dot{\xi}_n$ so that the normal coordinates and momenta ξ_n and p_n represent systems of unit mass.

The coefficients a_{in} are to be chosen so that when (10-60) is substituted in (10-57), the new Hamiltonian becomes

$$H = \sum_{n=1}^{N} (\tfrac{1}{2} p_n^2 + \tfrac{1}{2} \omega_n^2 \xi_n^2). \tag{10-61}$$

The new Hamiltonian is to represent a set of N *uncoupled* oscillators each of which has *unit mass* and a characteristic frequency ω_n. Furthermore, each oscillator or *normal mode* has one set of p_n and ξ_n assigned to it. Applying Hamilton's canonical equations to (10-61), we obtain an equation of motion for each ξ_n, whose solution is‡

$$\xi_n = A_n e^{-i\omega_n t}. \tag{10-62}$$

‡ In (10-62) and (10-63), only the real part is taken to represent physical motion.

A_n is a complex amplitude (amplitude and phase) to be determined from the initial conditions. Once the normal frequencies ω_n for each mode are known and the coefficients a_{in} determined, (10-60) may be used to establish the motion of the ith mass point as

$$\eta_i(t) = \sum_{n=1}^{N} a_{in} A_n e^{-i\omega_n t}. \qquad (10\text{-}63)$$

To evaluate ω_n and a_{in}, we substitute (10-63) into the equation of motion (10-59) and find

$$m\ddot{\eta}_i = K(\eta_{i+1} - 2\eta_i + \eta_{i-1})$$

or

$$\sum_{n=1}^{N} -m\omega_n^2 a_{in} A_n e^{-i\omega_n t} = \sum_{n=1}^{N} K(a_{i+1,n} - 2a_{in} + a_{i-1,n}) A_n e^{-i\omega_n t}. \qquad (10\text{-}64)$$

Since (10-64) is to be valid for any initial conditions, it must hold for all A_n. This is possible if and only if the corresponding coefficients of A_n on the right and left sides are equal, that is,

$$-m\omega_n^2 a_{in} = K(a_{i+1,n} - 2a_{in} + a_{i-1,n}) \qquad (i = 1, \ldots, N). \quad (10\text{-}65)$$

Equation (10-65) constitutes a set of N homogeneous linear algebraic equations for the unknowns a_{in}. Nontrivial $(a_{in} \neq 0)$ solutions exist only for certain values of ω_n^2. These values are found by requiring that the determinant of the coefficients of the a_{in} in (10-65) vanish. We are led to an algebraic equation of the Nth degree whose N roots give the allowed values of ω_n^2. Each value of ω_n is resubstituted in (10-65) and the corresponding set of coefficients a_{in} obtained.

The normal frequencies ω_n and the corresponding set of coefficients a_{in} are extremely difficult to find even for relatively small lattices. Oddly enough, for very large N, it becomes possible to find solutions to (10-65) using arguments based on the translational symmetry of the system. If $N \to \infty$, then a translation of the lattice by a multiple of the lattice constant b leaves the Hamiltonian unaffected. This suggests that we take the coefficients in (10-65) to be of the form

$$a_{in} = \frac{1}{(Nm)^{1/2}} \exp(ik_n x_{io}) \qquad (10\text{-}66)$$

where k_n is some constant (wave vector) which depends on the mode; equivalently, in a given mode the motions of the mass points differ at most by a phase. Substitution of (10-66) into (10-65) verifies that our "guess" was correct provided

$$-m\omega_n^2 = K\{\exp[ik_n(x_{(i+1)o} - x_{io})] - 2 + \exp[-ik_n(x_{io} - x_{(i-1)o})]\}$$

$$= 2K[\cos k_n b - 1]$$

or

$$\omega_n = \left(\frac{K}{m}\right)^{1/2} \sin \frac{1}{2} k_n b. \tag{10-67}$$

Once the mode wave vectors are known, the coefficients in (10-66) and the frequencies in (10-67) may be established. Equation (10-67) is an example of a *dispersion relation*, that is, a relation between ω_n and k_n for the lattice.

A rather interesting way of establishing the values k_n is to impose boundary conditions on the first and last masses on the chain. As $N \to \infty$, both end points may be thought of as merging at infinity. Equivalently, we can require that both exhibit identical motions (Born–von Karman cyclic conditions). Mathematically, we require

$$\eta_1(t) = \eta_N(t) \tag{10-68}$$

or from (10-60)

$$\sum_{n=1}^{N} a_{1,n} \xi_n(t) = \sum_{n=1}^{N} a_{N,n} \xi_n(t).$$

Since this relation is to be valid for all time, we must have

$$a_{1,n} = a_{N,n}, \qquad \exp(ik_n x_{1_0}) = \exp(ik_n x_{N_0})$$

or

$$\exp[ik_n(x_{N_0} - x_{1_0})] = e^{ik_n L} = 1 \qquad \text{for all} \quad n. \tag{10-69}$$

Equation (10-69) can be satisfied if and only if

$$k_n = \frac{2\pi n}{L} \qquad \text{where}\ddagger \quad n = 1, \ldots, N. \tag{10-70}$$

The normal frequencies and coefficients in (10-67) and (10-66) take the form

$$\omega_n = \left(\frac{K}{m}\right)^{1/2} \sin \frac{\pi b n}{L}$$

and

$$\tag{10-71}$$

$$a_{in} = \frac{1}{(Nm)^{1/2}} \exp\left(\frac{i2\pi x_{i_0} n}{L}\right).$$

---

‡ The particular choice for n is somewhat arbitrary and any set of N adjacent integers may be used.

The transition to quantum theory is accomplished making the replacements

$$\eta_i \to \hat{\eta}_i, \qquad P_i \to \hat{P}_i, \qquad \xi_n \to \hat{\xi}_n, \qquad p_n \to \hat{p}_n$$

and

$$\mathcal{H} \to \hat{\mathcal{H}} = \sum_{n=1}^{N} \frac{\hat{p}_n \hat{p}_n}{2} + \frac{1}{2}\omega_n^2 \hat{\xi}_n \hat{\xi}_n. \qquad (10\text{-}72)$$

The resulting Schroedinger equation is separable (since interactions are absent) and the total eigenvalues are given by the sum

$$E_{\text{tot}} = \sum_{n=1}^{N} \varepsilon_{v_n}^{(n)} = \sum_{n=1}^{N} (v_n + \tfrac{1}{2})\hbar\omega_n \qquad (v_n = 0, 1, 2, \ldots). \qquad (10\text{-}73)$$

The vibrational state of the lattice is indicated by giving the excitation integers v_n for each of the normal-mode oscillators. The ground state of the lattice is realized when all the v_n are zero. The corresponding "zero-point" lattice energy is

$$E_{\substack{\text{ground} \\ \text{state}}} = \sum_{i=1}^{N} \tfrac{1}{2}\hbar\omega_n. \qquad (10\text{-}74)$$

The equivalence between an elastic lattice and a set of independent normal-mode oscillators permits a rather simple theoretical formulation of lattice vibrations. This approach is quite useful in explaining the observed thermal and acoustical behavior of solids.

There is an interesting way of interpreting the energy spectrum of the lattice (10-73). We introduce the notion of a quasi-particle called a *phonon*. Each phonon has a frequency ω_n and an energy $\varepsilon_n = \hbar\omega_n$. With each normal mode we associate one such phonon with the same frequency. The excitation integers v_n are interpreted as the number of phonons of frequency ω_n present. When no phonons are present, that is, all $v_n = 0$, we have a "vacuum state" of energy

$$\sum_{n=1}^{N} \tfrac{1}{2}\hbar\omega_n.$$

Since the phonon number v_n may change, phonons may be created or destroyed. Furthermore, since v_n may be greater than unity, more than one phonon may have the same energy $\hbar\omega_n$. Phonons may therefore be regarded as indistinguishable bosons, obeying Bose–Einstein statistics. However, since phonons are not conserved in number, the chemical potential must be set equal to zero and the phonon distribution becomes

$$\bar{v}_n = \frac{1}{e^{\hbar\omega_n/k_B T} - 1} \qquad (n = 1, \ldots, N). \tag{10-75}$$

This is reminiscent of the Planck distribution for photons (1-32), the latter being quanta of the electromagnetic field.

The first two parts of this text have dealt mainly with the quantum theory of nonrelativistic particles. It is incorrect to assume that all relativistic corrections can be deduced from their classical counterparts. For example, of the relativistic fine structure terms in hydrogen only the mass variation term is derivable from a classical form. The spin–orbit and Darwin terms have no classical counterparts. While it is possible to explain the spin–orbit interaction using an analogy between $\hat{S}$ and $\hat{L}$, the Darwin term has no classical analog whatsoever and requires a relativistic quantum treatment. We shall see in Chapter 11 that all three fine-structure terms follow directly from the Dirac equation for the relativistic electron.

Suggested Reading

Bethe, H. A., and Jackiw, R. W., "Intermediate Quantum Mechanics." 2nd ed. Benjamin, New York, 1968.
Bethe, H. A., and Salpeter, E. E., "Quantum Mechanics of One and Two Electron Atoms." Springer-Verlag, Berlin, 1957.
Borowitz, S., "Fundamentals of Quantum Mechanics." Benjamin, New York, 1967.
Condon, E. U., and Shortley, G. H., "The Theory of Atomic Spectra." Cambridge Univ. Press, London and New York, 1963.
Eisberg, R. M., "Fundamentals of Modern Physics," Chapter 13. Wiley, New York, 1961.
Goldstein, H., "Classical Mechanics," Chapter 3. Addison-Wesley, Reading, Massachusetts, 1950.
Heitler, W., "Elementary Wave Mechanics," 2nd ed. Oxford Univ. Press, London and New York, 1956.
Stehle, P., "Quantum Mechanics." Holden-Day, San Francisco, 1966.

Problems

10-1. Find the shift in the first Balmer line in hydrogen when nuclear motion is taken into account. Compare this with the fine-structure corrections.

10-2. Consider a collection of hydrogen atoms in a large box, each characterized by $\psi = e^{i\mathbf{K}\cdot\mathbf{R}}\psi_i(\mathbf{r})$. In addition to internal factors (for example, fine-structure, reduced mass corrections, etc.) is there any classical effect, due to translation, which can affect the emission spectrum of the atoms? Write an expression for this effect.

10-3. (a) Show that when a moving mass collides elastically with a second equal mass initially at rest, the first mass can never scatter beyond 90° from incidence when viewed in the laboratory frame. (Hint: Use

$$\frac{\sin \theta}{\cos \theta + 1} = \tan \frac{\theta}{2}.)$$

(b) Using Eq. (10-5) and the relation following (10-8) show that the lab and c.m. cross sections are related by

$$\sigma_{\text{lab}} = \left\{ \frac{[1 - (m/M)^2 + 2(m/M) \cos \theta_{\text{c.m.}}]^{3/2}}{1 + (m/M) \cos \theta_{\text{c.m.}}} \right\} \sigma_{\text{c.m.}}.$$

10-4. The variational function for the helium atom as given in Eq. (10-15) can be written

$$\Psi_{\text{Var}}(1, 2) = \psi(1)\psi(2)$$

where

$$\psi(r) = \left(\frac{Z'}{\pi a^3}\right)^{1/2} e^{-Z'r/a}$$

and where the "effective charge" is $Z' = \frac{27}{16}$.

(a) Find the charge density about the nucleus. (Hint: Assume that the electron density is the sum of the probability densities associated with each electron function.)

(b) Verify that the electrostatic potential about the nucleus is

$$\phi(r) = 2e\left(\frac{1}{r} + \frac{Z'}{a}\right)e^{-2Z'r/a}.$$

(Hint: Use Poisson's equation.)

10-5. (a) Show that the average energy per particle of an ideal Fermi gas at $T = 0$ is

$$\bar{\varepsilon} = \frac{3}{5}\varepsilon_F.$$

(b) Show that the total kinetic energy of the atomic electron gas according to the Thomas–Fermi model is

$$T = \int \tau(r) \, d\mathbf{r} = \int \frac{\hbar^2 (3\pi^2)^{5/3}}{10m\pi^2} [\rho(r)]^{5/3} \, d\mathbf{r}.$$

(c) Write an expression for the *total* energy of the Thomas–Fermi atom in terms of $\rho(r)$.

10-6. Use Eq. (10-36) and the values of the quantum defects as given in the text to calculate the wavelength of sodium's characteristic yellow line. This line is emitted by the valence electron during the transition $3p \rightarrow 3s$.

10-7. Consider a Z-electron atom in which each electron moves independently in a screened electrostatic potential $V(\mathbf{r})$.

(a) Show that for each electron the following relation holds:

$$[\hat{z}, \mathcal{H}] = \frac{i\hbar}{m}\,\hat{p}_z.$$

$\mathcal{H} = (\hat{p}^2/2m) + V(\mathbf{r})$ is the Hamiltonian for each electron.

(b) Using the above result show that

$$(\varepsilon_i - \varepsilon_j)\langle\varepsilon_j|\hat{z}|\varepsilon_i\rangle = \frac{i\hbar}{m}\langle\varepsilon_j|\hat{p}_z|\varepsilon_i\rangle.$$

(c) Multiplying both sides by z_{ij} and summing over j, use the rules of matrix multiplication to show that

$$\sum_{j=1}^{\infty} (\varepsilon_i - \varepsilon_j)|z_{ji}|^2 = \frac{i\hbar}{m}(\hat{z}\hat{p}_z)_{ii}$$

or, conjugating both sides,

$$\sum_{j=1}^{\infty} (\varepsilon_i - \varepsilon_j)|z_{ji}|^2 = -\frac{i\hbar}{m}(\hat{p}_z\hat{z})_{ii}.$$

(d) Adding the above equations, show that

$$\sum_{j=1}^{\infty} (\varepsilon_j - \varepsilon_i)|z_{ji}|^2 = \frac{\hbar^2}{2m}$$

and that

$$\sum_{i=1}^{Z}\sum_{j=1}^{\infty} (\varepsilon_j - \varepsilon_i)|\mathbf{r}_{ji}|^2 = \frac{3\hbar^2}{2m}Z$$

where

$$|\mathbf{r}_{ji}|^2 = |x_{ji}|^2 + |y_{ji}|^2 + |z_{ji}|^2.$$

The above result is known as the *oscillator sum rule*.

10-8. Consider a helium atom whose ground (^1S) state function is

$$\Psi_{\text{He}}(1,2) = \psi_{100}^A(1)\psi_{100}^A(2)\left\{\frac{|\tfrac{1}{2}\rangle^{(1)}|-\tfrac{1}{2}\rangle^{(2)} - |\tfrac{1}{2}\rangle^{(2)}|-\tfrac{1}{2}\rangle^{(1)}}{\sqrt{2}}\right\}$$

where ψ^A is hydrogenic in form and is taken with respect to nucleus A.

(a) Write the total function for a system composed of the helium atom above and a hydrogen atom in its ground state, with nucleus B, when the two atoms are *far* apart.

(b) Show that when the two atoms are brought together the only properly symmetrized state is one in which exchange effects produce repulsion. Equivalently, you will have shown that a helium atom cannot form a stable bond with hydrogen. Generalize this result to show that exchange effects between noble gas atoms normally lead to repulsion.

10-9. In the *ionic* bond of the NaCl molecule, the valence electron of sodium is transferred to the chlorine atom. The Na^+ ion resembles a positively charged Ne atom whereas Cl^- is quite similar to a negatively charged argon atom. In terms of the results of Problem 10-8 explain the repulsive and attractive properties of the stable ionic bond.

10-10. Consider a molecular potential of the form

$$U(R) = -\frac{A}{R} + \frac{B}{R^2}.$$

(a) In terms of A and B find the equilibrium position and dissociation energy of the molecule.

(b) Using (10-47) find approximate expressions for the rotational and vibration spectra in terms of A and B.

10-11. Show that Eq. (10-56) follows from (10-55).

10-12. Find the appropriate dispersion relation $\omega = \omega(k)$ in the one-dimensional lattice when *next*-to-nearest neighbors are included.

10-13. Write the expression for the total energy of a vibrating one-dimensional lattice at a temperature T by assuming that the lattice is represented by an ideal phonon gas, the latter being bosons satisfying Bose–Einstein statistics with $\mu = 0$.

III RELATIVISTIC QUANTUM MECHANICS AND FIELD THEORY

11 | Relativistic Quantum Mechanics

In a rigorous approach to relativistic quantum theory a *covariant* formulation is used in which all laws are expressed in a four-dimensional space with time serving as the fourth dimension. The advantage to this formalism is that time and space variables are automatically treated in a symmetric manner. It also ensures that the resulting laws are invariant under a Lorentz transformation relating observables as seen by two observers moving at constant velocity with respect to each other.

To avoid the complexities of a covariant treatment, all relativistic quantities will be expressed in an ordinary three-dimensional space. The results will be assumed correct if

(a) they are consistent with their classical counterparts,

(b) they are symmetric in time and space variables,

and

(c) they reduce to the results of the Schroedinger theory when $v \ll c$.

I The Klein–Gordon Equation

The Schroedinger equation of motion can be developed by introducing the energy operator $\hat{\mathscr{E}} = i\hbar\,\partial/\partial t$. Setting this operator equal to the Hamiltonian operator, we obtain

$$\hat{\mathscr{H}}\Psi = \hat{\mathscr{E}}\Psi = \left\{-\frac{\hbar^2}{2m}\nabla^2 + V(\mathbf{r})\right\}\Psi = i\hbar\,\frac{\partial\Psi}{\partial t} \tag{11-1}$$

which is the required result. Inspection of (11-1) reveals that Schroedinger's equation is not acceptable as a relativistic equation since it involves second derivatives in space (that is, ∇^2) but only first derivatives in time. We are led to one of two obvious choices. Either we generate an equation which is second order in time (Klein–Gordon equation) or we linearize the Hamiltonian with respect to the momentum operator (Dirac equation). Both choices must reduce to the Schroedinger equation in the nonrelativistic limit, and in the case of a particle in an electromagnetic field both must be consistent with the relativistic classical relation‡

$$c^2\left(\mathbf{p} - \frac{q}{c}\mathbf{A}\right)^2 + m^2c^4 = (\varepsilon - q\Phi)^2. \tag{11-2}$$

Using the approach of Klein and Gordon, we set

$$\varepsilon \to \hat{\mathscr{E}} = i\hbar\,\frac{\partial}{\partial t} \qquad \text{and} \qquad \mathbf{p} \to \hat{\mathbf{p}} = \frac{\hbar}{i}\nabla$$

in (11-2) and generate the KG equation

$$\left\{c^2\left(\frac{\hbar}{i}\nabla - \frac{q}{c}\mathbf{A}\right)^2 + m^2c^4\right\}\Psi = \left\{\left(i\hbar\,\frac{\partial}{\partial t} - q\Phi\right)^2\right\}\Psi. \tag{11-3}$$

This equation was originally considered by Schroedinger himself, who disregarded it because it led to negative probabilities. The KG equation is consistent with conditions (a) and (b) above.

For static electric and magnetic fields

$$\left(\frac{\partial\mathbf{A}}{\partial t} = \frac{\partial\Phi}{\partial t} = 0\right)$$

the Klein–Gordon equation becomes

$$\left\{c^2\left(\frac{\hbar}{i}\nabla - \frac{q}{c}\mathbf{A}\right)^2 + m^2c^4\right\}\Psi = \left\{-\hbar^2\,\frac{\partial^2}{\partial t^2} - 2i\hbar q\Phi\,\frac{\partial}{\partial t} + q^2\Phi^2\right\}\Psi. \tag{11-4}$$

---

‡ We use m (rather than m_0) to denote rest mass. See Chapter 2, Eq. (2-86).

We shall verify next that the KG equation reduces to the Schroedinger equation as $v/c \to 0$. Recall that relativistically the total energy ε of a system includes a *rest* part mc^2 which must be reflected in the quantum-mechanical state function; we therefore separate out the dependence on mc^2 by expressing the latter in the form

$$\Psi = e^{-imc^2t/\hbar}\Psi_{kin}. \tag{11-5}$$

The kinetic part Ψ_{kin} is the term to be compared with the nonrelativistic Schroedinger state function. Substituting (11-5) into (11-4), we obtain the KG equation for the kinetic part, namely

$$\left\{c^2\left(\hat{\mathbf{p}} - \frac{q}{c}\mathbf{A}\right)^2 + m^2c^4\right\}\Psi_{kin} = \left\{\left(-\hbar^2\frac{\partial^2}{\partial t^2} + 2i\hbar(mc^2 - q\Phi)\frac{\partial}{\partial t}\right)\right.$$

$$\left. -[q\Phi(2mc^2 - q\Phi) + m^2c^4]\right\}\Psi_{kin}$$

$$\left\{c^2\left(\hat{\mathbf{p}} - \frac{q}{c}\mathbf{A}\right)^2\right\}\Psi_{kin} = \left\{-\hbar^2\frac{\partial^2}{\partial t^2} + 2i\hbar(mc^2 - q\Phi)\frac{\partial}{\partial t}\right.$$

$$\left. -q\Phi(2mc^2 - q\Phi)\right\}\Psi_{kin}. \tag{11-6}$$

However, if the particle is nonrelativistic, the rest energy is dominant and those terms not involving mc^2 on the right-hand side of (11-6) may be dropped. We then have, for $v \ll c$,

$$\left\{\frac{(\hat{\mathbf{p}} - (q/c)\mathbf{A})^2}{2m} + q\Phi\right\}\Psi_{kin} = i\hbar\frac{\partial}{\partial t}\Psi_{kin} \tag{11-7}$$

which will be recognized as the Schroedinger equation. Since spin properties do not arise in the KG equation, it governs the behavior of spinless bosons, for example, π mesons.

The stationary eigenstates of a KG particle may be assumed to evolve in time as

$$\Psi = e^{-i\varepsilon_it/\hbar}\psi_i(\mathbf{r}).$$

Substitution into (11-4) gives the KG *energy eigenvalue* equation

$$\left\{c^2\left(\hat{\mathbf{p}} - \frac{q}{c}\mathbf{A}\right)^2 + m^2c^4\right\}\psi_i(\mathbf{r}) = (\varepsilon_i - q\Phi)^2\psi_i(\mathbf{r}). \tag{11-8}$$

If we apply (11-8) to the case of a hydrogenic electron in which $q = -e$, $\mathbf{A} = 0$, and $\Phi = e/r$, we can show that spin effects do not appear. Since this is a central force problem, (11-8) is separable using

$$\psi_i = R_{nl}(r)Y_{lm_l}(\theta, \phi).$$

As in the nonrelativistic case, the radial equation determines the energy spectrum of the atom. We state without proof‡ that R_{nl} will satisfy bound-state requirements only for those energies given by

$$\varepsilon_{nl} = mc^2 \left[1 + \frac{\gamma^2}{\{n - (l + \frac{1}{2}) + [(l + \frac{1}{2})^2 - \gamma^2]^{1/2}\}^2} \right]^{-1/2}$$

$$(n = 1, 2, 3 \ldots; \quad l = 0, 1, \ldots, n - 1) \quad (11\text{-}9)$$

where $\gamma = e^2/\hbar c = 1/137$ is the fine-structure constant. In the nonrelativistic limit,§ using $\gamma \ll 1$, the "hydrogenic" levels become

$$\varepsilon_{nl} \simeq mc^2 \left\{ 1 - \frac{\gamma^2}{2n^2} - \frac{\gamma^4}{2n^4} \left(\frac{n}{l + \frac{1}{2}} - \frac{3}{4} \right) \right\} + \text{terms of order } \gamma^6. \quad (11\text{-}10)$$

The first term is just the rest energy of the particle. The next term is

$$-\frac{mc^2\gamma^2}{2n^2} = \frac{\varepsilon_1}{n^2} = \varepsilon_n$$

which will be recognized as the Bohr formula. The last terms are due to fine structure and take the form

$$\varepsilon_{\text{fs}} = \varepsilon_n \frac{\gamma^2}{n} \left[\frac{1}{(l + \frac{1}{2})} - \frac{3}{4n} \right]. \quad (11\text{-}11)$$

Comparing this result with (7-51) we see that (11-11) would be correct if l were half-integral (that is, $l = j$) instead of integral. Evidently, (11-11) represents fine structure without spin, as would be the case for a π-mesic atom (for example, a hydrogen atom with a π meson replacing the electron).

II The Dirac Equation

Dirac in a classic paper (1928) developed an alternative to the KG equation. While this equation satisfied all the relativistic requirements, it also implied the existence of intrinsic spin in the nonrelativistic limit. Not only was this spin equal to $\frac{1}{2}$ but the fine-structure terms obtained for a Dirac particle in a Coulomb field were in agreement with those observed in the spectrum

---

‡ See, for example, L. I. Schiff, "Quantum Mechanics," 3rd ed., p. 470. McGraw-Hill, New York, 1968.

§ Actually, it may be verified that the speed of the electron in the first Bohr orbit is $v = e^2/\hbar$. Thus the relation $\gamma = e^2/\hbar c = v/c = 1/137 \ll 1$ indicates generally that the electron in a bound state of hydrogen is highly nonrelativistic.

of hydrogen. There could be little doubt that the relativistic electron was a particle which obeyed the Dirac equation, that is, a Dirac particle.

Dirac chose to preserve the linear form of the Schroedinger equation, namely,

$$\mathcal{H}\Psi = \mathcal{E}\Psi = i\hbar \frac{\partial \Psi}{\partial t}. \tag{11-12}$$

Since $\mathcal{H}$ had to be linear in $\hat{\mathbf{p}}$, he suggested the form

$$\mathcal{H} = c\hat{\boldsymbol{\alpha}} \cdot \hat{\mathbf{p}} + \hat{\beta} m_0 c^2. \tag{11-13}$$

The nature of the Hermitian operators $\hat{\boldsymbol{\alpha}}$ and $\hat{\beta}$ will be determined below. Using the replacements ("minimal" substitution)

$$\hat{\mathbf{p}} \to \hat{\mathbf{p}} - \frac{q}{c}\mathbf{A} \quad \text{and} \quad \hat{\mathcal{E}} \to \hat{\mathcal{E}} - q\Phi \tag{11-14}$$

the Dirac equation of motion (11-12) for a particle in an electromagnetic field becomes

$$\left\{ c\hat{\boldsymbol{\alpha}} \cdot \left(\hat{\mathbf{p}} - \frac{q}{c}\mathbf{A} \right) + \hat{\beta} m_0 c^2 \right\}\Psi = \left\{ i\hbar \frac{\partial}{\partial t} - q\Phi \right\}\Psi. \tag{11-15}$$

"Squaring" both sides (operating a second time with the operators) in (11-15), we find

$$\left\{ c\hat{\boldsymbol{\alpha}} \cdot \left(\hat{\mathbf{p}} - \frac{q}{c}\mathbf{A} \right) + \hat{\beta} mc^2 \right\}^2 \Psi = \left\{ i\hbar \frac{\partial}{\partial t} - q\Phi \right\}^2 \Psi. \tag{11-16}$$

The right-hand side is the same as that in the KG equation. The left side will agree with (11-3) if and only if $\hat{\boldsymbol{\alpha}}$ and $\hat{\beta}$ commute with $\hat{\mathbf{p}}$ and $\mathbf{r}$ and if the following relations are satisfied:

$$[\hat{\alpha}_i, \hat{\alpha}_j]_+ = 2\delta_{ij}\hat{1}, \quad [\hat{\alpha}_i, \hat{\beta}]_+ = 0, \quad \text{and} \quad \hat{\beta}^2 = \hat{1}. \tag{11-17}$$

These commutation relations assure that the cross terms arising from the "squaring" of the operator on the left in (11-16) vanish.

The operators $\hat{\boldsymbol{\alpha}}$ and $\hat{\beta}$ have commutation properties quite similar to those of the Pauli matrices. This would suggest that we construct the 4×4 matrices

$$\hat{\alpha}_i = \begin{pmatrix} 0 & \hat{\sigma}_i \\ \hat{\sigma}_i & 0 \end{pmatrix} \quad \text{and} \quad \hat{\beta} = \begin{pmatrix} \hat{1} & 0 \\ 0 & -\hat{1} \end{pmatrix} \tag{11-18}$$

where the $\hat{\sigma}_i$ are the 2×2 matrices given by (6-7). Equations (11-18) are abbreviations for

$$\hat{\alpha}_x = \begin{pmatrix} 0 & 0 & 0 & 1 \\ 0 & 0 & 1 & 0 \\ 0 & 1 & 0 & 0 \\ 1 & 0 & 0 & 0 \end{pmatrix}, \qquad \hat{\alpha}_y = \begin{pmatrix} 0 & 0 & 0 & -i \\ 0 & 0 & i & 0 \\ 0 & -i & 0 & 0 \\ i & 0 & 0 & 0 \end{pmatrix},$$

$$\hat{\alpha}_z = \begin{pmatrix} 0 & 0 & 1 & 0 \\ 0 & 0 & 0 & -1 \\ 1 & 0 & 0 & 0 \\ 0 & -1 & 0 & 0 \end{pmatrix}, \qquad \hat{\beta} = \begin{pmatrix} 1 & 0 & 0 & 0 \\ 0 & 1 & 0 & 0 \\ 0 & 0 & -1 & 0 \\ 0 & 0 & 0 & -1 \end{pmatrix}. \qquad (11\text{-}19)$$

Direct substitution verifies that these matrices indeed satisfy the commutation relations in (11-17).‡ The Dirac equation (11-15) can therefore be regarded as a 4×4 *matrix* differential equation for the four-component state function (spinor)

$$\Psi = \begin{pmatrix} \Psi^{(1)} \\ \Psi^{(2)} \\ \Psi^{(3)} \\ \Psi^{(4)} \end{pmatrix}. \qquad (11\text{-}20)$$

Alternatively, it can be considered as a set of four coupled differential equations for the components of Ψ.

The stationary energy eigenstates are obtained by assuming

$$\Psi = e^{-i\varepsilon_i t/\hbar}\psi_i(\mathbf{r})$$

which when substituted into (11-15) gives the Dirac *energy eigenvalue* equation

$$\left\{ c\hat{\alpha} \cdot \left(\hat{\mathbf{p}} - \frac{q}{c} \mathbf{A} \right) + q\Phi + \hat{\beta}mc^2 \right\} \psi_i(\mathbf{r}) = \varepsilon_i \psi_i(\mathbf{r}). \qquad (11\text{-}21)$$

We are assuming that $\mathbf{A}$ and Φ represent *static* fields.

III   Free Dirac Particles

The Dirac (energy eigenvalue) equation for a free particle ($\mathbf{A} = \Phi = 0$) is

$$\left\{ c\frac{\hbar}{i}\hat{\alpha} \cdot \nabla + \hat{\beta}mc^2 \right\} \psi_\mathbf{p} = \varepsilon_\mathbf{p} \psi_\mathbf{p}. \qquad (11\text{-}22)$$

---

‡ This representation for $\hat{\alpha}$ and $\hat{\beta}$, although not unique, is used most commonly.

In analogy with the nonrelativistic case, we assume that the solution is a momentum eigenfunction of the form

$$\psi_\mathbf{p} = |u_\mathbf{p}\rangle e^{i\mathbf{p}\cdot\mathbf{r}/\hbar} \tag{11-23}$$

where the amplitude is the four-component *spinor*,

$$|u_\mathbf{p}\rangle = \begin{pmatrix} u_\mathbf{p}^{(1)} \\ u_\mathbf{p}^{(2)} \\ u_\mathbf{p}^{(3)} \\ u_\mathbf{p}^{(4)} \end{pmatrix}.$$

Substitution of (11-23) into (11-22) gives the 4×4 *matrix* eigenvalue equation

$$\{c\mathbf{p}\cdot\hat{\boldsymbol{\alpha}} + \hat{\beta}mc^2\}|u_\mathbf{p}\rangle = \varepsilon_\mathbf{p}|u_\mathbf{p}\rangle. \tag{11-24}$$

It is left as an exercise (see Problem 11-3) to show that the eigenvalues of (11-24) are

$$\varepsilon_\mathbf{p}^+ = (c^2p^2 + m^2c^4)^{1/2} \quad \text{and} \quad \varepsilon_\mathbf{p}^- = -(c^2p^2 + m^2c^4)^{1/2}. \tag{11-25}$$

We show below that the negative-energy states are not to be dismissed as extraneous. The eigenvectors for positive energy turn out to be

$$|u_{\mathbf{p}\uparrow}^+\rangle = N_p \begin{pmatrix} 1 \\ 0 \\ \dfrac{cp_z}{\varepsilon_p^+ + mc^2} \\ \dfrac{c(p_x + ip_y)}{\varepsilon_p^+ + mc^2} \end{pmatrix} \quad \text{and} \quad |u_{\mathbf{p}\downarrow}^+\rangle = N_p \begin{pmatrix} 0 \\ 1 \\ \dfrac{c(p_x - ip_y)}{\varepsilon_p^+ + mc^2} \\ \dfrac{-cp_z}{\varepsilon_p^+ + mc^2} \end{pmatrix} \tag{11-26a}$$

where N_p is a normalization constant. The arrows distinguish between the two types of states of positive energy, and as we shall see, designate the spin orientation (up or down) of the Dirac particle. The negative-energy states have the amplitudes

$$|u_{\mathbf{p}\uparrow}^-\rangle = N_p \begin{pmatrix} \dfrac{-cp_z}{-\varepsilon_p^- + mc^2} \\ \dfrac{-c(p_x + ip_y)}{-\varepsilon_p^- + mc^2} \\ 1 \\ 0 \end{pmatrix} \quad \text{and} \quad |u_{\mathbf{p}\downarrow}^-\rangle = N_p \begin{pmatrix} \dfrac{-c(p_x - ip_y)}{-\varepsilon_p^- + mc^2} \\ \dfrac{cp_z}{-\varepsilon_p^- + mc^2} \\ 0 \\ 1 \end{pmatrix}. \tag{11-26b}$$

All four of the above spinors may be normalized according to $\langle u_{\mathbf{p}} | u_{\mathbf{p}} \rangle = 1$ by setting

$$N_p = \left[\frac{\varepsilon_p^+ + mc^2}{2\varepsilon_p^+} \right]^{+1/2}.$$

The total state is specified by indicating its momentum $\mathbf{p}$, its spin (up or down), and its energy type (positive or negative). For example, a particle moving along the z axis with negative energy and spin up would be represented by

$$\psi_{\mathbf{p}\uparrow}^- = | u_{\mathbf{p}\uparrow}^- \rangle e^{ip_z z/\hbar} = N_p \begin{pmatrix} \dfrac{-cp_z}{-\varepsilon_p^- + mc^2} \\ 0 \\ 1 \\ 0 \end{pmatrix} e^{ip_z z/\hbar}$$

with energy

$$\varepsilon_{\mathbf{p}} = \varepsilon_{\mathbf{p}}^- = -(p_z^2 c^2 + m^2 c^4)^{1/2}.$$

It is left as an exercise (see Problem 11-5) to show that for positive-energy states, the lower two components in (11-26a) are of order v/c times the upper two components. Thus, in the nonrelativistic limit the lower components are small and can be ignored. As $v/c \to 0$, the positive-energy eigenfunctions correspond to the nonrelativistic states of the particle, and thus the properties of these states are determined by the upper two components of (11-26a). In the nonrelativistic limit, the relativistic state functions in (11-26a) reduce to the two-component spinors $\psi_{\mathbf{p}\uparrow} = \binom{1}{0} e^{i\mathbf{p}\cdot\mathbf{r}/\hbar}$ and $\psi_{\mathbf{p}\downarrow} = \binom{0}{1} e^{i\mathbf{p}\cdot\mathbf{r}/\hbar}$, representing a free particle of spin $\frac{1}{2}$ with spin "up" and spin "down" respectively.

IV Negative-Energy States

As was pointed out above, the negative-energy states cannot be dismissed as extraneous; we show below that they do have physical significance. The free-particle spectrum is continuous and ranges from plus to minus infinity except for a *forbidden* gap, $-mc^2 < \varepsilon < mc^2$ (Figure 11-1).

Dirac suggested that in nature there exists an infinity of electrons—more than enough to fill the negative-energy states. The Pauli principle restricts the occupancy of each free-particle state to a single electron. The negative-energy states are assumed to be completely filled. This infinite Fermi "sea" of negative-energy electrons produces a uniform charge density everywhere in space. Gauss's law implies that the field at the center of a uniform charged sphere is zero. As the radius of the sphere tends to infinity, any point in space can be regarded as the center of the sphere, and we conclude that the field

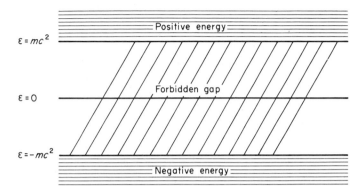

Figure 11-1   Energy level diagram for free Dirac particles.

of a uniform charge density of infinite extent produces no observable fields. The state in which there are exactly enough electrons to fill the negative levels is called the vacuum state. Any additional particles must enter positive-energy states and are observed as electrons.

Beginning with the vacuum state, it should be possible to excite a negative-energy electron to a positive level by absorption of a photon. The vacancy or "hole" left behind destroys the uniformity of the Fermi sea. The field produced by the hole is equivalent to that of a particle of opposite charge to the electron. This hole is observed as a positive electron—a positron—the anti-particle of the electron. The process in which a photon is absorbed and an electron–positron pair formed from the vacuum is called pair creation. The minimum photon energy required is

$$mc^2 - (-mc^2) = 2mc^2 = \hbar\omega \simeq 1 \quad \text{MeV.}$$

A photon of this energy corresponds to a gamma-ray quantum. It is left as an exercise (Problem 11-6) to show that it is impossible to conserve both momentum and energy when a gamma-ray photon disappears and an electron-positron pair is formed. Since both momentum and energy must be conserved, this process can only occur in the presence of an external field such as that of an atomic nucleus, the latter absorbing some of the momentum. The reverse process (pair annihilation) results when the electron returns to fill the hole and thereby liberates a *pair* of gamma-ray photons. Since *two* photons are emitted in this process, energy and momentum can simultaneously be conserved without any additional mass such as a nucleus. Note that this interpretation of negative-energy states cannot be applied to Klein–Gordon particles since they are bosons and do not obey the Pauli exclusion principle.

V A Dirac Particle in a Static Field

The Dirac energy-eigenvalue equation (11-21) for a particle in an electrostatic ($\mathbf{E} = -\nabla\Phi$) and a magnetostatic ($\mathbf{B} = \nabla \times \mathbf{A}$) field is

$$\left\{ c\left(\hat{\mathbf{p}} - \frac{q}{c}\mathbf{A}\right) \cdot \hat{\mathbf{\alpha}} + \hat{\beta}mc^2 + q\Phi \right\} \psi_i = \varepsilon_i \psi_i \tag{11-27}$$

where

$$\psi_i = \begin{pmatrix} \psi_i^{(1)} \\ \psi_i^{(2)} \\ \psi_i^{(3)} \\ \psi_i^{(4)} \end{pmatrix}.$$

Equation (11-27) is equivalent to the pair of coupled equations:

$$\left\{ c\left(\hat{\mathbf{p}} - \frac{q}{c}\mathbf{A}\right) \cdot \hat{\mathbf{\sigma}} \right\} \psi_i^{(B)} + (mc^2 + q\Phi)\psi_i^{(A)} = \varepsilon_i \psi_i^{(A)} \tag{11-28}$$

and

$$\left\{ c\left(\hat{\mathbf{p}} - \frac{q}{c}\mathbf{A}\right) \cdot \hat{\mathbf{\sigma}} \right\} \psi_i^{(A)} - (mc^2 - q\Phi)\psi_i^{(B)} = \varepsilon_i \psi_i^{(B)}$$

where

$$\psi_i^{(A)} = \begin{pmatrix} \psi_i^{(1)} \\ \psi_i^{(2)} \end{pmatrix} \quad \text{and} \quad \psi_i^{(B)} = \begin{pmatrix} \psi_i^{(3)} \\ \psi_i^{(4)} \end{pmatrix}$$

are the upper and lower components of the Dirac four-component eigenfunction. To decouple these equations, we first solve the second to give

$$\psi_i^{(B)} = \left\{ \frac{c(\hat{\mathbf{p}} - (q/c)\,\mathbf{A}) \cdot \hat{\mathbf{\sigma}}}{\varepsilon_i - q\Phi + mc^2} \right\} \psi_i^{(A)}. \tag{11-29}$$

Substituting this result into the first equation, we obtain

$$c^2 \left\{ \left(\hat{\mathbf{p}} - \frac{q}{c}\mathbf{A}\right) \cdot \hat{\mathbf{\sigma}}(\varepsilon_i - q\Phi + mc^2)^{-1}\left(\hat{\mathbf{p}} - \frac{q}{c}\mathbf{A}\right) \cdot \hat{\mathbf{\sigma}} \right\} \psi_i^{(A)} + (mc^2 + q\Phi)\psi_i^{(A)}$$

$$= \varepsilon_i \psi_i^{(A)}. \tag{11-30}$$

In the nonrelativistic limit, $\psi^{(B)}$ is negligible compared with $\psi^{(A)}$; in the limit $v/c \to 0$, the behavior of the particle is entirely determined by (11-30). Furthermore, in the nonrelativistic limit, the total nonrelativistic energy $\varepsilon_i' = \varepsilon_i - mc^2$ and the potential energy $q\Phi$ are much smaller than the rest energy. We can make the expansion

$$(\varepsilon_i - q\Phi + mc^2)^{-1} = (\varepsilon_i' - q\Phi + 2mc^2)^{-1}$$

$$\simeq \frac{1}{2mc^2}\left[1 - \frac{\varepsilon_i' - q\Phi}{2mc^2} + \cdots\right]. \tag{11-31}$$

In the *extreme* nonrelativistic limit, we retain only the first term on the right of (11-31) and (11-30) becomes

$$\left\{ \frac{[(\hat{\mathbf{p}} - (q/c)\,\mathbf{A}) \cdot \hat{\boldsymbol{\sigma}}]^2}{2m} + q\Phi \right\} \psi_i^{(A)} = \varepsilon_i' \psi_i^{(A)}. \qquad (11\text{-}32)$$

To show how spin arises as a natural consequence of the Dirac equation in the nonrelativistic limit, we simplify (11-31) by introducing the following operator indentities‡:

(a) $$\qquad\qquad (\hat{\boldsymbol{\sigma}} \cdot \hat{\mathbf{U}})(\hat{\boldsymbol{\sigma}} \cdot \hat{\mathbf{V}}) = \hat{\mathbf{U}} \cdot \hat{\mathbf{V}} + i\hat{\boldsymbol{\sigma}} \cdot (\hat{\mathbf{U}} \times \hat{\mathbf{V}}) \qquad\qquad (11\text{-}33a)$$

provided $$\qquad [\hat{\boldsymbol{\sigma}}, \hat{\mathbf{U}}] = [\hat{\boldsymbol{\sigma}}, \hat{\mathbf{V}}] = 0$$

and

(b) $$\left(\hat{\mathbf{p}} - \frac{q}{c}\mathbf{A} \right) \times \left(\hat{\mathbf{p}} - \frac{q}{c}\mathbf{A} \right) = \frac{iq\hbar}{c} \nabla \times \mathbf{A} = \frac{iq\hbar}{c} \mathbf{B} \qquad \left(\hat{\mathbf{p}} = \frac{\hbar}{i} \nabla \right).$$

$$(11\text{-}33b)$$

Setting $\hat{\mathbf{U}} = \hat{\mathbf{V}} = \hat{\mathbf{p}} - \dfrac{q}{c}\mathbf{A}$, (11-32) becomes

$$\left\{ \frac{1}{2m} \left(\hat{\mathbf{p}} - \frac{q}{c}\mathbf{A} \right)^2 - \frac{q}{mc}\mathbf{S} \cdot \mathbf{B} + q\Phi \right\} \psi_i^{(A)} = \varepsilon_i' \psi_i^{(A)} \qquad (11\text{-}34)$$

where $\hat{\mathbf{S}} = \frac{1}{2}\hbar\hat{\boldsymbol{\sigma}}$.

Equation (11-34) is identical to the nonrelativistic Schroedinger energy eigenvalue equation with Pauli's generalization. The first term in the Hamiltonian on the left side of (11-34) includes the kinetic energy and the orbital interaction with the **B** field. The second and third terms are the spin–magnetic and the electrostatic interactions respectively.

Thus, even in the nonrelativistic limit, a Dirac particle has spin–magnetic properties. It is quite remarkable that this feature, which originated from relativistic considerations, remains in the limit $v/c \to 0$! Note also that the spin dipole moment in (11-34) is correctly given as

$$\hat{\boldsymbol{\mu}}_s = \frac{g_s q \hat{\mathbf{S}}}{2mc} = \frac{q\hat{\mathbf{S}}}{mc}$$

with the spin gyromagnetic factor $g_s = 2$, as is experimentally observed for the electron.

---

‡ Note that the cross product of the two identical operators in (b) is not zero. This is because the components of $\hat{\mathbf{p}}$ and $\mathbf{A}$ do not commute.

It is interesting to consider the leading relativistic corrections to (11-30) in the case where $\mathbf{A} = \mathbf{0}$ and where $q\Phi = V(r)$ is some central force potential. This situation corresponds to a hydrogenic electron when we set $V = -e^2/r$. Using (11-31) in (11-30), we find

$$\left\{\left[\frac{\hat{\mathbf{p}} \cdot \hat{\boldsymbol{\sigma}}(1 - [\varepsilon_i' - V(r)]/2mc^2)\hat{\mathbf{p}} \cdot \hat{\boldsymbol{\sigma}}}{2m}\right] + V(r)\right\}\psi_i^{(A)} = \varepsilon_i'\psi_i^{(A)}. \quad (11\text{-}35)$$

In this case $T = \varepsilon_i' - V(r) = \hat{p}^2/2m$ is the nonrelativistic kinetic energy of the particle. Using the identities in (11-33) and the properties of the $\hat{\boldsymbol{\sigma}}$ and $\hat{\mathbf{p}} = (\hbar/i)\nabla$ operators, the following identity may be established:

$$\frac{1}{2m}(\hat{\mathbf{p}} \cdot \hat{\boldsymbol{\sigma}})\left[1 - \frac{(\varepsilon_i' - V)}{2mc^2}\right](\hat{\mathbf{p}} \cdot \hat{\boldsymbol{\sigma}})$$

$$= \frac{\hat{p}^2}{2m} - \frac{\hat{p}^4}{8m^3c^2} - \frac{\hbar^2}{4m^2c^2}\nabla V \cdot \nabla + \frac{\hbar}{4m^2c^2}\hat{\boldsymbol{\sigma}} \cdot \nabla V \times \hat{\mathbf{p}}. \quad (11\text{-}36)$$

Using this relation and noting that for central fields

$$\nabla V = \frac{dV}{dr}\frac{\mathbf{r}}{r}$$

(11-35) becomes, after regrouping terms,

$$\left\{\left(\frac{\hat{p}^2}{2m} + V\right) - \frac{\hat{p}^4}{8m^3c^2} + \frac{\hbar}{4m^2c^2}\frac{1}{r}\frac{dV}{dr}\hat{\boldsymbol{\sigma}} \cdot (\mathbf{r} \times \hat{\mathbf{p}}) - \frac{\hbar^2}{4m^2c^2}\frac{dV}{dr}\frac{\partial}{\partial r}\right\}\psi_i^{(A)}$$

$$= \varepsilon_i'\psi_i^{(A)}. \quad (11\text{-}37)$$

The first two terms on the right represent the nonrelativistic Hamiltonian. The third term is the mass-variation term which provides the leading relativistic corrections to the kinetic energy. The fourth term may be written

$$\Delta E(\text{spin–orbit}) = \frac{1}{2m^2c^2}\frac{1}{r}\frac{dV}{dr}\hat{\mathbf{S}} \cdot \hat{\mathbf{L}}$$

or in the case of hydrogen,

$$\Delta E(\text{spin–orbit}) = \frac{1}{2m^2c^2}\frac{e^2}{r^3}\hat{\mathbf{S}} \cdot \hat{\mathbf{L}}.$$

This will be recognized as the spin–orbit interaction term as in (6-35) with the Thomas precession factor of $\frac{1}{2}$ arising naturally.

The fifth term is the Darwin term alluded to in our discussion of fine structure in Chapter 7. In hydrogen, the Darwin term produces first-order energy corrections only in s ($l = 0$) states. These corrections are given by

$$\Delta E(\text{Darwin}) \propto \left\langle n, l \left| \frac{dV}{dr}\frac{\partial}{\partial r} \right| n, l \right\rangle \rightarrow \int_0^\infty R_{nl}\frac{dV}{dr}\frac{dR_{nl}}{dr} r^2 \, dr.$$

For hydrogen ($V = -e^2/r$), this becomes

$$\Delta E(\text{Darwin}) \propto e^2 \int_0^\infty R_{nl}\frac{1}{r^2}\frac{dR_{nl}}{dr} r^2 \, dr$$

$$= e^2 \int_0^\infty R_{nl}\, dR_{nl} = \tfrac{1}{2}e^2 \left| R_{nl}^2 \right|_0^\infty.$$

Recall that R_{nl} vanishes at infinity for all bound states. Since it also vanishes at the origin for all but s states, the energy corrections exist only for these states.

The fine-structure effects in hydrogen are thus first-order relativistic effects of the Dirac equation. We may therefore identify a Dirac particle with a relativistic electron.

VI The Dirac Particle in a Coulomb Potential— Fine Structure in Hydrogen

It is possible to calculate the fine-structure energy of the hydrogenic electron by applying first-order perturbation theory to the three fine-structure terms in (11-37). However, in the special case of a Coulomb potential, an exact solution to the Dirac equation is possible. The exact energy eigenvalues can be expressed in powers of the fine-structure constant, and the leading terms retained. Since the procedure is quite involved, we shall merely outline the steps and list the results.

The Dirac equation (11-27) for the Coulomb potential is

$$\left\{ c\hat{\boldsymbol{\alpha}} \cdot \hat{\mathbf{p}} + \hat{\beta}mc^2 - \frac{e^2}{r} \right\}\psi_i = \varepsilon_i\psi_i. \tag{11-38}$$

Although (11-38) appears to represent a central force problem, a separation cannot be accomplished using $R_{nl}(r)Y_{lm_l}(\theta, \phi)$ because the equation itself incorporates spin–orbit coupling and $\hat{L}_z$ is not conserved. Equivalently,

m_l is not a "good" quantum number. To effect a separation, we introduce the two classes of solutions

$$\psi_{\varepsilon l j \pm m_j} = \begin{pmatrix} u(r) \left[\dfrac{l \pm m_j + \frac{1}{2}}{2l+1} \right]^{1/2} Y_{l, m_j - \frac{1}{2}} \\[2ex] \mp u(r) \left[\dfrac{l \mp m_j + \frac{1}{2}}{2l+1} \right]^{1/2} Y_{l, m_j + \frac{1}{2}} \\[2ex] -iv(r) \left[\dfrac{(l \pm 1) \mp (m_j \mp \frac{1}{2})}{(2l+1) \pm 2} \right]^{1/2} Y_{l \pm 1, m_j - \frac{1}{2}} \\[2ex] \mp iv(r) \left[\dfrac{(l \pm 1) \pm (m_j \pm \frac{1}{2})}{(2l+1) \pm 2} \right]^{1/2} Y_{l \pm 1, m_j + \frac{1}{2}} \end{pmatrix} \qquad (11\text{-}39)$$

where $j^{\pm} = l \pm \frac{1}{2}$ and $|m_j| \leqslant j$.

Expressing (11-38) in spherical coordinates and substituting (11-39), we are led to the following equations for the radial parts, $u(r)$ and $v(r)$:

$$\frac{1}{\hbar c} \left\{ \varepsilon_i + \frac{e^2}{r} + mc^2 \right\} v = \frac{du}{dr} + \frac{(1+k)u}{r}$$

$$\frac{1}{\hbar c} \left\{ \varepsilon_i + \frac{e^2}{r} - mc^2 \right\} u = -\frac{dv}{dr} - \frac{(1-k)v}{r} \qquad (11\text{-}40)$$

where

$$k = -(l+1) \qquad \text{for} \quad \psi_{j+}$$
$$k = l \qquad\qquad \text{for} \quad \psi_{j-}.$$

These equations are treated in a manner analogous to nonrelativistic radial equations. Bound-state requirements[‡] on u and v can be satisfied if and only if ε_i is restricted to the values

$$\varepsilon_i = \varepsilon_{n'k} = mc^2 \left\{ 1 + \frac{\gamma^2}{[(k^2 - \gamma^2)^{1/2} + n']^2} \right\}^{-1/2}, \qquad \left(\gamma = \frac{e^2}{\hbar c} \right) \qquad (11\text{-}41)$$

where $n' = 0, 1, 2, \ldots$. This result is exact and gives the quantized energies of a relativistic hydrogenic electron. Since $\gamma^2 \simeq (\frac{1}{137})^2 \simeq 10^{-4}$, we may expand (11-41) in powers of γ^2. The leading terms are

$$\varepsilon_i \simeq mc^2 \left[1 - \frac{\gamma^2}{2n^2} - \frac{\gamma^4}{2n^4} \left(\frac{n}{|k|} - \frac{3}{4} \right) + \cdots \right] \qquad (11\text{-}42)$$

---

‡ For details, see L. I. Schiff, "Quantum Mechanics," 3rd ed., p. 485, McGraw-Hill, New York, 1968, or H. A. Bethe and E. E. Salpeter, "Quantum Mechanics of One and Two Electron Atoms," p. 65, Springer-Verlag, Berlin, 1957.

where $n = n' + |k| = 1, 2, 3, \ldots$. Recalling that $|k| = l + 1$ for $j = l + \frac{1}{2}$ and that $|k| = l$ for $j = l - \frac{1}{2}$, it follows that for either case $|k| = j + \frac{1}{2}$ so that (11-42) may be expressed as

$$\varepsilon_{n,\,j} = mc^2 - \frac{me^4}{2\hbar^2 n^2} - \frac{\gamma^2 me^4}{2\hbar^2 n^3}\left(\frac{1}{j+\frac{1}{2}} - \frac{3}{4n}\right)$$

$$= mc^2 + \varepsilon_n + \frac{\gamma^2 \varepsilon_n}{n}\left(\frac{1}{j+\frac{1}{2}} - \frac{3}{4n}\right). \qquad (11\text{-}43)$$

These terms are respectively the rest energy, the "Bohr" energy, and the fine-structure energy as given by (7-51). The last term is in agreement with observed spectroscopic data in hydrogen.

The fine-structure terms in hydrogen as given by (11-43) provide corrections of order 10^{-4} with respect to the electrostatic Bohr levels. Additional factors exist which produce much smaller corrections than those due to fine structure. For example, the proton, which has its own magnetic spin dipole moment, can interact with the dipole associated with the electron's spin. These interactions produce hyperfine structure in hydrogen's spectrum. Also, the hydrogenic electron experiences a Coulombic potential only outside the proton. Inside the proton ($r < 10^{-13}$ cm), the potential is definitely not a $1/r$ potential. This deviation produces a minor shift in the hydrogenic energy spectrum.

A third factor of importance is related to the interaction between an electron and the radiation vacuum. In the absence of external radiation, the average values of $\mathbf{E}$ and $\mathbf{B}$ vanish, as expected. However, the nonvanishing fluctuations or uncertainties of the quantized field can still interact with the electron to produce an anomalous spin g factor, namely, $g_s \simeq 2.00232$. This effect is responsible for the small shift in the spectrum of hydrogen known as the *Lamb* shift and produces a slight separation between the $^2S_{\frac{1}{2}}$ and $^2P_{\frac{1}{2}}$ levels in hydrogen (Figure 11-2). The fluctuations of the quantized radiation field are also responsible for spontaneous transitions in the absence of observable $\mathbf{E}$ and $\mathbf{B}$ fields.

Quantization of the electromagnetic field not only ensures consistency with Heisenberg's uncertainty principle but also justifies the notion of a photon as a field quantum. We shall address ourselves to the general problem of field quantization in Chapter 12.

While the Dirac equation appears to govern the quantum behavior of the relativistic electron, its solutions are not always physically obvious. For one thing, the notion of position and velocity of a Dirac particle remains somewhat vague. In fact, a careful analysis of the Dirac equation would show that a Dirac particle traveling in a given direction actually executes two motions at once. The first is a translation of the particle's mean position along the

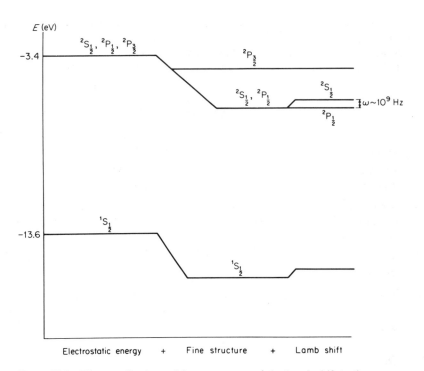

Figure 11-2 The contributions of fine structure and the Lamb shift to the energy spectrum of hydrogen. The level separations are in the ratio of electrostatic: fine structure: Lamb shift $\simeq 1:10^{-4}:10^{-5}$.

direction of classical motion. The second is an erratic, high-frequency ($\omega \simeq 2mc^2/\hbar = 2 \times 10^{21}$ Hz) oscillation of the particle about its mean position. The oscillatory frequency of this *Zitterbewegung* ("trembling" motion) is much too high to be observed directly. The Darwin "fluctuation" term in the fine–structure energy can be explained, physically, in terms of the electron's Zitterbewegung.‡

Suggested Reading

Berestetskii, V. B., Lifshitz, E. M., and Pitaevskii, L. P., "Relativistic Quantum Theory." Addison-Wesley, Reading Massachusetts, 1971.

Bethe, H. A., and Salpeter, E. E., "Quantum Mechanics of One and Two Electron Atoms." Springer-Verlag, Berlin, 1957.

‡ See, for example, J. D. Bjorken and S. D. Drell, "Relativistic Quantum Mechanics," p. 52. McGraw-Hill, New York, 1964.

Bjorken, J. D., and Drell, S. D., "Relativistic Quantum Mechanics." McGraw-Hill, New York, 1964.
Messiah, A., "Quantum Mechanics," Volume II. Wiley, New York, 1962.
Schiff, L. I., "Quantum Mechanics," 3rd ed. McGraw-Hill, New York, 1968.

Problems

11-1. Show that the commutation relations (11-17) ensure that the Dirac and Klein–Gordon equations are consistent with each other.

11-2. Show that the Dirac matrices (11-19) satisfy (11-17).

11-3. Set up and solve the matrix eigenvalue equation (11-24) and verify that the eigenvectors and eigenvalues are correctly given by (11-25) and (11-26).

11-4. Verify that the free-particle eigenspinors in (11-26) are normalized to $\langle u|u \rangle = 1$.

11-5. Show that in the nonrelativistic limit the lower two components of (11-26a) are v/c times the upper components.

11-6. Show that the reaction $\gamma \to e^+ + e^-$ (pair creation) cannot conserve both energy and momentum in free space.

11-7. Derive the identities given in (11-33) and (11-36).

11-8. (a) Show that if the probability density associated with a Dirac spinor is $\rho = \Psi^\dagger \Psi$, then the probability current is given by

$$\mathbf{J} = c\Psi^\dagger \hat{\boldsymbol{\alpha}}\Psi.$$

[Hint: Put Eq. (11.15) into the form $\nabla \cdot \mathbf{J} + (\partial\rho/\partial t) = 0$.]

(b) Consider a free Dirac particle of positive energy, moving along the positive z axis with momentum p. Using the definition of $\mathbf{J}$ above, find the velocity at which the probability is flowing. Explain the result.

11-9. Using first-order perturbation theory, calculate the shift in energy of the ground state of hydrogen if the proton is regarded as a uniformly charged sphere of radius $R = 10^{-13}$ cm.

12 | Quantum Field Theory

A classical field is mathematically represented by a function $\psi(\mathbf{r}, t)$ characterizing the state of the field at the point $\mathbf{r}$ and time t. For example, ψ could represent the density fluctuation associated with a sound wave in air or the displacement of a point on a vibrating string. These fields are essentially granular fields since they represent motions of particles.

A field representation often simplifies a many-particle problem. For example, transverse waves on a stretched string satisfy the one-dimensional wave equation

$$\frac{\partial^2 y}{\partial x^2} - \frac{1}{v^2} \frac{\partial^2 y}{\partial t^2} = 0 \tag{12-1}$$

where $y(x, t)$ is the displacement of the string and $v = (\text{tension/linear density})^{1/2}$ is the phase velocity of waves along the string. It is simpler to treat this single field equation than the $\sim 10^{23}$ equations of motion for the individual molecules of the string. Approximating a microscopic many-body system by a classical field is useful when macroscopic properties are considered. The electromagnetic field on the other hand is fundamentally a nongranular field and is dynamically equivalent to a many-body problem involving an infinite number of particles.

Equation (12-1) is essentially a Newtonian formulation of continuum dynamics. The quantum theory of fields must be developed from a canonical Lagrangian–Hamiltonian formalism. We shall develop a canonical theory by considering the transition from a discrete elastic chain to a smooth continuous string.

I Classical Theory of Fields

The Lagrangian for transverse vibrations of an N-particle linear elastic chain is‡

$$L = \sum_{i=1}^{N} \tfrac{1}{2}m\dot{y}_i{}^2 - \sum_{i=1}^{N} \tfrac{1}{2}K(y_{i+1} - y_i)^2 \tag{12-2}$$

where m is the mass of each particle, K is the elastic constant of each spring, and y_i is the transverse displacement of the ith mass point. We will use a to represent the equilibrium separation of the masses (lattice constant) and Na will therefore be the total length of the chain. The transition from the discrete to the continuous chain is initiated by removing half of each mass and translating it an amount $a/2$ to the right. Repeating this operation an infinite number of times, we generate a chain with a continuous mass distribution. Furthermore, at each "halving" of the lattice constant, the elastic constant of each spring doubles. We are thus led to the following mathematical relations:

(a) $\displaystyle \lim_{a \to 0} \sum a \to \int dx$

(b) $\displaystyle \lim_{a \to 0} \frac{m}{a} \to \frac{dm}{dx} = \mu = $ linear mass density

(c) $\displaystyle \lim_{\substack{a \to 0 \\ K \to \infty}} Ka \to \tau = $ elastic tension.

The limiting form of the Lagrangian in (12-2) is

$$L = \lim_{a \to 0} \sum a \left\{ \frac{1}{2}\frac{m}{a}\dot{y}_i{}^2 - \frac{1}{2}Ka\left(\frac{y_{i+1} - y_i}{a}\right)^2 \right\} \to \int dx \left\{ \frac{1}{2}\mu\dot{y}^2 - \frac{1}{2}\tau y'^2 \right\}$$

$$\tag{12-3}$$

where $\dot{y} = \partial y/\partial t$ and $y' = \partial y/\partial x$. Equation (12-3) may be written   as

---

‡ In this chapter we will use L and H to represent the Lagrangian and Hamiltonian and $\mathscr{L}$ and $\mathscr{H}$ to represent their respective densities.

$L = \int \mathscr{L}(\dot{y}, y') \, dx$; $\mathscr{L}$ is the *Lagrangian density* (per unit length) and is expressible in terms of the kinetic and potential energy densities as $\mathscr{L} = \mathscr{T} - \mathscr{V}$, where $\mathscr{T} = \frac{1}{2}\mu\dot{y}^2$ and $\mathscr{V} = \frac{1}{2}\tau y'^2$.

The Lagrangian density is a function of the field velocity $\dot{y}(x, t)$ and the field coordinate $y(x, t)$ (and possibly its derivative y'). It should be stressed that x, which represents the point at which the field is considered, is not a field coordinate; rather it is a variable originating with the subscript i in (12-3). We have yet to establish how the equation of motion for the field, (12-1), is derived from $\mathscr{L}$. The three-dimensional generalization of (12-3) is obtained by going from the field function $y(x, t)$ to $\psi(\mathbf{r}, t)$ where ψ is a scalar field. The Lagrangian for this field has the general form

$$L = \int \mathscr{L}(\dot{\psi}, \psi, \nabla\psi) \, d\mathbf{r} = \int \mathscr{L}\left(\dot{\psi}, \psi, \frac{\partial\psi}{\partial x}, \frac{\partial\psi}{\partial y}, \frac{\partial\psi}{\partial z}\right) d\mathbf{r}.$$

We *define* the "functional" or "variational" derivative of L with respect to ψ as

$$\frac{\delta L}{\delta\psi} = \frac{\partial\mathscr{L}}{\partial\psi} - \sum_{i=1}^{3} \frac{\partial}{\partial x_i} \frac{\partial\mathscr{L}}{\partial(\partial\psi/\partial x_i)} \qquad (x_1 = x, \quad x_2 = y, \quad x_3 = z). \quad (12\text{-}4)$$

Similarly, the derivative with respect to the field velocity (12-4), $\delta L/\delta\dot{\psi}$, is taken by replacing ψ by $\dot{\psi}$ above. Using this definition of the derivative, we state without proof that the equation of motion for the field is derivable using the familiar scheme‡ of chapter 2

$$\frac{d}{dt}\frac{\delta L}{\delta\dot{\psi}} - \frac{\delta L}{\delta\psi} = 0. \qquad (12\text{-}5)$$

These are Lagrange's equations for a *field*. Applying (12-5) to the Lagrangian density of the string, (12-3), we obtain§

$$\frac{\delta L}{\delta y} = \frac{\partial\mathscr{L}}{\partial y} - \frac{\partial}{\partial x}\frac{\partial\mathscr{L}}{\partial y'} = \tau\frac{\partial^2 y}{\partial x^2}$$

$$\frac{d}{dt}\frac{\delta L}{\delta\dot{y}} = \frac{\partial}{\partial t}\left(\frac{\partial\mathscr{L}}{\partial\dot{y}} - \frac{\partial}{\partial x}\frac{\partial\mathscr{L}}{\partial\dot{y}'}\right) = \mu\frac{\partial^2 y}{\partial t^2}$$

---

‡ The derivative with respect to time is evaluated as

$$\frac{d}{dt}\frac{\delta L}{\delta\dot{\psi}} = \frac{\partial}{\partial t}\left(\frac{\partial\mathscr{L}}{\partial\dot{\psi}} - \sum_i \frac{\partial}{\partial x_i} \frac{\partial\mathscr{L}}{\partial(\partial\dot{\psi}/\partial x_i)}\right).$$

§ Since $\dot{y}'$ will not enter in $\mathscr{L}$ we may set $\partial\mathscr{L}/\partial\dot{y}' = 0$.

so that (12-5) gives

$$\mu \frac{\partial^2 y}{\partial t^2} - \tau \frac{\partial^2 y}{\partial x^2} = 0$$

as required by (12-1).

For a vector field $\psi(\mathbf{r}, t)$, we obtain a field equation of the form (12-5) for each component ψ_i provided these components are independent.

II The Hamiltonian Density

The Hamiltonian for a field may be expressed in terms of its density as $H = \int \mathscr{H}\, d\mathbf{r}$. The density $\mathscr{H}$ is a function of the field coordinates ψ_i and their canonically conjugate momentum densities defined by

$$\pi_i = \frac{\delta L}{\delta \dot{\psi}_i}. \tag{12-6}$$

Performing a Legendre transformation similar to (2-59), we obtain

$$\mathscr{H} = \sum_{i=1}^{3} \dot{\psi}_i \pi_i - \mathscr{L}. \tag{12-7}$$

It is understood that $\mathscr{H}$ must be expressed as a function of π_i, ψ_i, and $\nabla \psi_i$ but cannot contain $\dot{\psi}_i$. Using the definition in (12-4), the Hamilton equations for a field follow as

$$\frac{\delta H}{\delta \pi_j} = \dot{\psi}_j \quad \text{and} \quad \frac{\delta H}{\delta \psi_j} = -\dot{\pi}_j. \tag{12-8}$$

These are equivalent to Lagrange's field equations.

In particular, the Hamiltonian density for a string may be shown to be of the form

$$\mathscr{H} = \frac{\pi_y^2}{2\mu} + \frac{\tau y'^2}{2}. \tag{12-9}$$

This function is numerically equal to the energy density associated with the vibrating string. An application of Hamilton's equations (12-8) to (12-9) gives

$$\dot{y} = \frac{\pi_y}{\mu} \quad \text{and} \quad \dot{\pi}_y = \tau y''$$

which when combined give $\mu \ddot{y} - \tau y'' = 0$. This result is identical to (12-1) as expected.

III   Field Quantization

The quantization of a field is accomplished by first making the Hermitian operator replacements for the canonical field variables

$$\pi_j \to \hat{\pi}_j \qquad \text{and} \qquad \psi_j \to \hat{\psi}_j$$

and then postulating commutation relations between the field coordinates and momenta in analogy with those for a system of particles, namely,

$$[\hat{\pi}_j(\mathbf{r}'), \hat{\pi}_k(\mathbf{r})] = [\hat{\psi}_j(\mathbf{r}'), \hat{\psi}_k(\mathbf{r})] = 0$$

and   (12-10)

$$[\hat{\psi}_j(\mathbf{r}'), \hat{\pi}_k(\mathbf{r})] = i\hbar\delta_{jk}\delta(\mathbf{r} - \mathbf{r}').$$

These commutation relations imply that simultaneous measurements of two different field coordinates or two different field momenta may be performed with perfect precision at the same time. On the other hand, perfect *simultaneous* measurements of $\hat{\psi}_j$ and $\hat{\pi}_j$ cannot be made at the same point in space. In certain situations (for example, with relativistic fields) retardation effects are important. The commutation relations in (12-10) must then refer to measurements taken at different times. However, (12-10) adequately illustrates the general approach to quantization.

The Hamiltonian for the field is represented by the Hermitian operator

$$\hat{H}_{\text{field}} = \int \mathscr{H}(\hat{\pi}_j, \hat{\psi}_j, \nabla\hat{\psi}_j)\, d\mathbf{r}.$$

The energy eigenstates of $\hat{H}_{\text{field}}$ determine the stationary states of the quantized field. Equivalently, we seek the eigenvectors and eigenvalues of the equation

$$\hat{H}_{\text{field}}|\text{field}\rangle = \varepsilon_{\text{field}}|\text{field}\rangle.$$

Since $\hat{\mathbf{r}}$ is not a relevant observable for the field, this eigenvalue equation cannot be formulated in the differential language of wave mechanics.

IV   Classical Electrodynamics

Ironically, one of the most difficult classical fields to quantize is the one associated with electromagnetic radiation. Since the radiation field by its very nature is relativistic, care must be taken when applying commutation relations. As we shall see, the components of $\mathbf{E}$ and $\mathbf{B}$ describing the electromagnetic field are not independent. A reduction to independent field coordinates is therefore necessary before applying quantization.

We consider a system of charged particles in the presence of electromagnetic fields. Newton's equations of motion‡ for the particles are determined by the Lorentz force law,

$$m \frac{d^2 \mathbf{r}_i}{dt^2} = \mathbf{F}_i = q\mathbf{E}_i + \frac{q}{c} \mathbf{v}_i \times \mathbf{B}_i \qquad (12\text{-}11)$$

where $\mathbf{E}_i$ and $\mathbf{B}_i$ are the electric and magnetic fields evaluated at the position of the ith particle. The equations for the fields are given by Maxwell's equations

(1) $$\nabla \cdot \mathbf{E} = 4\pi\rho$$

(2) $$\nabla \cdot \mathbf{B} = 0$$

(3) $$\nabla \times \mathbf{E} = -\frac{1}{c}\dot{\mathbf{B}} \qquad (12\text{-}12)$$

(4) $$\nabla \times \mathbf{B} = \frac{1}{c}\dot{\mathbf{E}} + \frac{4\pi}{c}\mathbf{J};$$

the charge and current densities are related to the particle coordinates by

$$\rho = \sum_i q\delta(\mathbf{r} - \mathbf{r}_i) \quad \text{and} \quad \mathbf{J} = \sum_i q\mathbf{v}_i \delta(\mathbf{r} - \mathbf{r}_i). \qquad (12\text{-}13)$$

It is pointless to construct a Lagrangian in terms of $\mathbf{E}$ and $\mathbf{B}$ as field coordinates since they are not independent. The six components (of $\mathbf{E}$ and $\mathbf{B}$) can be reduced to four using the second and third of Maxwell's equations. The second, $\nabla \cdot \mathbf{B} = 0$, implies that $\mathbf{B}$ is derivable from a vector potential according to

$$\mathbf{B} = \nabla \times \mathbf{A}. \qquad (12\text{-}14)$$

The third equation may be written

$$\nabla \times \left(\mathbf{E} + \frac{1}{c}\dot{\mathbf{A}}\right) = \mathbf{0}$$

which allows us to define a scalar potential Φ such that

$$\mathbf{E} + \frac{1}{c}\dot{\mathbf{A}} = -\nabla\Phi \quad \text{or} \quad \mathbf{E} = -\nabla\Phi - \frac{1}{c}\dot{\mathbf{A}}. \qquad (12\text{-}15)$$

The three components of $\mathbf{A}$ along with Φ represent four field coordinates. Although these four are still not quite independent, we shall perform the final

---

‡ In what follows, we assume the particles to be nonrelativistic.

reduction after the Lagrangian and Hamiltonian have been constructed. The equations of motion (12-11) and (12-12) may now be written by expressing **E** and **B** in terms of **A** and Φ.

The Lagrangian for the entire electrodynamic system may be written as

$$L = \int (\mathscr{L}_{\text{part}} + \mathscr{L}_{\text{rad}} + \mathscr{L}_{\text{int}}) \, d\mathbf{r} \qquad (12\text{-}16)$$

where

$$\mathscr{L}_{\text{part}} = \sum_i \tfrac{1}{2} m \delta(\mathbf{r} - \mathbf{r}_i) v_i^2$$

$$\mathscr{L}_{\text{rad}} = \frac{\left(-\nabla\Phi - \dfrac{1}{c}\dot{\mathbf{A}}\right)^2 - (\nabla \times \mathbf{A})^2}{8\pi} = \frac{E^2 - B^2}{8\pi}$$

$$\mathscr{L}_{\text{int}} = -\rho\Phi + \frac{\mathbf{J} \cdot \mathbf{A}}{c}$$

and where ρ and $\mathbf{J}$ are given by (12-13). It is a lengthy but straightforward procedure to verify that an application of Lagrange's equations to (12-16) reproduces both the Lorentz force law for the particles and Maxwell's equations for the fields.‡

The momentum density conjugate to **A** is, from (12-6),

$$\Pi = \frac{\delta L}{\delta \dot{\mathbf{A}}} = \frac{\partial \mathscr{L}}{\partial \dot{\mathbf{A}}} = \frac{1}{4\pi c}\left(-\nabla\Phi - \frac{1}{c}\dot{\mathbf{A}}\right) = \frac{\mathbf{E}}{4\pi c}.$$
$$(12\text{-}17)$$

Since $\dot{\Phi}$ does not appear in $\mathscr{L}$, its conjugate momentum density is zero. The Legendre transformation for constructing H is straightforward, with the result being

$$H = \int \left\{ \frac{1}{8\pi}(E^2 + B^2) + \frac{\mathbf{E} \cdot \nabla\Phi}{4\pi} \right\} + \left\{ \sum_i \left[\frac{(\mathbf{p}_i - (q/c)\,\mathbf{A})^2}{2m} + q\Phi \right] \delta(\mathbf{r} - \mathbf{r}_i) \right\} d\mathbf{r}$$
$$(12\text{-}18)$$

where $\mathbf{E} = 4\pi c \Pi$ and $\mathbf{B} = \nabla \times \mathbf{A}$.

We observe that **A** and Φ are physical only to the extent that **E** and **B** can be derived from them. In particular, the *gauge* transformation,

$$\mathbf{A} \to \mathbf{A} + \nabla\chi, \qquad \Phi \to \Phi - \frac{\dot{\chi}}{c}$$

---

‡ See, for example, H. Goldstein, "Classical Mechanics," Chapter 11. Addison-Wesley, Reading, Massachusetts, 1950.

where χ is an arbitrary scalar, leaves the **E** and **B** fields in (12-14) and (12-15) unaffected (Problem 12-3). This can be seen by direct substitution, noting that the curl of a gradient of any function χ is zero. We are therefore at liberty to restrict **A** and Φ by an appropriate choice of χ. The *Lorentz* gauge in which we make the restriction $\mathbf{V} \cdot \mathbf{A} = \dot{\Phi}/c$ lends itself to a relativistically covariant formulation but leads to overwhelming complications in the quantum theory. A simpler gauge for our purposes is the *radiation gauge* which requires that

$$\mathbf{V} \cdot \mathbf{A} = 0. \tag{12-19}$$

In choosing this gauge, we are sacrificing manifest covariance for the sake of simplicity; the results remain valid nevertheless.

Substituting (12-15) into Maxwell's first equation, we obtain

$$\nabla^2 \Phi + \frac{1}{c} \frac{\partial}{\partial t} (\mathbf{V} \cdot \mathbf{A}) = -4\pi\rho$$

or using (12-19)

$$\nabla^2 \Phi = -4\pi\rho \qquad \text{(Poisson's equation)}. \tag{12-20}$$

The solution to (12-20) is given by Coulomb's law, that is,

$$\Phi = \int \frac{\rho(\mathbf{r}')d\mathbf{r}'}{|\mathbf{r} - \mathbf{r}'|} = \int \frac{\sum_i q\delta(\mathbf{r}' - \mathbf{r}_i) \, d\mathbf{r}'}{|\mathbf{r} - \mathbf{r}'|} .$$

Understandably, the radiation gauge is often referred to as the *Coulomb gauge*. In this gauge, Φ is determined by the instantaneous positions of the charged particles and vanishes in their absence. We shall see below that Φ is not a coordinate of the radiation field in the radiation gauge.

Any vector field can always be decomposed into a longitudinal (irrotational) and a transverse (solenoidal) part according to

$$\mathbf{A} = \mathbf{A}_L + \mathbf{A}_T. \tag{12-21a}$$

The two parts are *defined* by the properties

$$\mathbf{V} \cdot \mathbf{A}_T = 0, \qquad \mathbf{V} \times \mathbf{A}_L = 0. \tag{12-21b}$$

In the radiation gauge (12-19), it follows that **A** is purely transverse. The radiation gauge is also referred to as the *transverse gauge*. From Maxwell's second equation, $\mathbf{V} \cdot \mathbf{B} = 0$, it follows that **B** is always transverse regardless of the gauge.

The electric field is, from (12-17),

$$\mathbf{E} = 4\pi c\mathbf{\Pi} = -\nabla\Phi - \frac{1}{c} \dot{\mathbf{A}}.$$

It follows that

$$\mathbf{E} = \mathbf{E_L} + \mathbf{E_T}$$

where

$$\mathbf{E_L} = -\nabla\Phi = 4\pi c \mathbf{\Pi_L} \quad \text{and} \quad \mathbf{E_T} = -\frac{1}{c}\dot{\mathbf{A}} = 4\pi c \mathbf{\Pi_T}. \quad (12\text{-}22)$$

This decomposition of $\mathbf{E}$ in (12-22) is valid in the radiation gauge since only then will $\dot{\mathbf{A}}$ be transverse. We obtain from (12-22)

$$E^2 = E_T^{\ 2} + E_L^{\ 2} + 2\mathbf{E_T} \cdot \mathbf{E_L} = E_T^{\ 2} + E_L^{\ 2} - 2\mathbf{E_T} \cdot \nabla\Phi$$

and

$$\mathbf{E} \cdot \nabla\Phi = \mathbf{E_T} \cdot \nabla\Phi + \mathbf{E_L} \cdot \nabla\Phi = \mathbf{E_T} \cdot \nabla\Phi - E_L^{\ 2}.$$

Using these relations, the Hamiltonian in (12-18) becomes

$$H = \int \left\{ \frac{1}{8\pi}(E_T^{\ 2} + B^2) + \frac{\mathbf{E} \cdot \nabla\Phi}{8\pi} - \frac{\mathbf{E_T} \cdot \nabla\Phi}{8\pi} \right\}$$

$$+ \left\{ \sum_i \left[\frac{(\mathbf{p}_i - (q/c)\,\mathbf{A})^2}{2m} + q\Phi \right] \delta(\mathbf{r} - \mathbf{r}_i) \right\} d\mathbf{r}. \quad (12\text{-}23)$$

Integrating by parts, neglecting the surface term which for a physical field vanishes at infinity, and using

$$\nabla \cdot \mathbf{E} = 4\pi\rho = 4\pi\Sigma_i q\delta(\mathbf{r} - \mathbf{r}_i)$$

we find

$$\int \frac{\mathbf{E} \cdot \nabla\Phi}{8\pi}\,d\mathbf{r} = -\frac{1}{2}\int \frac{\nabla \cdot \mathbf{E}}{4\pi}\,\Phi\,d\mathbf{r} = -\frac{1}{2}\int \sum q\delta(\mathbf{r} - \mathbf{r}_i)\,d\mathbf{r}$$

and

$$\int \mathbf{E_T} \cdot \nabla\Phi\,d\mathbf{r} = -\int \Phi\nabla \cdot \mathbf{E_T}\,d\mathbf{r} = 0.$$

Substituting this into (12-23) and using the properties of the delta function, the electrodynamic Hamiltonian becomes

$$H = \int \frac{1}{8\pi}(E_T^{\ 2} + B^2)\,d\mathbf{r} + \sum_i \left[\frac{(\mathbf{p}_i - (q/c)\,\mathbf{A}_i)^2}{2m} + \frac{1}{2}q\Phi_i \right].$$

Using Coulomb's law for Φ_i, the electrodynamic Hamiltonian takes the final form

$$H = H_{\text{rad}} + H_{\text{part}} + H_{\text{int}}$$

where

$$H_{\text{rad}} = \int \frac{1}{8\pi}(E_T{}^2 + B^2)\,d\mathbf{r} = \int \frac{1}{8\pi}[(4\pi c\mathbf{\Pi}_T)^2 + (\nabla \mathbf{x}A)^2]\,d\mathbf{r}$$

$$H_{\text{part}} = \sum_i \frac{p_i{}^2}{2m} + \frac{1}{2}\sum_i\sum_j \frac{q^2}{|\mathbf{r}_j - \mathbf{r}_i|} \qquad (12\text{-}24)$$

$$H_{\text{int}} = \sum_i \left[-\frac{q}{mc}\,\mathbf{p}_i \cdot \mathbf{A}_i + \frac{q^2}{2mc^2}\,A_i{}^2 \right].$$

In the radiation gauge, $\mathbf{E}_L$ and Φ do not enter in the free radiation Hamiltonian. Rather, they contribute to the electrostatic Coulomb interactions between the particles. Unfortunately the Coulomb term in (12-24) includes the self-energies of the particles. This difficulty arises because we have assumed point particles. In practice, if the charges are conserved in number, this self-energy does not change and cannot affect the dynamics of the particles. We shall therefore ignore the terms in the Coulomb interactions that have $i = j$.

The radiation field involves transverse components only. We next consider quantization of this free radiation field, that is, the situation in which no particles are present $(q = p_i = 0)$.

V   The Equivalence between Free Radiation and Oscillators

While it should be possible to quantize H_{rad} in (12-24) directly, it is convenient to perform first a Fourier expansion of the free radiation field. This avoids the ambiguities associated with retardation effects in the commutation relations for the fields. We assume that the radiation is confined to a large box of volume $\mathscr{V}$ and take the characteristic modes of the cavity to be plane waves of the form

$$\mathbf{u}_{\mathbf{k}\lambda} = \frac{(4\pi)^{1/2}c}{\sqrt{\mathscr{V}}}\,\mathbf{e}_{\mathbf{k}\lambda}\,e^{i\mathbf{k}\cdot\mathbf{r}}. \qquad (12\text{-}25)$$

Each mode is characterized by a wave vector $\mathbf{k}$ and a polarization index λ. The longitudinal polarization vector $\mathbf{e}_{\mathbf{k},3}$ is a unit vector along $\mathbf{k}$, while the transverse vectors $\mathbf{e}_{\mathbf{k},1}$ and $\mathbf{e}_{\mathbf{k},2}$ are mutually orthogonal and perpendicular to $\mathbf{k}$. The modes of the field have the orthogonality property

$$\int_{\mathscr{V}} \mathbf{u}_{\mathbf{k}\lambda}^* \cdot \mathbf{u}_{\mathbf{k}'\lambda'}\,d\mathbf{r} = 4\pi c^2 \delta_{\mathbf{k}\mathbf{k}'}\,\delta_{\lambda\lambda'}. \qquad (12\text{-}26)$$

The general state of the radiation in the cavity is given by a superposition of these modes and we may make the expansion‡

$$\mathbf{A}(\mathbf{r}, t) = \sum_{\mathbf{k}\lambda} q_{\mathbf{k}\lambda}(t)\mathbf{u}_{\mathbf{k}\lambda}(\mathbf{r})$$

$$\mathbf{\Pi}_{\mathrm{T}}(\mathbf{r}, t) = \frac{1}{4\pi c^2} \sum_{\mathbf{k}\lambda} p_{\mathbf{k}\lambda}(t)\mathbf{u}_{\mathbf{k}\lambda}(\mathbf{r})$$

(12-27)

where $q_{\mathbf{k}\lambda}$ and $p_{\mathbf{k}\lambda}$ are real expansion coefficients determined by the initial conditions. (Note that the coefficients $q_{\mathbf{k}\lambda}$ and $p_{\mathbf{k}\lambda}$ are functions of time alone and not of position.) The allowed values of $\mathbf{k}$ are determined by the boundary conditions imposed at the walls of the cavity. As the volume of the enclosure becomes large, the quantum number $\mathbf{k}$ becomes quasi-continuous. The sums contain only transverse polarizations ($\lambda = 1, 2$) since the free radiation Hamiltonian involves purely transverse fields.

The curl operation on the plane wave $e^{i\mathbf{k}\cdot\mathbf{r}}$ amounts to multiplication by $i\mathbf{k} \times$. We therefore have

$$\nabla \times \mathbf{A} = \sum_{\mathbf{k}, \lambda = 1, 2} q_{\mathbf{k}\lambda}\, i\mathbf{k} \times \mathbf{u}_{\mathbf{k}\lambda}.$$

(12-28)

Using the relations [see Fig. 12-1)] $\mathbf{k} \times \mathbf{u}_{\mathbf{k}1} = k\,\mathbf{u}_{\mathbf{k}2}$ and $\mathbf{k} \times \mathbf{u}_{\mathbf{k}2} = -k\,\mathbf{u}_{\mathbf{k}1}$ along with (12-26), we obtain after some simplification

$$\int |\nabla \times \mathbf{A}|^2\, d\mathbf{r} = \sum_{\mathbf{k}\lambda = 1, 2} q_{\mathbf{k}\lambda}^2 k^2 4\pi c^2.$$

(12-29)

Similarly, referring to the definition of $\mathbf{\Pi}_{\mathrm{T}}$ in (12-27), we find

$$|\mathbf{\Pi}_{\mathrm{T}}|^2 = \frac{1}{(4\pi c^2)^2} \sum_{\mathbf{k}\mathbf{k}'\lambda\lambda'} p_{\mathbf{k}'\lambda'} p_{\mathbf{k}\lambda}\, \mathbf{u}_{\mathbf{k}'\lambda'}^* \cdot \mathbf{u}_{\mathbf{k}\lambda}.$$

(12-30)

Substituting these results into (12-24), integrating, and using the orthonormality relations (12-26), we obtain

$$H_{\mathrm{rad}} = \int \left[\frac{|4\pi c\mathbf{\Pi}_{\mathrm{T}}|^2 + |\nabla \times \mathbf{A}|^2}{8\pi} \right] d\mathbf{r} = \sum_{\mathbf{k}} \sum_{\lambda = 1, 2} \left[\frac{p_{\mathbf{k}\lambda}^2}{2} + \frac{1}{2}\omega_k^2 q_{\mathbf{k}\lambda}^2 \right]$$

(12-31)

where $\omega_k = ck$. *The free radiation Hamiltonian is therefore equivalent to one representing an infinite collection of oscillators of unit mass each of which is associated with one of the characteristic modes.* The expansion coefficients

‡ Strictly speaking, the fields for a physical system in (12-27) must be real and should vanish at infinity. While the modes in (12-25) do not satisfy these criteria, they do simplify the calculations considerably and the results are nevertheless valid.

$q_{k\lambda}$ and $p_{k\lambda}$ play the role of canonical variables for each of the mode oscillators. Each oscillator is characterized by a wave vector $\mathbf{k}$ and a polarization index $\lambda = 1, 2$. The infinity of oscillators occurs here (in contrast to the N oscillators of the discrete chain in Chapter 10) because of the nongranularity of the field.

VI   Quantization of the Free Radiation (Transverse) Field

The quantization of the radiation field follows the familiar pattern, making the associations

$$\mathbf{A} \to \hat{\mathbf{A}} = \sum_{k\lambda} \hat{q}_{k\lambda} \mathbf{u}_{k\lambda} = \sum_{k\lambda} \hat{\mathbf{A}}_{k\lambda} \tag{12-32}$$

and

$$\mathbf{\Pi}_t \to \hat{\mathbf{\Pi}}_T = \frac{1}{4\pi c^2} \sum_{k\lambda} \hat{p}_{k\lambda} \mathbf{u}_{k\lambda} = \sum_{k\lambda} \hat{\mathbf{\Pi}}_{T\,k\lambda} \tag{12-33}$$

where

$$\hat{\mathbf{A}}_{k\lambda} = \hat{q}_{k\lambda} \mathbf{u}_{k\lambda} \qquad \text{and} \qquad \hat{\mathbf{\Pi}}_{T\,k\lambda} = \frac{1}{4\pi c^2} \hat{p}_{k\lambda} \mathbf{u}_{k\lambda}.$$

Using (12-10), the field commutation relations for the transverse Fourier components become

$$[\hat{\mathbf{A}}_{k'\lambda'}(\mathbf{r}'), \hat{\mathbf{A}}_{k\lambda}(\mathbf{r})] = [\hat{\mathbf{\Pi}}_{T\,k'\lambda'}(\mathbf{r}'), \hat{\mathbf{\Pi}}_{T\,k\lambda}(\mathbf{r})] = 0$$

and

$$[\hat{\mathbf{A}}_{k'\lambda'}(\mathbf{r}'), \hat{\mathbf{\Pi}}_{T\,k\lambda}(\mathbf{r})] = i\hbar \delta_{kk'} \delta_{\lambda\lambda'} \delta(\mathbf{r} - \mathbf{r}'). \tag{12-34}$$

In accordance with Heisenberg's uncertainty principle, we are limited in our ability to accurately measure the same transverse Fourier components of $\hat{\mathbf{A}}$ and $\hat{\mathbf{\Pi}}_T$ at the same point in space at the *same time*. Integrating (12-34) over all space and using (12-26), we obtain

$$[\hat{q}_{k'\lambda'}, \hat{q}_{k\lambda}] = [\hat{p}_{k'\lambda'}, \hat{p}_{k\lambda}] = 0$$

and

$$[\hat{q}_{k'\lambda'}, \hat{p}_{k\lambda}] = i\hbar \delta_{kk'} \delta_{\lambda\lambda'}. \tag{12-35}$$

The quantization of the Hamiltonian,

$$\hat{H}_{\text{rad}} = \sum_{k\lambda} \tfrac{1}{2} \hat{p}_{k\lambda}^2 + \tfrac{1}{2} \omega_k^2 \hat{q}_{k\lambda}^2 \tag{12-36}$$

follows a pattern characteristic of the one for oscillators. The state of the radiation is represented by a direct product of oscillator states, each of which is associated with a characteristic mode, that is,

$$|\text{rad}\rangle = |v_{\mathbf{k}_1 \lambda_1}\rangle \otimes |v_{\mathbf{k}_2 \lambda_2}\rangle \otimes \cdots = |\{v_{\mathbf{k}\lambda}\}\rangle \qquad (12\text{-}37)$$

where $v_{\mathbf{k}\lambda}$ are the excitation integers of the mode $\mathbf{k}, \lambda$. The radiation state is known when the sequence of excitation integers $\{v_{\mathbf{k}\lambda}\}$ is indicated. The energy of the state is

$$\varepsilon(\text{rad}) = \sum_{\mathbf{k}\lambda} (v_{\mathbf{k}\lambda} + \tfrac{1}{2})\hbar\omega_k. \qquad (12\text{-}38)$$

The resemblance between (12-38) and (10-73) suggests that we introduce the *photon* as the quantum of the electromagnetic field. With each mode we associate a photon of energy $\varepsilon_k = \hbar\omega_k$. The photon is a completely relativistic particle of zero rest mass moving at the speed of light. Setting $m = 0$ in the relation $\varepsilon = (P^2 c^2 + m^2 c^4)^{1/2}$, we find the photon's momentum to be $P_k = \varepsilon/c = \hbar\omega_k/c$ or in vector form $\mathbf{P_k} = \hbar\mathbf{k}$. The integers of the sequence for a given state, that is, $v_{\mathbf{k}\lambda}$, are interpreted as the numbers of photons of each type present. The vacuum state (all $v_{\mathbf{k}\lambda} = 0$) has energy

$$\varepsilon_{\text{vac}} = \sum_{\mathbf{k}\lambda}^{\infty} \tfrac{1}{2}\hbar\omega_k \to \infty.$$

The infinite energy of the vacuum is characteristic of the electromagnetic field. However, since this energy is constant, we shall ignore it as we did in the case of the infinite self-energy of the Coulomb interaction in (12-24).

VII Quantum Electrodynamics—Radiative Transitions

The stationary eigenstates of an electrodynamical system are represented by the eigenvectors of $\hat{H} = \hat{H}_{\text{rad}} + \hat{H}_{\text{part}} + \hat{H}_{\text{int}}$. The interaction term is, from (12-24),

$$\hat{H}_{\text{int}} = -\sum_i \frac{q}{mc} \,\hat{\mathbf{p}}^{(i)} \cdot \hat{\mathbf{A}}(\mathbf{r}^{(i)}) + \sum_i \frac{q^2}{2mc^2} \,\hat{A}^2(\mathbf{r}^{(i)}). \qquad (12\text{-}39)$$

For electrons, the perturbation constant, in cgs units, is taken to be

$$\lambda = \frac{q}{c} = \frac{4.8 \times 10^{-10}}{3 \times 10^{10}} \sim 10^{-20}.$$

Neglecting the second term (since it involves λ^2), the interaction may be regarded as a first-order perturbation and written

$$\hat{H}_{\text{int}} \simeq -\lambda \sum_i \frac{\hat{\mathbf{p}}^{(i)} \cdot \hat{\mathbf{A}}(\mathbf{r}^{(i)})}{m}. \qquad (12\text{-}40)$$

Taking $\hat{H}_0 = \hat{H}_{\text{rad}} + \hat{H}_{\text{part}}$ as the unperturbed Hamiltonian, its eigenstates may be written as a direct product

$$|\alpha\rangle = |\text{rad} + \text{part}\rangle = |\text{rad}\rangle \otimes |\text{part}\rangle = |\{v_{k\lambda}\}\rangle \otimes |i\rangle \quad (12\text{-}41)$$

where $|i\rangle$ represents an eigenstate of the particle system. The unperturbed state is known when both the radiation and particle states have been specified. The unperturbed energy of the state (neglecting the vacuum energy) is

$$\varepsilon_\alpha = \varepsilon_{\text{rad} + \text{part}} = \varepsilon_i(\text{part}) + \sum_{k\lambda} v_{k\lambda} \hbar\omega_k. \quad (12\text{-}42)$$

It is understood that radiation and particle operators operate only on their respective kets in (12-41).

We shall assume that the perturbation in (12-40) induces transitions between unperturbed states in (12-41) and take as our initial state

$$|\alpha\rangle = |\{v_{k\lambda}\}\rangle |i\rangle;$$

the sequence $\{v_{k\lambda}\}$ represents the initial number of photons of each type present and $|i\rangle$ represents the initial state of the particle system. The final state is written

$$|\beta\rangle = |\{v'_{k\lambda}\}\rangle |f\rangle.$$

The initial and final radiation states have a different distribution of photons (that is, $\{v_{k\lambda}\} \neq \{v'_{k\lambda}\}$). The particle system has made a transition from $|i\rangle$ to some final state $|f\rangle$. The first-order transition rate is given by Fermi's Golden Rule (7-71)

$$R_{\alpha \to \beta} = \frac{2\pi}{\hbar} \lambda^2 \left[\langle \alpha | \sum_i \frac{\hat{\mathbf{p}}^{(i)}}{m} \cdot \hat{\mathbf{A}}(\mathbf{r}^{(i)}) | \beta \rangle \right]^2 \rho_\beta \Bigg|_{\varepsilon_\alpha = \varepsilon_\beta} \quad (12\text{-}43)$$

where $\lambda = q/c$.

We shall apply (12-43) to a particle system composed of a single hydrogenic electron. The evaluation of (12-43) is simplified by making the following observations:

(1) Since the operators

$$\hat{\mathbf{p}} = \frac{\hbar}{i} \nabla$$

$$\hat{\mathbf{A}} = \frac{(4\pi c^2)^{1/2}}{\sqrt{\mathscr{V}}} \sum_{k\lambda} \hat{q}_{k\lambda} \mathbf{e}_{k\lambda} e^{i\mathbf{k} \cdot \mathbf{r}}$$

operate only on the particle and radiation kets respectively, the matrix elements in (12-43) may be written

$$\left| \langle \alpha | \frac{\hat{\mathbf{p}} \cdot \hat{\mathbf{A}}(\mathbf{r})}{m} | \beta \rangle \right|^2 = \frac{4\pi c^2}{\mathscr{V}} \left| \sum_{\mathbf{k}, \lambda} \langle \{v_{\mathbf{k}\lambda}\} | \hat{q}_{\mathbf{k}\lambda} | \{v'_{\mathbf{k}\lambda}\} \rangle \mathbf{e}_{\mathbf{k}\lambda} \cdot \langle i | \frac{\hat{\mathbf{p}} e^{i\mathbf{k} \cdot \mathbf{r}}}{m} | f \rangle \right|^2.$$

(12-44)

(2)   The Golden Rule involves only transitions between states of the same energy and we require

$$\varepsilon_\alpha = \varepsilon_\beta$$

$$\varepsilon_i + \sum_{\mathbf{k}\lambda} v_{\mathbf{k}\lambda} \hbar \omega_k = \varepsilon_f + \sum_{\mathbf{k}\lambda} v'_{\mathbf{k}\lambda} \hbar \omega_k.$$

(12-45)

(3)   In accordance with the results of Chapter 4 (4-53), we may write the oscillator-like radiation matrix elements in (12-44) as

$$\langle \{v_{\mathbf{k}\lambda}\} | \sum_{\mathbf{k}\lambda} \hat{q}_{\mathbf{k}\lambda} | \{v'_{\mathbf{k}\lambda}\} \rangle = \sum_{\mathbf{k}\lambda} \langle v_{\mathbf{k}\lambda} | \hat{q}_{\mathbf{k}\lambda} | v'_{\mathbf{k}\lambda} \rangle$$

$$= \sum_{\mathbf{k}\lambda} \left[\frac{(v_{\mathbf{k}\lambda} + 1)\hbar}{2\omega_k} \right]^{1/2} \delta_{v_{\mathbf{k}\lambda} + 1, \, v'_{\mathbf{k}\lambda}}$$

$$+ \left[\frac{v_{\mathbf{k}\lambda} \hbar}{2\omega_k} \right]^{1/2} \delta_{v_{\mathbf{k}\lambda} - 1, \, v'_{\mathbf{k}\lambda}}.$$

(12-46)

Equivalently, the number of photons of each type may increase (emission) or decrease (absorption) by no more than one. In higher-order perturbation theory, multiple photon processes are possible.

We shall assume that the particle system, in this case the hydrogen atom, is making a transition in which $\varepsilon_i > \varepsilon_f$. In order to satisfy the principle of conservation of energy, (12-45), the photon energy must increase by $\varepsilon_i - \varepsilon_f$. However, the one-photon rule can be satisfied if and only if photons of frequency

$$\omega_{if} = |\mathbf{k}| c = \frac{\varepsilon_i - \varepsilon_f}{\hbar}$$

are emitted. This suggests that only the first Kronecker delta in (12-46) is to be retained (for the emission process) and that the sum be over photon states with

$$|\mathbf{k}| = \frac{\omega_{if}}{c} = \frac{\varepsilon_i - \varepsilon_f}{\hbar c}.$$

(4)   The density of states (neglecting polarization) for a photon of wave vector **k** is obtained using

$$\frac{\mathscr{V} \, d\mathbf{k}}{(2\pi)^3} = \rho(\omega) \, d\varepsilon = \rho(\omega)\hbar \, d\omega$$

$$\frac{\mathscr{V} k^2 \, dk}{(2\pi)^3} \, d\Omega_\mathbf{k} = \frac{\mathscr{V} \omega^2 \, d\omega}{(2\pi c)^3} \, d\Omega_\mathbf{k} = \rho(\omega)\hbar \, d\omega$$

or

$$\rho(\omega) = \frac{\mathscr{V} \omega^2}{(2\pi c)^3 \hbar} \, d\Omega_\mathbf{k}. \tag{12-47}$$

(5)   The optical transitions in hydrogen involve photons whose wavelengths are much longer than the extent of the atom. We may therefore write $\lambda \gg r$ or $\mathbf{k} \cdot \mathbf{r} \ll 1$, so that the plane wave in the particle matrix element may be approximated by $e^{i\mathbf{k} \cdot \mathbf{r}} \simeq 1$. The element becomes

$$\langle f | \hat{\mathbf{p}} e^{i\mathbf{k} \cdot \mathbf{r}} | i \rangle \simeq \langle f | \hat{\mathbf{p}} | i \rangle. \tag{12-48}$$

This is, as we shall see, the *dipole* approximation used in Chapter 7.

Putting the results above into the Golden Rule (12-43), and replacing the sum over **k** by an integral, we find the emission rate to be

$$R_{\alpha \to \beta} = \frac{2\pi}{\hbar} \frac{e^2}{c^2} \int \sum_\lambda \frac{4\pi c^2}{\mathscr{V}} \frac{\hbar(v_{\mathbf{k}\lambda} + 1)}{2\omega_{fi}} \left| \mathbf{e}_{\mathbf{k}\lambda} \cdot \langle f | \frac{\hat{\mathbf{p}}}{m} | i \rangle \right|^2 \frac{\mathscr{V} \omega_{fi}^2}{(2\pi c)^3 \hbar} \, d\Omega_\mathbf{k}$$

$$= \frac{e^2 \omega_{fi}}{2\pi c^3 \hbar} \int \sum_\lambda (v_{\mathbf{k}\lambda} + 1) \left| \mathbf{e}_{\mathbf{k}\lambda} \cdot \langle f | \frac{\hat{\mathbf{p}}}{m} | i \rangle \right|^2 \, d\Omega_\mathbf{k}.$$

Using the identity

$$\frac{\hat{\mathbf{p}}}{m} = \frac{[\hat{\mathbf{r}}, \hat{H}]}{i\hbar}$$

we may write

$$\langle i | \frac{\hat{\mathbf{p}}}{m} | f \rangle = \frac{1}{i\hbar} \langle i | \hat{\mathbf{r}} \hat{H} - \hat{H} \hat{\mathbf{r}} | f \rangle$$

$$= \frac{\varepsilon_i - \varepsilon_f}{i\hbar} \langle i | \hat{\mathbf{r}} | f \rangle = \frac{\omega_{if}}{i} \langle i | \hat{\mathbf{r}} | f \rangle.$$

The emission rate now becomes

$$R_{\alpha \to \beta} = \frac{e^2 \omega_{if}^3}{2\pi \hbar c^3} \int d\Omega_\mathbf{k} \sum_\lambda (v_{\mathbf{k}\lambda} + 1) |\mathbf{e}_{\mathbf{k}\lambda} \cdot \langle i | \hat{\mathbf{r}} | f \rangle|^2. \tag{12-49}$$

The absorption rate is similar except that the second Kronecker delta in (12-46) is retained; equivalently, the term $(v_{k\lambda} + 1)$ in (12-49) is replaced by $v_{k\lambda}$ for absorption. Note that the integral in (12-49) is over a sphere of radius $k = \omega_{if}/c$.

If the initial radiation state is unpolarized, then the occupation integers are independent of λ and we set $v_{k\lambda} = v_k$. The polarization sum in (12-49) becomes

$$\sum_{\lambda} (v_{k\lambda} + 1) |\mathbf{e}_{k\lambda} \cdot \langle i|\hat{\mathbf{r}}|f\rangle|^2 = (v_k + 1) \sum_{\lambda = 1, 2} \langle i|\hat{\mathbf{r}}|f\rangle|^2 \cos^2 \Theta_{k\lambda} \quad (12\text{-}50)$$

where $\Theta_{k\lambda}$ is the angle between $\langle \mathbf{r} \rangle_{if}$ and $\mathbf{e}_{k\lambda}$ (Figure 12-1). However, since the

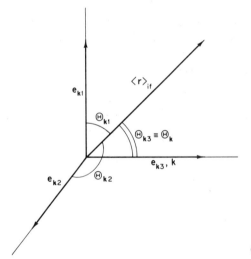

Figure 12-1 Illustrating the angles Θ_{k1}, Θ_{k2}, and Θ_{k3}.

sum of the squares of three direction cosines is unity, we have

$$\sum_{\lambda = 1, 2} \cos^2 \Theta_{k\lambda} = \cos^2 \Theta_{k1} + \cos^2 \Theta_{k2} = 1 - \cos^2 \Theta_{k3} = \sin^2 \Theta_k$$

$$(12\text{-}51)$$

where Θ_k is the angle between $\langle \mathbf{r} \rangle_{if}$ and $\mathbf{k}$ (that is, $\mathbf{e}_{k3}$). If the radiation is isotropic (the photon number is the same for all directions of $\mathbf{k}$), then v_k may be set equal to v_k and taken out of the integral. The transition rate finally becomes

$$R_{\alpha \to \beta} \begin{pmatrix} \text{emission} \\ \text{absorption} \end{pmatrix} = \frac{e^2 \omega_{if}^3}{2\pi \hbar c^3} \begin{Bmatrix} v_k + 1 \\ v_k \end{Bmatrix} \left[\int d\Omega_k \sin^2 \Theta_k = \frac{8\pi}{3} \right] |\langle \mathbf{r} \rangle_{if}|^2$$

$$\left(k = \frac{\omega_{if}}{c} \right). \quad (12\text{-}52)$$

The emission rate (upper term) is composed of two terms. The first involves the initial radiation state (that is, v_k) and corresponds to induced emission. This term is identical to the induced absorption rate as expected from the principle of microscopic reversibility. The second term in the emission rate is independent of the initial radiation state and occurs even in the absence of radiation. This spontaneous rate follows from (12-52) as

$$R_{\alpha \to \beta}^{\text{spont}} = \frac{4e^2 \omega_{if}^3}{3\hbar c^3} |\langle i|\hat{\mathbf{r}}|f\rangle|^2 \tag{12-53}$$

and is identical to the Einstein "A" coefficient calculated in (7-91). Note that this derivation is purely quantum-mechanical and does not rely on statistical detailed balancing.

The radiation energy density $\rho_{\text{rad}}(\omega)$ per frequency interval $d\omega$ is defined by the relation

$$2v_k \, \hbar \omega \, \frac{d\mathbf{k}}{(2\pi)^3} = \rho_{\text{rad}}(\omega) \, d\omega.$$

Using $\omega = ck$, this becomes

$$2v_k \, \hbar \omega \, \frac{k^2 \, dk \, d\Omega_\mathbf{k}}{(2\pi)^3} = 2v_k \, \hbar \omega \, \frac{\omega^2 \, d\omega}{(2\pi c)^3} \, d\Omega_\mathbf{k} = \rho_{\text{rad}}(\omega) \, d\omega. \tag{12-54}$$

The factor of two results from the two polarization states. The isotropy of the radiation permits the replacement $d\Omega_\mathbf{k} \to 4\pi$. Solving for v_k, we find

$$v_k = \frac{\pi^2 c^3}{\omega^3 \hbar} \rho_{\text{rad}}(\omega) \qquad \left(k = \frac{\omega}{c}\right). \tag{12-55}$$

Substituting (12-55) into (12-52) gives

$$R_{\alpha \to \beta}^{(\text{induced absorption or emission})} = \frac{4\pi^2 e^2}{3\hbar^2} |\langle i|\hat{\mathbf{r}}|f\rangle|^2 \rho_{\text{rad}}(\omega_{if}) \tag{12-56}$$

which leads to an Einstein "B" coefficient in agreement with (7-83).

The transition rates in (12-53) and (12-56) are valid provided the radiation field is not very intense; otherwise, first-order theory is inadequate. Also, the dipole approximation is valid for transitions in which the wavelength of the emitted or absorbed photons are large compared to the atomic system. In optical transitions we find $\lambda \simeq 5000$ Å, whereas the atomic size is only $d \sim 1$ Å. For X-ray emission ($\lambda \sim 1$ Å), the dipole approximation is clearly inadequate.

The quantization of the field justifies the notion of a particle-like photon with energy $\varepsilon = \hbar\omega$ and momentum $\mathbf{P} = \hbar\mathbf{k}$. The polarization may be related to the spin state of the photon. The photon is a boson of unit spin with the up and down spin orientations corresponding to the two degrees of transverse

polarization. The zero spin orientation is absent as it corresponds to longitudinal polarization which does not exist in free radiation (transverse gauge).

Finally, since photons need not be conserved in number, the thermal distribution is given by the Bose–Einstein formula with the chemical potential set
equal to zero, namely,

$$\bar{v}_k = \frac{1}{e^{+(\hbar\omega_k/k_BT)} - 1}.$$

Using (12-55), the energy contained by these thermal photons of frequency
ω is

$$\rho_{\text{rad}}(\omega) = \frac{\hbar}{\pi^2 c^3} \frac{\omega^3}{e^{\hbar\omega/k_BT} - 1}$$

which is exactly Planck's formula (7-90).

VIII Broadening of Spectral Lines— The Energy–Time Uncertainty Relation

Throughout our discussions we have assumed the existence of true
stationary bound states and the ability to observe accurately their characteristics. There exists another uncertainty relation independent of and in addition
to the position–momentum uncertainty relation introduced in Chapter 3.
This new principle limits the accuracy with which the energy of a stationary
state can be measured when the measurement is taken over a *finite* time. It
should be understood that the notion of a stationary bound state, in fact,
subtly implies that from a physical standpoint the characteristics of the state
are invariant with respect to translations in time. Any measurement process
involving a probe which is energy-coupled to the system and which takes place
over a finite interval τ destroys this invariance. We shall show that this leads
to an uncertainty of the measurement of the energy $\Delta\varepsilon$ given by

$$\Delta\varepsilon \cdot \tau \simeq \hbar.$$

In the derivation of the Golden Rule in Chapter 7 [see (7-68)] we were
led to a time-dependent transition probability of the form

$$\mathscr{P}_{ij}(\tau) \propto \frac{(1 - \cos \omega_{ij}\tau)}{\tau\omega_{ij}^2} = \frac{\left(1 - \cos\left(\frac{\varepsilon_j - \varepsilon_i}{\hbar}\right)\tau\right)}{\tau\left[\frac{\varepsilon_j - \varepsilon_i}{\hbar}\right]^2}.$$

This relation gives the probability of transition between the levels ε_i and ε_j induced by a static perturbation. This probability implies the existence of transitions to final states whose energies differ from the initial energy on the average by

$$|\varepsilon_j - \varepsilon_i| = \Delta\varepsilon \simeq \frac{\hbar}{\tau}.$$

We may interpret this result by saying that the static perturbation acts as a device for measuring the energy of the unperturbed system. Further, if the device acts for a finite time τ, the result of the measurement leads to an uncertainty in the energy $\Delta\varepsilon$. If $\tau \to \infty$, the effective measurement of energy becomes precise. It is worthwhile to note that this uncertainty relation depends only on the duration τ and not on the strength of the perturbation.

Consider a system composed of two weakly interacting subsystems, for example, a box composed of atoms and electromagnetic radiation. Suppose that the atomic system is being measured energetically by the interaction with the radiation. Since the radiation induces transitions in the atomic system, the atoms remain in their initial states a finite time known as the lifetime, given by

$$\tau_i = \frac{1}{\gamma_i}$$

where γ_i represents the total rate of depopulation of the level ε_i induced by the photons. In fact, as we have seen, even in the absence of photons (radiation vacuum), there exists a finite lifetime due to spontaneous emission. Thus the energy of a level of an atomic system cannot be observed with greater accuracy than

$$\Delta\varepsilon_i \sim \frac{\hbar}{\tau_i}$$

where τ_i is the lifetime due to spontaneous emission. This uncertainty leads to a *natural level width* and means that at best an atomic level is quasi-discrete.

Transitions between naturally broadened states lead to the emission (or absorption) of photons with a spectral width

$$\Delta\omega_{ij} = \frac{\Delta\varepsilon_i + \Delta\varepsilon_j}{\hbar} = \frac{1}{\tau_i} + \frac{1}{\tau_j} = \gamma_i + \gamma_j.$$

Thus the spectral lines of an atom are *naturally broadened*, with the spectral width of each line determined by the lifetime of the initial and final states associated with the transition. This broadening underlies the fact that radiation cannot be observed in a monochromatic state when it is measured in a finite amount of time.

Suggested Reading

Berestetskii, V. B., Lifshitz, E. M., and Pitaevskii, L. P., "Relativistic Quantum Theory."
    Addison-Wesley, Reading, Massachusetts, 1971.
Bjorken, J. D., and Drell, S. D., "Relativistic Quantum Fields." McGraw-Hill, New York,
    1965.
Goldstein, H., "Classical Mechanics." Addison-Wesley, Reading, Massachusetts, 1950.
Heitler, W., "Quantum Theory of Radiation," 3rd ed. Oxford Univ. Press, London and
    New York, 1954.
Messiah, A., "Quantum Mechanics," Vol. II. Wiley, New York, 1962.
Schiff, L. I., "Quantum Mechanics," 3rd ed. McGraw-Hill, New York, 1968.
Yourgrau, W., and Mandelstam, S., "Variational Principles in Dynamics and Quantum
    Theory," 3rd ed. Saunders, Philadelphia, 1968.

Problems

12-1. Take the Lagrangian density for the vibrations of an ideal fluid to be

$$\mathscr{L} = \tfrac{1}{2}\rho_0 \dot{\boldsymbol{\eta}}^2 + P_0 \nabla \cdot \boldsymbol{\eta} - \tfrac{1}{2}\gamma P_0 (\nabla \cdot \boldsymbol{\eta})^2$$

where ρ_0 is the equilibrium density, P_0 the equilibrium pressure, and γ the ratio of the specific heat at constant pressure to that at constant volume of the fluid. Let $\boldsymbol{\eta}(\mathbf{r}, t)$ represent the field coordinate describing the displacement of a point in the fluid at the point $\mathbf{r}$ and the time t.
(a)   Find the conjugate momentum density π_j.
(b)   Find the equation of motion for $\eta_j(\mathbf{r}, t)$.
(c)   Show that the local density of the fluid is given by

$$\rho(\mathbf{r}, t) = \rho_0(1 + \xi(\mathbf{r}, t))$$

where

$$\xi = -\nabla \cdot \boldsymbol{\eta}.$$

(d)   Express part (b) in terms of $\xi(\mathbf{r}, t)$.
(e)   Using a Legendre transformation, find the Hamiltonian density for the fluid.

12-2. Verify that the Lagrangian in (12-16) leads to the Hamiltonian in (12-18).

12-3. Show that the gauge transformation $\mathbf{A} \rightarrow \mathbf{A} + \nabla \chi$ and $\Phi \rightarrow \Phi - \dot{\chi}/c$ leaves the $\mathbf{E}$ and $\mathbf{B}$ fields unaffected.

12-4. Establish the identity

$$\frac{\hat{\mathbf{p}}}{m} = \frac{[\hat{\mathbf{r}}, \hat{H}]}{i\hbar} \qquad \text{where} \quad \hat{H} = \frac{\hat{p}^2}{2m} + \hat{V}(\hat{\mathbf{r}}).$$

12-5. Calculate the photoionization cross section for electron emission from the ground state of hydrogen induced by an X ray in the form of a plane electromagnetic wave of frequency ω propagating along the z axis and polarized along the y axis. Assume that $\hbar\omega \gg 13.6$ eV and ignore the spin of the electron. [Hint: Assume that the initial state is composed of a hydrogenic electron in the state

$$\psi_i = \frac{1}{(\pi a^3)^{1/2}} e^{-r/a}$$

along with $\nu_{\mathbf{k}\lambda}$ photons with their momentum vectors ($\mathbf{k}$) along the z axis and with their polarization vectors ($\mathbf{e}_{\mathbf{k}\lambda}$) along the y axis. The final state is composed of one less photon (absorption) and a free electron in the state

$$\psi_f = \frac{1}{\sqrt{\mathscr{V}}} e^{i\mathbf{K}\cdot\mathbf{r}} \qquad \text{where} \qquad \mathbf{P} = \hbar\mathbf{K}$$

is the momentum of the electron. Take the density of final electron states to be (neglecting spin)

$$\rho_{\mathbf{K}} = \frac{\mathscr{V}}{(2\pi)^3} \frac{mK}{\hbar^2} d\Omega_{\mathbf{K}}.$$

Note that $\hbar\omega = p^2/2m + 13.6$ eV.]

12-6. Using the results above, show that the photoelectric effect (photoabsorption) cannot take place if the electron is initially free. [Hint: Set $\psi_i = e^{i\mathbf{K}\cdot\mathbf{r}}/\sqrt{\mathscr{V}}.$]

12-7. The operator in (12-40) is a product of noncommuting Hermitian operators and is therefore not in general Hermitian. Show that this does not affect the results in (12-44). [Hint: Show that in the transverse gauge, the terms in (12-44) are unaffected by an interchange of $\hat{\mathbf{p}}$ and $e^{i\mathbf{k}\cdot\mathbf{r}}$.]

A | The Wentzel-Kramers-Brillouin (WKB or "Phase Integral") Approximation

Consider the one-dimensional Schroedinger equation for a particle in a potential $V(x)$ with energy $\varepsilon > V(x)$,

$$\left\{-\frac{\hbar^2}{2m}\frac{d^2}{dx^2} + V(x)\right\}\psi = \varepsilon\psi$$

or

$$\left\{\frac{d^2}{dx^2} + k^2(x)\right\}\psi = 0 \qquad (A\text{-}1)$$

where

$$k(x) = \left[\frac{2m(\varepsilon - V(x))}{\hbar^2}\right]^{1/2}.$$

If $V(x)$ is equal to some constant, let us say V_0, then the solution to (A-1) is $\psi = e^{\pm ik_0 x}$ where

$$k_0 = \left[\frac{2m(\varepsilon - V_0)}{\hbar^2}\right]^{1/2}.$$

Let us now consider the case where $V(x)$ is slowly varying and try a solution of the form

$$\psi = e^{iu(x)}. \tag{A-2}$$

Substitution into (A-1) gives

$$iu'' - (u')^2 + k^2 = 0 \quad \text{or} \quad u' = \pm(k^2 + iu'')^{1/2}. \tag{A-3}$$

If V is constant, then from (A-2) we expect $u \to \pm k_0 x$, $u' \to \pm k_0$, and $u'' = 0$. This is consistent with (A-3). However, if V is slowly varying, then to a first approximation we set $u'' = 0$ in (A-3) and find

$$u' = \pm k(x)$$

which yields upon integration

$$u = \pm \int k(x)\, dx + C.$$

We obtain a better approximation by substituting this result, that is, $u'' = \pm k'(x)$, into the right side of (A-3) giving

$$u' = \pm(k^2 \pm ik')^{1/2}$$

or

$$u = \pm \int (k^2 \pm ik')^{1/2}\, dx + C.$$

Using (A-2), the approximate solution to the Schroedinger equation becomes

$$\psi \simeq \exp[\pm i \int (k^2 \pm ik')^{1/2}\, dx]. \tag{A-4}$$

The constant e^C has been dropped since, at most, it affects the normalization of ψ. Let us now assume that the potential is slowly varying so that

$$|k'| \ll k^2. \tag{A-5}$$

In that case, the radical in (A-4) can be expanded, giving

$$\psi \simeq \exp \pm i \int \left(k \pm \frac{i}{2}\frac{k'}{k}\right) dx = \exp[-\tfrac{1}{2}\ln k]\exp[\pm i \int k\, dx]$$

or

$$\psi = \frac{1}{\sqrt{k}}\exp[\pm i \int k\, dx] \tag{A-6}$$

where

$$k(x) = \left[\frac{2m(\varepsilon - V)}{\hbar^2}\right]^{1/2}.$$

Equation (A-6) is known as the *WKB approximation* to the solution of (A-1).

In the case where $V(x) > \varepsilon$, it may be shown that the approximate function takes the form

$$\psi \simeq \frac{1}{\sqrt{K}} \exp(\pm \int K\, dx) \tag{A-7}$$

where

$$K(x) = \left[\frac{2m(V - \varepsilon)}{\hbar^2}\right]^{1/2}.$$

Both (A-6) and (A-7) are valid in regions where

$$|k'| \ll k^2 \quad \text{or} \quad |K'| \ll K^2. \tag{A-8}$$

Suppose we are dealing with a potential barrier $V(x)$ where the energy of our particle is less than the barrier's peak (Figure A-1).

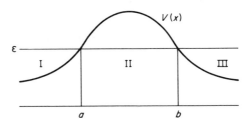

Figure A-1

Note that at the classical turning points ($x = a$ and $x = b$), k (or K) vanishes. Hence (A-5) *cannot* be satisfied and the WKB approximation *fails* at these points.

Let us assume that we know the general solution to the Schroedinger equation (A-1) which satisfies the continuity requirements at $x = a$ and $x = b$. In region I sufficiently far from $x = a$, this solution can be approximated by the WKB function

$$\psi_1 \simeq \frac{A}{\sqrt{k}} \exp\left(i \int_a^x k\, dx\right) + \frac{B}{\sqrt{k}} \exp\left(-i \int_a^x k\, dx\right)$$

$$\left(k = \left[\frac{2m(\varepsilon - V(x))}{\hbar^2}\right]^{1/2}\right). \tag{A-9a}$$

The choices of the lower limits in the integrals are unimportant since they merely affect the coefficients A and B which at this point are arbitrary. For regions II and III we have

$$\psi_{II} \simeq \frac{C}{\sqrt{K}} \exp\left(\int_a^x K \, dx\right) + \frac{D}{\sqrt{K}} \exp\left(-\int_a^x K \, dx\right) \qquad \text{(A-9b)}$$

and

$$\psi_{III} \simeq \frac{E}{\sqrt{K}} \exp\left(i \int_a^x k \, dx\right) + \frac{F}{\sqrt{K}} \exp\left(-i \int_a^x k \, dx\right) \qquad \text{(A-9c)}$$

where

$$K = \left[\frac{2m(V(x) - \varepsilon)}{\hbar^2}\right]^{1/2}.$$

These approximations are acceptable as long as we are not in the vicinity of $x = a$ or $x = b$. The constants A, B, C, D, E, and F remain to be determined. It is tempting to generate equations for these constants by imposing continuity requirements on (A-9) at $x = a$ and $x = b$. This procedure is entirely *incorrect* since the solutions in (A-9) are only valid away from the classical turning points.

In a sense, the WKB solutions describe the asymptotic behavior of the exact solution on both sides but far from the classical turning points. In the immediate neighborhood of these points an entirely different approximation is required. What we seek are *connection formulas* which link the WKB solutions to each other smoothly through these "neighborhood" functions. The mathematics required to establish these formulas is beyond the scope of this text. The results are as follows[‡]:

Case (a)   Turning point to the *left* of a classically *forbidden* region ($x = a$ in Figure A-1).

$$\qquad x < a \qquad\qquad\qquad x > a$$

$$\frac{2}{\sqrt{k}} \cos\left[\int_x^a k \, dx - \frac{1}{4}\pi\right] \leftarrow \frac{1}{\sqrt{K}} \exp\left(-\int_a^x K \, dx\right) \qquad \text{(A-10a)}$$

$$\frac{1}{\sqrt{k}} \sin\left[\int_x^a k \, dx - \frac{1}{4}\pi\right] \rightarrow \frac{-1}{\sqrt{K}} \exp\left(\int_a^x K \, dx\right) \qquad \text{(A-10b)}$$

[‡] See, for example, P. Stehle, "Quantum Mechanics," Holden-Day, San Francisco, 1966, or S. Borowitz, "Fundamentals of Quantum Mechanics," Benjamin. New York, 1967.

Case (b) Turning point to the *right* of a classically *forbidden* region ($x = b$ in Figure A-1).

$$x < b \qquad\qquad\qquad\qquad x > b$$

$$\frac{1}{\sqrt{K}} \exp\left(-\int_x^b K\,dx\right) \to \frac{2}{\sqrt{k}} \cos\left(\int_b^x k\,dx - \frac{1}{4}\pi\right) \qquad \text{(A-10c)}$$

$$\frac{-1}{\sqrt{K}} \exp\left(\int_x^b K\,dx\right) \leftarrow \frac{1}{\sqrt{k}} \sin\left(\int_b^x k\,dx - \frac{1}{4}\pi\right) \qquad \text{(A-10d)}$$

These connection formulas connect the exponential to the oscillatory solutions at the turning points. The arrows indicate the direction of the connection although we shall be unconcerned with the direction.

The ratios of the constants associated with the solutions in the three regions as given by (A-9) can now be obtained using the connection formulas as "boundary" conditions. The mathematics is straightforward but lengthy. The transmission coefficient in the WKB approximation turns out to be

$$\mathcal{T} = \left|\frac{E}{A}\right|^2 = \left[\exp\left(\int_a^b K\,dx\right) + \frac{1}{4}\exp\left(-\int_a^b K\,dx\right)\right]^{-2},$$

For high and wide barriers, the negative exponential in the brackets may be dropped and we find

$$\mathcal{T} \simeq \exp\left(-2\int_a^b K\,dx\right) \qquad \left(K = \left[\frac{2m(V(x) - \varepsilon)}{\hbar^2}\right]^{1/2}\right). \qquad \text{(A-11)}$$

The WKB approximation can also be applied to the bound states of an attractive potential as in Figure A-2. Here the classical turning point $x = a$

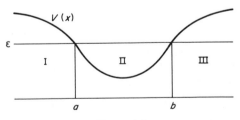

Figure A-2

is to the *right* of the classically forbidden region I while $x = b$ is to the *left* of the classically forbidden region III. The acceptable functions (that is, those that vanish at $x = \pm\infty$) have the form

$$\psi_I \simeq \frac{A}{\sqrt{K}} \exp\left(\int_a^x K\, dx\right)$$

$$\psi_{II} \simeq \frac{B}{\sqrt{k}} \exp\left(i \int_b^x k\, dx\right) + \frac{C}{\sqrt{k}} \exp\left(-i \int_b^x k\, dx\right)$$

$$\psi_{III} \simeq \frac{D}{\sqrt{K}} \exp\left(-\int_b^x K\, dx\right).$$

We apply the connection formula (A-10c) to ψ_I and ψ_{II} at $x = a$. Writing

$$\psi_I = \frac{A}{\sqrt{K}} \exp\left(-\int_x^a K\, dx\right)$$

we note that this must connect, using (A-10c), to

$$\psi_{II} \simeq \frac{2A}{\sqrt{k}} \cos\left(\int_a^x k\, dx - \frac{1}{4}\pi\right)$$

which can also be written, using $\cos\theta = \sin(\theta + \frac{1}{2}\pi)$, as

$$\psi_{II} \simeq \frac{2A}{\sqrt{k}} \cos\left(\int_a^b k\, dx - \int_x^b k\, dx - \frac{1}{4}\pi\right)$$

$$= \frac{2A}{\sqrt{k}} \sin\left(\int_a^b k\, dx - \int_x^b k\, dx + \frac{1}{4}\pi\right).$$

Using the identity $\sin(A - B) = \sin A \cos B - \cos A \sin B$, this takes the form

$$\psi_{II} \simeq \left[\sin\int_a^b k\, dx\right]\left[\frac{2A}{\sqrt{k}} \cos\left(\int_x^b k\, dx - \frac{1}{4}\pi\right)\right]$$

$$- \left[\cos\int_a^b k\, dx\right]\left[\frac{2A}{\sqrt{k}} \sin\left(\int_x^b k\, dx - \frac{1}{4}\pi\right)\right]. \tag{A-12}$$

However, at $x = b$, the second term (according to (A-10b)) connects to an increasing exponential in region III. Since the state is bound such a term must be absent and we have acceptable solutions only if the coefficient of the sine term in (A-12) vanishes, that is,

$$\cos\int_a^b k\, dx = 0 \quad \text{or} \quad \int_a^b k_n\, dx = \left(n + \frac{1}{2}\right)\pi. \tag{A-13}$$

Since

$$k_n = \left[\frac{2m(\varepsilon_n - V)}{\hbar^2}\right]^{1/2}$$

(A-13) is actually a quantization condition for the energy. Condition (A-13) can be written

$$\frac{2}{\hbar} \int_a^b [2m(\varepsilon_n - V)]^{1/2}\, dx = (n + \tfrac{1}{2})2\pi$$

or

$$2 \int_a^b p_n(x)\, dx = (n + \tfrac{1}{2})h.$$

The integral on the left is an integral over the classical motion from a to b and then back to a and can be written as

$$J_n = \oint p_n\, dx = (n + \tfrac{1}{2})h \qquad\qquad \text{(A-14)}$$

where J is known as the *phase integral*. Equation (A-14) is a generalization of what are known as the Wilson–Sommerfeld quantization rules.

Applying (A-13) to the oscillator, we set $V = \tfrac{1}{2}m\omega^2 x^2$ and find

$$\int_a^b \left[\frac{2m\varepsilon_n - m^2\omega^2 x^2}{\hbar^2}\right]^{1/2} dx = (n + \tfrac{1}{2})\pi.$$

The classical turning points are at $a = -(2\varepsilon/m\omega^2)^{1/2}$ and $b = (2\varepsilon/m\omega^2)^{1/2}$. The integral on the left is evaluated as

$$\frac{\pi \varepsilon_n}{\hbar\omega} = (n + \tfrac{1}{2})\pi \qquad \text{or} \qquad \varepsilon_n = (n + \tfrac{1}{2})\hbar\omega$$

which is the well-known result for the oscillator.

The WKB solution (A-2) can be written in three dimensions [setting $u(\mathbf{r}) = W(\mathbf{r})/\hbar$] as

$$\psi(\mathbf{r}) = e^{iW(\mathbf{r})/\hbar}. \qquad\qquad \text{(A-15)}$$

Substituting this solution into the Schroedinger energy-eigenvalue equation

$$\left\{\frac{-\hbar^2}{2m} \nabla^2 + V(\mathbf{r})\right\} \psi(\mathbf{r}) = \varepsilon\psi(\mathbf{r}) \qquad\qquad \text{(A-16)}$$

we obtain

$$\frac{1}{2m} (\nabla W)^2 - [\varepsilon - V(\mathbf{r})] - \frac{i\hbar}{2m} \nabla^2 W = 0. \qquad\qquad \text{(A-17)}$$

Note that Planck's constant h enters (A-17) only in the last term on the left. In the classical limit, $h \to 0$, we obtain

$$\frac{1}{2m} (\nabla W)^2 + V(r) = \varepsilon. \qquad\qquad \text{(A-18)}$$

Equation (A-18) is known as the time-independent *Hamilton–Jacobi equation* for Hamilton's characteristic function W. A solution of this equation amounts to a solution of the classical problem.‡ The WKB method provides corrections to this classical result in powers of h.§

An interesting analogy exists between quantum mechanics and optics. In optics, we seek solutions to the equation

$$\nabla^2 \phi + k^2 \phi = 0 \qquad\qquad (A\text{-}19)$$

where $k(\mathbf{r})$ is the wave vector and $\phi(\mathbf{r})$ is the wave "function." The wave vector may be expressed as

$$k(\mathbf{r}) = k_0\, n(\mathbf{r})$$

where

$$k_0 = \frac{\omega}{c} = \frac{2\pi}{\lambda}$$

and $n(\mathbf{r})$ is the local index of refraction. Equation (A-19) is the analog of (A-16) provided we identify the quantum-mechanical "index of refraction" with

$$n_{\text{quant}}(\mathbf{r}) = \sqrt{\frac{\varepsilon - V(\mathbf{r})}{\varepsilon}}.$$

Substituting a trial solution of the form

$$\phi = e^{ik_0 L(\mathbf{r})} \qquad\qquad (A\text{-}20)$$

into (A-19), we are led to an equation for L of the form

$$ik_0 \nabla^2 L - k_0^2 (\nabla L)^2 + n^2 k_0^2 = 0. \qquad\qquad (A\text{-}21)$$

Equation (A-21) is the analog of (A-17). In the limit of geometrical optics ($\lambda \to 0$) we find k_0 to be large and the first term in (A-21) can be dropped giving

$$(\nabla L)^2 = n^2. \qquad\qquad (A\text{-}22)$$

Equation (A-22) is known as the *eikonal equation* and is a fundamental equation of geometrical optics; it is the analog of the Hamilton–Jacobi equation (A-18). The function L is termed the *optical path length* or *eikonal*.

The analogy can be summarized briefly by saying that classical mechanics (Hamilton–Jacobi equation) represents the geometrical limit of quantum mechanics (Schroedinger equation) as $h \to 0$.

---

‡ See H. Goldstein, "Classical Mechanics." Addison-Wesley, Reading, Massachusetts, 1950.

§ See S. Borowitz, "Fundamentals of Quantum Mechanics." Benjamin, New York, 1967.

B | The Heisenberg and Interaction Pictures

In the *Schroedinger picture*, operators can evolve only according to an explicit time dependence, that is,

$$\hat{A}_\mathrm{S} = \hat{A}(t).$$

For convenience, we take $t_0 = 0$ as our initial time. State vectors in the Schroedinger picture evolve as

$$|\beta, t\rangle_\mathrm{S} = e^{-i\mathscr{H}t/\hbar}|\beta, 0\rangle$$

where we have assumed the Hamiltonian $\mathscr{H}$ to be time independent. The time dependence of expectation values is given by

$$\langle A(t)\rangle = \langle \beta, t|_\mathrm{S}\, \hat{A}_\mathrm{S}|\beta, t\rangle_\mathrm{S} = \langle \beta, 0|\, e^{i\mathscr{H}t/\hbar}\hat{A}(t)e^{-i\mathscr{H}t/\hbar}|\beta, 0\rangle. \qquad \text{(B-1)}$$

We can obtain the same time dependence as in (B-1) if we keep the state vectors fixed at their values at $t = 0$ and allow the operators to evolve as

$$\hat{A}_\mathrm{H} = e^{i\mathscr{H}t/\hbar}\hat{A}_\mathrm{S}\, e^{-i\mathscr{H}t/\hbar}. \qquad \text{(B-2)}$$

This prescription is known as the *Heisenberg picture*.

We are assuming that $\mathcal{H}$ is explicitly independent of time. Note that (B-2) implies $\mathcal{H}_H = \mathcal{H}_S$, so that a subscript on the Hamiltonian is unnecessary. Also at $t = 0$, the vectors and operators in both pictures coincide.

In the Schroedinger picture a state vector rotates; consequently its projections on the eigenvectors of the operator observables vary in time producing a time dependence of the expectation values as given by (B-1). In the Heisenberg picture, the state vector remains fixed but the eigenvectors of the Heisenberg operators rotate; the effect is to produce the same overall time dependence as given by (B-1).

The state vectors in the Schroedinger picture satisfy the Schroedinger equation of motion, that is,

$$\mathcal{H} \mid \beta, t\rangle_S = i\hbar \frac{d}{dt} \mid \beta, t\rangle_S \tag{B-3a}$$

while in the Heisenberg picture, operators satisfy the Heisenberg equation of motion

$$\frac{d\hat{A}_H}{dt} = \frac{[\hat{A}_H, \mathcal{H}]}{i\hbar} + \frac{\partial \hat{A}_H}{\partial t}. \tag{B-3b}$$

Equation (B-3b) is obtained upon differentiation of (B-2). The last term on the right of (B-3b) involves any explicit time dependence and is defined by

$$\frac{\partial \hat{A}_H}{\partial t} = e^{i\mathcal{H}t/\hbar} \frac{\partial \hat{A}_S}{\partial t} e^{-i\mathcal{H}t/\hbar}.$$

The two pictures are summarized below:

Schroedinger picture	Heisenberg picture
$\mid \beta, t\rangle_S = e^{-i\mathcal{H}t/\hbar} \mid \beta, 0\rangle$	$\mid \beta, t\rangle_H = \mid \beta, 0\rangle$
$\hat{A}_S = \hat{A}(t)$	$\hat{A}_H = e^{i\mathcal{H}t/\hbar} \hat{A}_S e^{-i\mathcal{H}t/\hbar}$
$\mathcal{H} \mid \beta, t\rangle_S = i\hbar \dfrac{d}{dt} \mid \beta, t\rangle_S$	$\dfrac{d\hat{A}_H}{dt} = \dfrac{[\hat{A}_H, \mathcal{H}]}{i\hbar} + \dfrac{\partial \hat{A}_H}{\partial t}.$

Intermediate pictures between those of Schroedinger and Heisenberg can also be generated. Consider a Hamiltonian of the form $\mathcal{H} = \mathcal{H}_0 + \hat{V}$. In practice, $\mathcal{H}_0$ usually represents the Hamiltonian for ideal many-body systems whereas $\hat{V}$ corresponds to the interaction between the particles. Hence the picture to be developed is known as the *interaction* picture. Imagine a picture in which operator observables evolve in a manner similar to Heisenberg operators, that is,

$$\hat{A}_I = e^{i\mathcal{H}_0 t/\hbar} \hat{A}_S e^{-i\mathcal{H}_0 t/\hbar}. \tag{B-4a}$$

This is not the Heisenberg picture since $\mathscr{H}_0$ and *not* $\mathscr{H}$ appears in the relation. Now we ask, what should the evolution of the state vector be like in order that the results be consistent with (B-1)? It is simple to verify that the behavior must be

$$|\beta, t\rangle_{\mathrm{I}} = e^{i\mathscr{H}_{0}t/\hbar} e^{-i\mathscr{H}t/\hbar} |\beta, 0\rangle; \qquad \text{(B-4b)}$$

in that case, we have

$$\langle A \rangle = \langle \beta, t|_{\mathrm{I}} \hat{A}_{\mathrm{I}} |\beta, t\rangle_{\mathrm{I}}$$

which agrees with (B-1). Note that (B-4b) can be written as

$$|\beta, t\rangle_{\mathrm{I}} = e^{i\mathscr{H}_{0}t/\hbar} |\beta, t\rangle_{\mathrm{S}}. \qquad \text{(B-4c)}$$

The equation of motion for the operators in the interaction picture follows the Heisenberg form, that is,

$$\frac{d\hat{A}_{\mathrm{I}}}{dt} = \frac{[\hat{A}_{\mathrm{I}}, \mathscr{H}_{0}]}{i\hbar} + \frac{\partial \hat{A}_{\mathrm{I}}}{\partial t} \qquad \text{(B-5)}$$

as can be seen by differentiating (B-4a).

The equation of motion for the state vectors in this picture is obtained by differentiating (B-4c) and using (B-3a). This gives

$$\hat{V}_{\mathrm{I}} |\beta, t\rangle_{\mathrm{I}} = i\hbar \frac{d}{dt} |\beta, t\rangle_{\mathrm{I}} \qquad \text{(B-6a)}$$

where

$$\hat{V}_{\mathrm{I}} = e^{i\mathscr{H}_{0}t/\hbar} \hat{V}_{\mathrm{S}} e^{-i\mathscr{H}_{0}t/\hbar}. \qquad \text{(B-6b)}$$

In the interaction picture operators evolve according to $\mathscr{H}_0$ while states evolve according to $\hat{V}_{\mathrm{I}}$. However the solution to (B-6a) cannot be found easily since $\hat{V}_{\mathrm{I}}$ depends on time in (B-6b) even when $\hat{V}_{\mathrm{S}}$ is time independent. When interactions vanish the interaction picture reduces to the Heisenberg picture. The interaction picture is valid even when $\hat{V}_{\mathrm{S}}$ depends on time explicitly.

Although (B-6a) cannot be solved easily, it can be transformed to an integral equation by multiplying both sides by $dt/i\hbar$, integrating, and adding an initial-value constant. Performing these operations, we find

$$|\beta, t\rangle_{\mathrm{I}} = |\beta, 0\rangle + \frac{1}{i\hbar} \int_{0}^{t} \hat{V}_{\mathrm{I}}(t') |\beta, t'\rangle_{\mathrm{I}} \, dt'. \qquad \text{(B-7)}$$

At this point, it is convenient to express (B-7) in the Schroedinger picture using (B-4c) and (B-4a); we thus obtain

$$e^{i\mathscr{H}_{0}t/\hbar} |\beta, t\rangle_{\mathrm{S}} = |\beta, 0\rangle + \frac{1}{i\hbar} \int_{0}^{t} e^{i\mathscr{H}_{0}t'/\hbar} \hat{V}_{\mathrm{S}} \, e^{-i\mathscr{H}_{0}t'/\hbar} e^{i\mathscr{H}_{0}t'/\hbar} |\beta, t'\rangle_{\mathrm{S}} \, dt'$$

or simplifying,

$$|\beta, t\rangle_S = e^{-i\mathcal{H}_0 t/\hbar}|\beta, 0\rangle + \frac{1}{i\hbar}\int_0^t e^{-i\mathcal{H}_0(t-t')/\hbar}\hat{V}_S(t')|\beta, t'\rangle_S \, dt'. \quad \text{(B-8)}$$

Henceforth, we drop the subscript S for the Schroedinger picture. The notation $\hat{V}(t)$ refers to any *explicit* time dependence in the interaction potential. If $\hat{V}$ is small, that is, $\hat{V} \to \lambda\hat{V}$, then (B-8) can be solved by successive iterations. The first-order approximation is obtained by inserting the zeroth-order result, $|\beta, t\rangle = e^{-i\mathcal{H}_0 t/\hbar}|\beta, 0\rangle$, into the integral giving

$$|\beta, t\rangle \simeq e^{-i\mathcal{H}_0 t/\hbar}|\beta, 0\rangle + \frac{\lambda}{i\hbar}\int_0^t e^{-i\mathcal{H}_0(t-t')/\hbar}\hat{V}(t')e^{-i\mathcal{H}_0 t'/\hbar}|\beta, 0\rangle \, dt'$$

or

$$|\beta, t\rangle = [\hat{U}^{(0)}(t, 0) + \hat{U}^{(1)}(t, 0)]|\beta, 0\rangle$$

where

$$\hat{U}^{(0)}(t, 0) = e^{-i\mathcal{H}_0 t/\hbar}$$

and

$$\hat{U}^{(1)}(t, 0) = \frac{\lambda}{i\hbar}\int_0^t e^{-i\mathcal{H}_0(t-t')/\hbar}\hat{V}(t')e^{-i\mathcal{H}_0 t'/\hbar} \, dt'.$$

This is exactly the result for the first-order correction to the evolution operator as given in Chapter 7 except that t_0 has been set equal to zero here. Further iterations of (B-8) generate higher-order corrections $\hat{U}^{(n)}$ to the evolution operator.

תושלב״ע.

Index